Advances in Industrial Control

Advances in Industrial Control is a series of monographs and contributed titles focusing on the applications of advanced and novel control methods within applied settings. This series has worldwide distribution to engineers, researchers and libraries.

The series promotes the exchange of information between academia and industry, to which end the books all demonstrate some theoretical aspect of an advanced or new control method and show how it can be applied either in a pilot plant or in some real industrial situation. The books are distinguished by the combination of the type of theory used and the type of application exemplified. Note that "industrial" here has a very broad interpretation; it applies not merely to the processes employed in industrial plants but to systems such as avionics and automotive brakes and drivetrain. This series complements the theoretical and more mathematical approach of Communications and Control Engineering.

Indexed by SCOPUS and Engineering Index.

Proposals for this series, composed of a proposal form (please ask the in-house editor below), a draft Contents, at least two sample chapters and an author cv (with a synopsis of the whole project, if possible) can be submitted to either of the:

Series Editor

Professor **Michael J. Grimble**:
Department of Electronic and Electrical Engineering, Royal College Building, 204 George Street, Glasgow G1 1XW, United Kingdom;
e-mail: m.j.grimble@strath.ac.uk

or the

In-house Editor

Mr. **Oliver Jackson**:
Springer London, 4 Crinan Street, London, N1 9XW, United Kingdom;
e-mail: oliver.jackson@springer.com

Proposals are peer-reviewed.

Publishing Ethics

Researchers should conduct their research from research proposal to publication in line with best practices and codes of conduct of relevant professional bodies and/or national and international regulatory bodies. For more details on individual ethics matters please see: https://www.springer.com/gp/authors-editors/journal-author/journal-author-helpdesk/publishing-ethics/14214.

Bing Chu · David H. Owens

Optimal Iterative Learning Control

A Practitioner's Guide

Springer

Bing Chu
School of Electronics and Computer Science
University of Southampton
Southampton, UK

David H. Owens
Department of Automatic Control and Systems Engineering
University of Sheffield
Sheffield, UK

ISSN 1430-9491 ISSN 2193-1577 (electronic)
Advances in Industrial Control
ISBN 978-3-031-80235-5 ISBN 978-3-031-80236-2 (eBook)
https://doi.org/10.1007/978-3-031-80236-2

Mathematics Subject Classification: 49J15, 49N05, 58E25, 93C40

This Springer imprint is published by the registered company Springer Nature Switzerland AG
The registered company address is: Gewerbestrasse 11, 6330 Cham, Switzerland

This text is dedicated to our wives, Rose and Zhi, with thanks for their patience and support during the many hours needed for the preparation of the manuscript.

Series Editor's Foreword

Control engineering is viewed differently by researchers who produce general theories and engineers who must design, calibrate, implement and maintain control systems. Researchers very often develop algorithms for control problems with a well-defined mathematical basis, whilst engineers may have more immediate concerns over the limitations of equipment, quality of control, safety, security and system downtime. The monograph series *Advances in Industrial Control* (AIC) attempts to bridge this divide by encouraging the consideration of advanced control techniques when they offer genuine real-world benefits.

The rapid development of new control theory, techniques and technology has an impact on all areas of engineering. This series focuses on applications that may stimulate the development of new and more general control paradigms. This is desirable if the different aspects of the "control design" problem are to be explored with the same dedication that "analysis and synthesis" problems have received. The series enables researchers to introduce new ideas motivated by the challenging problems in the applications of interest. It raises awareness of the various benefits that advanced control can provide at the same time as explaining the challenges that can arise.

This AIC monograph by authors Bing Chu and David Owens covers both the interesting theoretical aspects of Iterative Learning Control (ILC) and the practical issues involved. The authors have provided a very clear exposition of the topic with useful hints and pointers to the nature and value of chapters. Interest in the use of "Learning" has been greatly stimulated by the data-driven, Artificial-Intelligence (AI)-based developments in Machine Learning (ML). There are many links between the ideas presented here and those in ML systems. Both, of course, use the data collected in the previous iteration or pass of the process. In fact, it may be argued that it is beneficial to understand ILC before turning to AI-inspired concepts, and this text provides an excellent introduction to ILC. The text covers the use of feed-forward and feedback control action in ILC and issues of robustness in the presence of uncertainties.

Chapter 1 provides a simple introduction to ILC in which error dynamics are introduced and the important property of the convergence of the error is discussed. A quadratic optimal control problem is defined that provides a formal solution to the problem with well-defined properties. It also mentions the problem of uncertainties, including variations in disturbances that may occur between iterations.

Chapter 2 provides a straightforward introduction to mostly linear system concepts, such as state-space models, optimal control, and the mathematics used in the remainder of the text. The inverse-based ILC design approach, which can provide a benchmark for some systems is introduced in Chap. 3. Chapter 4 uses a general operator form approach and a gradient-based ILC solution. This is attractive since gradient optimization is so well understood and results are appealing intuitively. The problem of systems with non-minimum-phase zeros is not so serious as in the inverse ILC situation but difficulties still arise and are discussed.

In Chap. 5, on norm optimal ILC, the operator representation and Hilbert-space setting provides a rigorous solution and yet the main strategy and results are relatively easy to follow. Constraints are often problematic with systems designed using linear system concepts so Chap. 6 considers norm optimal ILC with hard and soft constraints. This is a very practical problem in real systems that does deserve a proper solution but is often treated with empirical fixes. The chapter offers more soundly based solutions to such problems.

Chapter 7 considers methods to accelerate the convergence of norm optimal ILC algorithms, such as moving some computations off-line where possible. Several useful practical measures to improve the algorithms convergence rate are suggested. For well over two decades one of the more popular topics has been the use of prediction in advanced controls (such as model predictive control) and Chap. 8 uses prediction for ILC problems which also provides useful benefits.

The final chapters of the text are concerned with applications. Chapter 9 describes probably the most likely area for application of ILC methods, namely applications in robotics. The details of the test rigs and experimental results are very helpful in illustrating the ideas and methods. Chapter 10 considers servo types of application by using a rotary electro-mechanical system problem. This is a non-minimum-phase plant that can introduce problems in the convergence rate and robustness. However, the chapter illustrates how such problems can be circumvented.

Chapter 11, on robot-assisted upper-limb stroke rehabilitation, illustrates the wider point about advanced controls that they can provide significant advances in health and wellbeing. Applications that increase performance in industry are of course important but the very strong trend to use advanced or indeed classical controls in health care is heartening.

The text includes a valuable set of examples and references that include seminal contributions to the subject by the authors. The monograph should be helpful to control-design engineers, researchers in dynamics and control, and to engineering/ IT course students interested in learning systems. The authors have provided a very readable text on an increasingly important topic, and it is a welcome addition to the AIC series.

Glasgow, UK
July 2024

Michael J. Grimble

Preface

Iterative Learning Control is an established part of the subject area of Control Engineering. It has an increasing number of applications in sectors including robotics, mechanical test systems, and medical rehabilitation. It differs from classical control problem as it is characterized by the need to repeat an identical operation on a system many times, and to do so with an accuracy that feedback and feedforward systems cannot achieve, particularly in the presence of uncertainties in the precise details of the system dynamics.

Unfortunately, the control input signal required to produce the desired output exactly is not known. Iterative Learning Control (ILC) uses the repetitive nature of the problem in the initial phases of operation of the plant, when errors are present, to find that input to a desired accuracy. It does this by using the first set of repetitions to systematically improve on an initial estimate using data accumulated from the current and previous repetitions (often called *iterations*). As a consequence, the core of an ILC design is an *algorithm* for *learning* that input over a number of *iterations*.

The central elements of the subject are a representation of the plant (typically in the form of a model), the accumulation of measured input and output data from iteration to iteration, and an algorithm for using this data for sequentially (iteratively) improving the input estimate. The design of the algorithm needs an understanding of model dynamics and the construction of a procedure for systematic input updating. An approach to such a design procedure is the central theme of this text.

The approach taken in this text is to focus on linear systems and to use *optimization procedures* as the basis of input updating. Such an approach has the advantage of providing a natural link to the established topic of *Optimal Control Theory* and will be seen to be a natural mechanism for ensuring that each iteration leads to an improvement in completing the task. The content of this text is strongly related to that of a companion text[1] which explores a wide range of possibilities for optimization-based ILC. The content and style used here has overlap with the companion volume

---

[1] Owens D. H. (2015) Iterative Learning Control—An optimization paradigm. Springer.

but, as well as including more recent results, it places emphasis on a subset of algorithms with well understood performance potential whilst providing a theoretical basis for the choice of the parameters available to the user to influence algorithm performance and to achieve the required control objective. It aims to provide the algorithms and understanding needed by students and practitioners to develop their own implementation strategy.

The text has many, simulation-based, examples of algorithms and their performance but, necessarily, has a significant mathematical content. In particular, it requires a reasonable background in classical control design, state-space methods, and Optimal Control Theory as it applies to linear systems with quadratic performance indices. It should be accessible to graduate students and users working on research projects or applications of iterative methods to repetitive systems. To support these readers, some of the background material needed from these areas is included.

It is worthwhile noting that the text uses some mathematical techniques that may be new to some readers. They are strongly linked to the methods of Optimal Control but, on the surface, have a structural similarity to the use of matrix or transfer function methods. The formal mathematical subject is that of Functional Analysis, but, in reality, the approach used is that of an operator representation of the system dynamics, the use of what is called the adjoint operator, and a way of manipulating signals and their magnitudes using inner products and norms for time series and functions of time. Ultimately, these play no role in the practical application of the algorithms but are an essential components of the underlying theory that allow the various algorithms and their properties to be developed.

After two initial chapters introducing the subject area, and providing essential mathematical background, the text introduces the simplest ILC algorithm that attempts to use the inverse system to reconstruct the required input. The issues that make this approach problematic are identified, and the optimization-based approach is introduced in Chap. 4 using a generalization of the familiar, iterative, gradient algorithm for solving sets of algebraic equations. The resultant algorithm can reduce a measure of error each iteration and, for linear systems has a well-defined parameter range within which success is guaranteed. Following this, Chap. 5 introduces the idea of Norm Optimal Iterative Learning Control (NOILC) which forms the basis for the remainder of the text. This algorithm has strong connections to the Gradient ILC Algorithm but uses an Optimal Control formulation to generate input updates. The approach is extended to include convex constraints in Chap. 6, as well as to optimize some additional performance objectives of interest.

One of the central themes in ILC research is ensuring convergence of the algorithms. In addition, the rate of convergence is important as each iteration involves an on-line experiment with the implied commitments in time and other resources. The number of iterations required for satisfactory convergence can be controlled, in part, using a design parameter in the NOILC Objective Function. Other useful approaches to accelerating NOILC are described in Chap. 7 with detailed explanations of the mechanism used and benefits achievable. This chapter is followed by Chap. 8 which explains how a predictive formulation of NOILC can accelerate the algorithm using an objective function that incorporates predicted decisions and performance on future iterations. Emphasis is placed on the most practical case to minimize the algorithm complexity whilst retaining its benefits.

The final chapters describe case studies that illustrate the application and potential performance of the algorithms. The success of the methodologies is evident from the material and some of the challenges in applications are evident. Rate of convergence predictions is verified and acceleration mechanisms are shown to be successful, but the effect of non-minimum-phase (NMP) zeros on certain types of problem is evident. The issue that arises in NMP systems is summarized as algorithm behaviours where initially rapid convergence rates slow down to become infinitesimally slow even though the error in achieving the task objective may still be significant. Chapters 4 and 5 include a detailed discussion of the mechanism, a calculation of the relevant observed characteristics in the simplest case, and ways are suggested for alleviating the issue in practice.

A number of other issues that add understanding and guide design are addressed in some or all of the chapters. These include the following:

1. Eigenvalue/eigenvector methods are used to link convergence to eigenvalue values of the system operator. These are linked closely to the frequency response characteristics for discrete-time, linear, time-invariant, state-space systems and can provide some guidance in the choice of parameters (as in Chap. 7 and the companion volume [1]).
2. Models are never exact descriptions of real systems so the robustness of the property of convergence is an important characteristic of any ILC algorithm. Modelling errors can take many forms, the analysis of many being highly complex and possibly impossible. The text examines a specific form of linear modelling error for all but the constrained algorithms. The analysis produces robustness criteria in a very similar form in each case. These criteria characterize the modelling errors that can be tolerated and, in some cases of interest, have a frequency domain interpretation in terms of the transfer function (matrix) of the modelling error. This provides considerable insight into what is needed and also some confidence that the algorithms can be robust in practice.

It is hoped that the readers get a clear picture of the nature of ILC and the benefits of the optimization-based approach. They should see the conceptual and mathematical foundations of the problem of algorithm construction and gain considerable insight into the reasons for algorithm convergence and the impact of the available parameters that make acceleration of that convergence possible. They should also see that, as in most design problems, there is always a balance to be made between, for example, convergence rates and robustness of the algorithm and gain some insight into how that balance might be controlled.

Sheffield, UK
Southampton, UK
June 2024

David H. Owens
Bing Chu

Acknowledgments The authors wish to acknowledge the contributions, over many years, of their friends, collaborators, and research students, without whom much of the related work and experimental studies would not have been possible.

Contents

Abbreviations and Notation

Abbreviations

FES	Functional Electrical Stimulation
ILC	Iterative Learning Control
MIMO	Multi-Input–Multi-Output
NMP	Non-Minimum Phase
NOILC	Norm Optimal Iterative Learning Control
PID	Proportional–Integral–Differential
PNOILC	Predictive Norm Optimal Iterative Learning Control
POILC	Parameter Optimal Iterative Learning Control
SISO	Single-Input–Single-Output
TFM	Transfer Function Matrix
TPBVP	Two-Point Boundary Value Problem

Notation

$\mathcal{R}^n$	Euclidean n-space
$L_2[0, T]$	Linear vector space of real valued functions on $[0, T]$ whose integral of 2-norm square over $[0, T]$ is well-defined and finite
$L_2^p[0, T]$	Linear vector space of real valued $p \times 1$ vector functions on $[0, T]$ whose integral of 2-norm square over $[0, T]$ is well-defined and finite
$\mathcal{L}$ and $\mathcal{L}^{-1}$	Laplace Transform and Inverse Laplace Transform
$\mathcal{Z}$ and $\mathcal{Z}^{-1}$	$\mathcal{Z}$-Transform and Inverse $\mathcal{Z}$-Transform
$\mathcal{R}[L]$	Range of operator L
$ker[L]$	Kernel of operator L
L^*	Adjoint of operator L
$P_S[\cdot]$	Projection operator onto set S

$M^{\perp}$	Orthogonal complement of M
M^T	Transpose of matrix M
$M^{\dagger}$	Hermitian transpose of matrix M
$\lvert \cdot \rvert$	Modulus of a number
$\lVert \cdot \rVert$	Norm of a vector, matrix, or operator

Chapter 1
Introduction

The main objective of the text is to describe and evaluate control algorithms and to introduce guides to control parameter choice. The topic presents the normal problems of control systems design with the added ingredient of repetition. Because of this, the design decisions made necessary by repetition are added to those met in more traditional control design and algorithm development. This “expansion” requires a change in our way of thinking. As with the ubiquitous Proportional plus Integral Control concept, any iterative algorithm ideally has a structure that applies quite generally and gives rise to similar issues in control performance. Some of the main issues arising in the following chapters are anticipated using matrix iteration as a simple model of the problem. Matrix iteration sets the scene for future chapters but leaves many unanswered questions and, significantly, leaves out much of the detail necessary in the iterative control of dynamical systems. Despite this, it aims to provide important pointers to the ideas and techniques used later.

1.1 Control Systems

Iterative Learning Control sits on the shoulders of the subject of Control Engineering, now a long established discipline that has worked with other areas of Engineering and Science in the design and working efficiency of almost every machine or process used by mankind including household devices, the utilities and aerospace, and satellite systems. The basis of the control operation is algorithms embedded in electro-mechanical devices, computer control systems, or simpler rule-based operator guidelines. Much of the hardware and software of control technology is hidden from the user who, as a consequence, can be forgiven for taking the beneficial consequences of good control for granted. In reality the algorithms are an essential part of the overall system without which control objectives cannot be met.

B. Chu and D. H. Owens, *Optimal Iterative Learning Control*, Advances in Industrial Control, https://doi.org/10.1007/978-3-031-80236-2_1

Control systems are essentially objects driven by input data streams and creating output data streams constructed to achieve a desired objective to a satisfactory accuracy. This is because effective control requires measurement or estimation of relevant outputs and other signals as the basis for computation of a suitable intervention using available actuators. Although simply stated, the reality is that control systems take a wide variety of forms and design is a skilled process. The complexities of control systems design are reflected in a number of challenges as follows:

1. The system to be controlled, often called the *plant*, typically has its own temporal and/or spatial dynamics. Dynamics can be linear or nonlinear. Behaviours of apparently complex plant can be quite simple to characterize and control whilst those of even apparently simple systems can exhibit more complexity ranging from instability to oscillation, to counter-intuitive non-minimum phase characteristics, and to time variation or structural change.
2. The accuracy and capabilities of sensors and actuators available to the designer may be limited by cost, availability, and/or the need for a natural simplicity to make commissioning and maintenance easy.
3. The form and structure of the control system have to be appropriate to the task and implementable in hardware and software form. In particular, there should be sufficient flexibility within the controller to permit the design engineer and, ultimately, the user to adjust parameters to reach an acceptable compromise/tradeoff that meets the real-world multi-objective performance specifications.

As a consequence, the need for general theoretical principles and mathematical tools to underpin and guide the design process has been essential.

1.1.1 Classical Feedback and Feedforward Control

Traditionally, control objectives have been expressed as tracking problems where the objective is to track a specified reference signal $r(t), t \geq 0$. The two guiding design principles have been the use of *feedback* and *feedforward* loops as illustrated in Fig. 1.1 which is a block diagram representing the situation where a process G produces an output time signal y using an input u.

The signal u is constructed from three sources as follows:

1. a feedback element K operating on the observed tracking error signal $e = r - y,\ t \geq 0$,
2. a feedforward element F operating on the desired reference r, and
3. a minor feedback loop element M operating on a measured or estimated additional plant signal z.

The structure of this control system arises from the simple intuition that effective control intervention at a given time can only be undertaken based on, and with a knowledge of, current and past behaviour of the system (i.e., data feedback) and a knowledge of the desired behaviour (i.e., feedforward elements). The general design

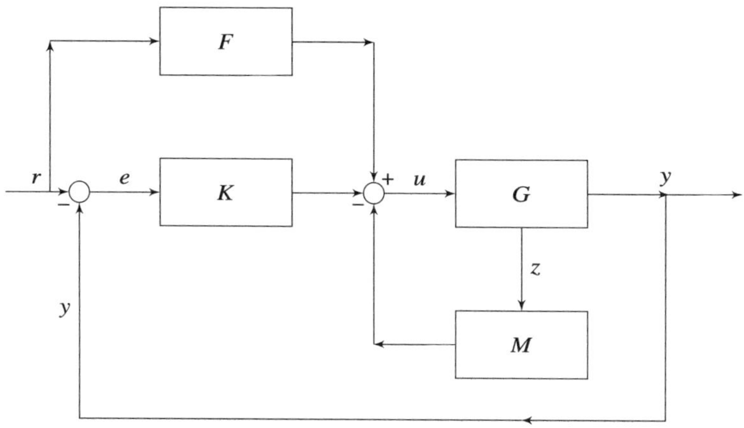

Fig. 1.1 Feedback and feedforward control structure

challenge is to choose a suitable control structure and control elements and parameters within that structure to achieve the desired plant behaviour. That is, the design process is essentially the choice of K, F, and M to meet performance needs and practical limitations.

These classical ideas are embedded in the ideas of Iterative Learning Control with the added dimension of repetition.

1.2 Repetition and Iteration

The idea of repetition is familiar in everyday life, often, but not solely, associated with learning a new skill, repeating an exercise to rehabilitate damaged muscles or limbs or, more generally, an industrial process going through a procedure many times to produce a large number of identical outcomes or products.

1.2.1 Examples of Repetition in Control Situations

1.2.1.1 Repetition and Robotics

The simplest visualization of repetition in Control can be found in the area of robotic assembly where a robotic device is used to complete a specified task such as "pick and place" as shown in Fig. 1.2. In such situations, the objective is to transfer a sequence of identical objects (delivered to the robot by conveyor) from a specified position to a new position after travelling along a well-defined spatial trajectory. The time taken to complete each task is assumed to be fixed at a specified value. It is important to

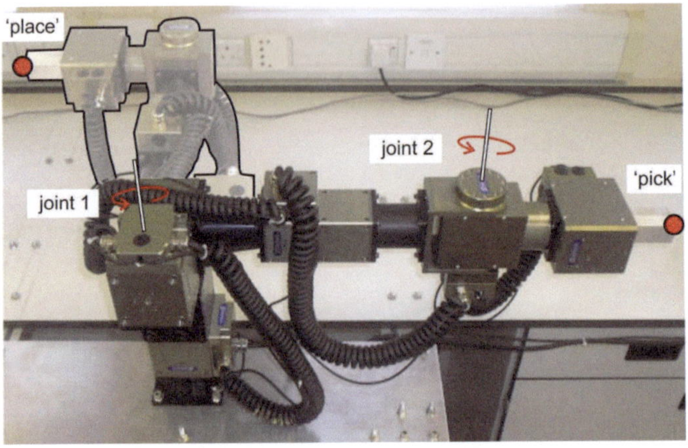

Fig. 1.2 Robot arm "pick-and-place" task

note, however, that the tracking of the trajectory during each transfer is essentially a classical control problem that can be approached using feedback and feedforward methodologies. This approach inevitably leads to non-zero tracking errors due to the very nature of feedback control and the imperfections in the success of feedforward elements arising from the uncertainties in the model of the system. It is hence natural to ask the following general questions:

1. *Can the repetitive nature of the task provide a means of improving, significantly, the tracking accuracy despite the model uncertainties?*
2. *How can this objective be achieved and what are the properties of the plant dynamics that aid or inhibit the process?*

The questions are quite general and have many potential answers. However, the general approach to sequential improvement in tracking accuracy taken here is to make use of the data recorded during each execution of the task.

Conceptual Use of Data: *Let k be the index of the repetition. Conceptually, the use, on repetition $k + 1$, of data recorded on previous repetitions with index $k' \leq k$ will contain data on actual input–output behaviour and the success of previously used control signals. If used intelligently, this "intelligence" could provide a means of error reduction in future repetitions so that, as k increases indefinitely, the tracking error will, ideally, be reduced to zero. The combination of any such control scheme with the robot is an example of an* **Iterative Learning Control scheme**.

Note 1.1 The historical development of the subject has seen the word repetition replaced by "pass", "trial", "task", "experiment", or, more generally, "iteration". This text will follow what is now normal practice and use "iteration" but may use the alternatives if the situation permits.

The control procedure has the conceptual form summarized below as an "algorithm".

Algorithm 1.1 (Conceptual Algorithm) The conceptual algorithm has the following five steps:

Step 1 Given the reference trajectory $r(t)$ over the desired task completion time $[0, T]$, choose an initial control input $u_0(t),\ 0 \le t \le T$. Set up the system in the desired initial state and undertake a complete pick-and-place attempt on the interval $0 \le t \le T$ to find the observed robot response $y_0(t),\ 0 \le t \le T$. Compute the tracking error $e_0(t),\ 0 \le t \le T$. Store the signal data $\mathcal{D}_0 = (e_0, y_0, u_0)$ and set $k = 0$.

Step 2 Reset the system to the desired initial state and use the information in the data sets $\mathcal{D}_j = (e_j, y_j, u_j),\ j \le k$ to construct a new input $u_{k+1}(t),\ 0 \le t \le T$.

Step 3 Undertake a complete pick-and-place attempt to find the observed robot response $y_{k+1}(t),\ 0 \le t \le T$. Compute the tracking error e_{k+1} and add the signal data $\mathcal{D}_{k+1} = (e_{k+1}, y_{k+1}, u_{k+1})$ to the store.

Step 4 If the error e_{k+1} is satisfactory, terminate the iterative procedure. Otherwise continue.

Step 5 Replace k by $k + 1$ and return to Step 2.

The key to the success of the procedure is the algorithm for constructing the new input in Step 2 from past data sets. In general, there will be many requirements placed on the control scheme, for example, the following:

1. A good Iterative Learning Control scheme will ensure the *reasonable trajectory tracking behaviour during initial iterations* but, ultimately,
2. it will also ensure the *convergence property* that this tracking accuracy will become perfect (i.e., exact) as $k \to \infty$. This statement also leads to the notion of *rate of convergence* as the designer will naturally only tolerate a small number of iterations (pick-and-place operations) where large tracking errors will be observed.
3. Consideration should also be given to the *initialization of the iterative process*. That is, the choice of the control signals used on the "zeroth" iteration when the robot is set up.
4. The Iterative Learning Control scheme behaviour should also be *insensitive (robust) to modelling errors* and/or changes in robot parameters such as a change in the weight of the objects to be lifted.

1.2.1.2 Iteration and Mechanical Test

Mechanical test procedures have a structure very similar to the robotic example described above. Test procedures are widely used as a means of either

1. exposing a mechanical specimen to a repeated sequence of defined force profiles over a specified time interval or
2. replicating, in the laboratory, either user-defined conditions representing typical operational events or physical conditions recorded during actual operation of the plant.

For example, a four-poster automotive test system (as shown in Fig. 1.3) is specifically designed for the testing of vehicles (cars, trucks). Such test systems consist of four hydraulic actuators on top of which the wheels of the vehicle are placed. Movements of the actuators aim to simulate movement of the vehicle over the road surface, the forces exerted by the road on the wheels, and the physical displacements of components or subassemblies. To achieve this, the movements of the system need to be tightly controlled by a digital test controller. During the development phase, a four-poster system is often used to test newly designed suspension systems and their durability. Different kinds of testing are possible including sine tests (including frequency sweeps) and tests aimed at reproducing the conditions met by the vehicle in an actual road environment (represented by data coming from an outdoor acquisition session/road test).

Successful control during mechanical test is not as straightforward as might initially appear. For visualization purposes, consider the control of a test facility that is required to take a road vehicle through a laboratory test that simulates *very accurately* the conditions recorded on a chosen road surface. The vehicle is assumed to have been fully instrumented and driven at the desired speed over the surface with measurements of useful variables made and recorded as time series over a specified time period. Measurements could include measured forces, accelerations, and displacements at chosen points of the vehicle including points on the suspension and subassembly. These signals form the reference signal for the following process of iteration in the laboratory.

For commercial or research reasons, suppose that it is required that the vehicle be subjected to the same "road" conditions but in a laboratory setting. Typically, high accuracy replication of road conditions is required. The forces are recreated by suitable controlled operation of the rig hydraulics but, as the rig is operating dynamically,

Fig. 1.3 Tractor on a four-poster rig. Reprinted from [6], Copyright 2007, with permission from Elsevier

it interacts with the vehicle and responds to its motions. As a consequence, errors in reproducing the desired vehicle response are inevitable in any one test. For this reason, it is essential that the process is repeated through a sequence of repetitions or iterations designed to reduce errors by changing the hydraulic actuation signals. Physically, the iterations consist of repeated operation of the rig using improved actuation signals. The aim of the iterative process is to reduce the errors between signals observed on road tests and those observed in the laboratory tests. When sufficiently small to satisfy the needs of the test, the road conditions have been faithfully replicated and the user is able to take the vehicle through the same conditions, whenever it is required, without the time and expense of more road tests.

The technological key to achieving improved replication of road conditions in the laboratory is some form of Iterative Learning Control which, in a similar manner to the robotics example outlined above, uses previously recorded test data to improve accuracy from iteration to iteration. More precisely, data recorded on previous iterations with index $k' \leq k$ will contain data on actual vehicle behaviour whilst on the rig and, if used intelligently, this could provide a means of generating actuation signal updates that guarantee error reductions. As k increases indefinitely, the tracking error, representing the mismatch between road and laboratory conditions, will hopefully be reduced, asymptotically, to zero.

1.2.2 Repetition and Iterative Learning Control

The examples illustrate the nature and benefits of and/or the need for repetition in some control problems that repeat a task over a fixed and specified time period. Iteration adds an extra dimension to the control problem, namely the *iteration index* $k \geq 0$ describing the place in the sequence of repetitions of a given data set. A measured or estimated signal S on the kth iteration will be denoted by S_k. With this notation, the iterative control process sequentially generates a sequence $\{ S_0, S_1, S_2, \cdots \}$. On the $(k+1)$th iteration, the data set $\{ S_0, S_1, S_2, \cdots, S_k \}$ could be made available to advise the decision making and the construction of the new control signal u_{k+1}. This decision making is the core of the algorithm design and should ensure that not only should each signal S_k have acceptable characteristics, but also the sequence should have desired characteristics.

A schematic description of a conceptual Iterative Learning Control Scheme is shown in Fig. 1.4 with reference signal r, an input signal u, and an output signal y on the kth iteration. At this level of generality, there are three blocks shown, namely the plant G and the iterative controller K plus a block S representing stored data obtained on iterations $k' \leq k$. The control block K will contain feedback loops together with processes that feed data forward, including data obtained on previous iterations.

Of course, the figure hides many complexities and a whole field of study that considers the nature and design of the processes within the control blocks. That field of study is *Iterative Learning Control* (commonly abbreviated to *ILC*). The use of

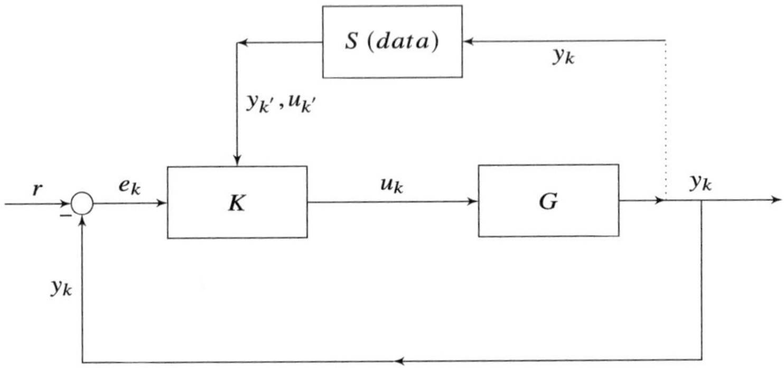

Fig. 1.4 Conceptual ILC scheme with computational, K, and stored data, S, blocks

the word "Iterative" is natural as the reader will come to recognize that ILC has strong conceptual links to Computational Mathematics (Numerical Analysis) and the iterative solution of sets of equations. The use of the word "Learning" is perhaps more surprising—

Important Note

The use of the word "learning" in ILC is now established in the published literature and has a precise interpretation. It reflects the perception that the iterative structure creates a desired and previously unknown control signal based on the observed behaviour of previous trials. In this sense, the iteration "learns" the desired signal using "past experience".

The following section discusses some of the issues that will arise in ILC by exploring the iterative solution of a set of linear matrix equations.

1.3 Dynamical Properties of Iteration: A Review of Ideas

In the context of the material in this text, the examples of the previous section indicate that Iterative Learning Control has two dynamical characteristics, namely the following:

1. *the dynamics within each iteration, as exemplified by the time response of the vehicle to hydraulic actuation forces during each iteration of the mechanical test procedure, and*
2. *the dynamical behaviour introduced by the design process through control computations, between iterations, based on past measurement data. This interaction between iterations produces a systematic dynamic behaviour in the sequence*

of outputs from the iterations. These dynamics are dependent on both the plant dynamics and the chosen data transfer mechanisms between and within iterations.

In general, the dynamics within each iteration will interact with the dynamics induced by data transfer between iterations. This interaction can be positive (that is, beneficial) or negative (that is, performance degrading) and it is certainly not the case that good control within each iteration implies that error reduction will be achieved from iteration to iteration. These ideas will form part of the following chapters. The following discussion uses a simple matrix iteration to provide an insight into some of the issues that can arise.

1.3.1 Matrix Iteration

1.3.1.1 A Matrix Model of Iterative Learning Control

The dynamic nature of iteration can be illustrated by the analysis of iteration in matrix theory in the form of iterative solution of linear simultaneous sets of algebraic equations. In what follows, some of the issues and problems that arise are noted and will be further developed in later chapters.

Iterative Control is analogous to the iterative computation of the solution vector u, for a given choice of reference/target r, of the matrix equations

$$\begin{array}{lll} & r = y & (The \text{ ``}Control\ \ Objective\text{''}) \\ where & y = Gu + d & (The \text{ ``}Input/Output\ \ Relation\text{''}). \end{array} \tag{1.1}$$

The dimensions of the column vectors r, u, d and matrix G are not relevant to the discussion but are required to be consistent with the matrix operations indicated. For any choice of u a corresponding y can be computed, the problem being to find a "feasible" u such that y is equal (or, in practice, very close) to the desired value r.

1.3.1.2 A Feedforward/Feedback Approach to Iterative Solution

For any sequence $\{u_0, u_1, u_2, \cdots\}$ of estimated solutions of the equation, let $\{y_0, y_1, \cdots\}$ be the corresponding values of y. Choose an initial guess u_0, compute y_0, and construct the remainder of the sequence by the update relation

$$\underbrace{u_{k+1} = u_k + K_0 e_k + K_1 e_{k+1}, \quad k \geq 0,}_{Input\ Update\ Rule} \tag{1.2}$$

where, for any $j \geq 0$, $e_j = r - y_j = r - Gu_j - d$ represents the error in satisfying the equation $r = y$ using u_j. The matrices K_0, K_1 are chosen by the user and $K_0 e_k + K_1 e_{k+1}$ represents a proposed correction to the iterate u_k to produce a new iterate

u_{k+1}. The iterates change every iteration stopping only if the user decides or if $u_{k+1} = u_k$ (when, necessarily, $e_{k+1} = e_k$).

Remark 1.1 The update relationship (1.2) is an algebraic parallel of the use of feedback (represented by $K_1 e_{k+1}$) and feedforward (represented by $K_0 e_k$) control in ILC. The term "feedback" is used as e_{k+1} is a consequence of the choice of u_{k+1}. The term "feedforward" is used to represent the fact that e_k is known data that is independent of u_{k+1}, i.e., the algorithm feeds data forward from previous iterations.

1.3.1.3 Error Dynamics

The update relationship is implicit as e_{k+1} depends on u_{k+1}. An explicit formula is obtained by noting that

$$e_{k+1} = e_k - GK_0 e_k - GK_1 e_{k+1}, \quad k \geq 0, \tag{1.3}$$

so that, denoting the unit matrix by I and assuming that $I + GK_1$ is nonsingular,

$$e_{k+1} = Le_k \quad where \quad L = (I + GK_1)^{-1}(I - GK_0). \tag{1.4}$$

It follows that the iteration update formula can be written in the equivalent "feedforward form"

$$u_{k+1} = u_k + (K_0 + K_1 L)e_k \tag{1.5}$$

and the evolution of the iteration errors has the dynamic behaviour

$$e_k = L^k e_0, \quad for \quad k \geq 0. \tag{1.6}$$

That is, once the "control elements" K_0 and K_1 have been chosen, the errors observed in the iterative process evolve in a dynamic way dependent on the initial error e_0 and the matrix G. A schematic block diagram of the algorithm is shown in Fig. 1.5 which reaffirms the link with the ideas of ILC.

Note 1.2 The reduction of the feedback plus feedforward form of control update Eq. (1.2) to the feedforward form (1.5) is a methodology that will be used from time to time in later chapters. As e_k is available after the kth iteration but before the $(k + 1)$th iteration starts, it makes possible the off-line computations of control updates and reveals methodologies for robustness assessment.

1.3.1.4 Error Convergence to Zero

Error convergence to zero is the primary objective of the iterative process as when $e = 0$ we have $y = r$. It clearly depends upon the nature of G, K_0, K_1, and e_0. In

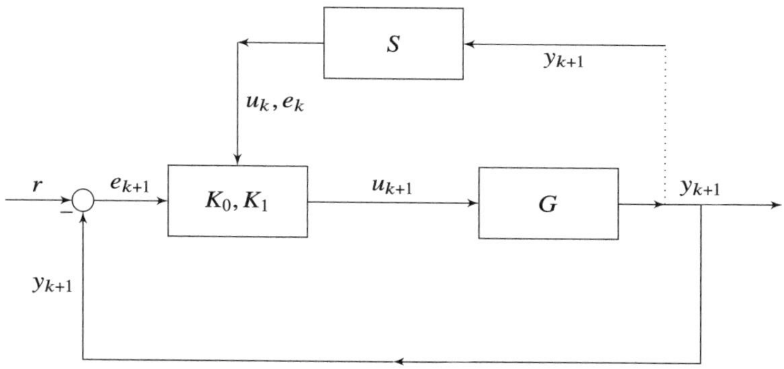

Fig. 1.5 Schematic representation of the matrix iteration

particular, e_k converges to zero for all possible initial errors e_0 if, and only if, all eigenvalues of L have modulus strictly less than unity. This condition, in particular, means that, after a sufficiently large number of iterations, the error in satisfying the equation is arbitrarily small so that, even if the iteration is terminated, an arbitrarily accurate solution can be achieved.

Note 1.3 (Technical Note) This analysis can be refined to describe the situation when L has eigenvalues less than or equal to unity. These details are left for later in the text but, in essence, the error sequence does converge but possibly to a non-zero value.

1.3.1.5 Nature of the Convergence

The eigenvalue condition describes asymptotic convergence precisely and, in simple terms, the closer the eigenvalues are to zero, the faster the convergence to the solution will be. This simplicity is misleading however as the precise nature of this convergence depends critically on the detail of the matrices involved. For example, if

$$L = \begin{bmatrix} 0.5 & \alpha \\ 0 & 0.7 \end{bmatrix}, \tag{1.7}$$

then errors will always converge to zero as the eigenvalues are 0.5 and 0.7. In particular, error magnitudes will reduce from iteration to iteration if $\alpha = 0$. In contrast, if $\alpha \neq 0$, errors from the defined initial condition take the form

$$e_0 = \begin{bmatrix} 0 \\ 1 \end{bmatrix}, \quad e_1 = \begin{bmatrix} \alpha \\ 0.7 \end{bmatrix}, \quad e_2 = \begin{bmatrix} 1.2\alpha \\ (0.7)^2 \end{bmatrix},$$
$$\textit{and} \quad e_3 = \begin{bmatrix} (0.6 + (0.7)^2)\alpha \\ (0.7)^3 \end{bmatrix}, \quad \cdots \tag{1.8}$$

which will ultimately reduce to zero but, before that happens, can take arbitrarily large values (dependent on the parameter α) as the iteration index k increases. As a consequence, it can be expected that, *in any iterative process, convergence to a desired solution will not necessarily imply that the convergence will be accompanied by acceptable magnitudes and forms of the iterates.*

Note 1.4 In Iterative Learning Control involving experimental work on the physical system, the magnitudes of signals are normally subject to constraints due to safety boundaries and/or the physical range of actuators. Ensuring acceptable signal magnitudes is crucial in this case.

1.3.1.6 Sensitivity of the Convergence

If the matrix G is such that L only has eigenvalues with modulus less than unity, then convergence of iterations will be guaranteed. However, if G is perturbed by a small amount, convergence may be lost. In general terms, the effect of a small perturbation will be to perturb the eigenvalues also by a small amount. It follows that those eigenvalues with modulus closest to unity will be the most sensitive to perturbations to G as perturbations with only small magnitudes may be all that is required to move the eigenvalue to a position with modulus greater than unity. Divergence of the iterates will then occur.

The Sign of the Modelling Error: It is interesting to note that the nature of the perturbation is crucial to the consequences for convergence. More precisely, for an eigenvalue close to unity, small perturbations could increase the eigenvalue modulus (potentially causing divergence) whilst the same perturbation but with opposite sign will tend to reduce the modulus (hence increasing convergence speeds). The ability of an iterative process to tolerate perturbations and retain convergence therefore can be expected to depend on its eigenvalue magnitudes and some concept of the "sign" of the perturbation. This will be reflected in the following text where, for dynamical systems, the concept of *positive real* operators plays a role in understanding this issue more fully.

1.3.1.7 Design Factors

The factors K_0 and K_1 are chosen as part of the update relationship by the user whilst the factor G is analogous to the plant dynamical model. The interplay between these three objects is represented mathematically by the form of L which is the key object in ensuring algorithm convergence. Clearly the primary requirement is convergence and hence eigenvalue positioning. However, an important secondary objective is to control the magnitudes of the iterates as k increases. To achieve this requires a careful design of K_0 and K_1, an issue that dominates this text and the area of Iterative Control. Last but not least, the choices made will have an impact on the sensitivity/robustness of the convergence properties, and choices that provide good robustness might be preferred.

An insight into possible guidelines to design choices is suggested by the form of L. More precisely, "ideal" K_0 and K_1 can be linked, respectively, to

(a) *approximate inversion of* G and (b) *high gain feedback.*

In more detail:

1. If K_0 is a right inverse of G in the sense that $GK_0 = I$, then $L = 0$ and the iterations converge in one step with $e_1 = 0$ independent of the starting condition e_0. This choice requires matrix inversion which may not be regarded as suitable or possible, but, using some notion of K_0 as an approximate right inverse might have the effect of reducing eigenvalues of L.
2. The factor $(I + GK_1)^{-1}$ in L may tend to reduce in magnitude if K_1 is "large" in some sense. It may therefore tend to reduce the eigenvalue magnitudes. For example, if GK_1 has only strictly positive eigenvalues and K_1 is replaced by the scaled version gK_1 where $g > 0$ is a scalar "gain", then it is easily seen that $(I + GK_1)^{-1} \to 0$ (and hence $L \to 0$) as $g \to \infty$. That is, in theory, extremely rapid convergence of the error to zero is possible under high gain conditions.

Note 1.5 Although mathematically true, in control situations the use of high gains can be problematic so, in practice, the range of gains will be limited.

The observations made above have a great similarity to the use of inverse systems in classical feedforward control and the notion of high gain feedback being beneficial for control performance. The discussion does not prove that the ideas carry over to Iterative Learning Control without change but they do suggest that inverse systems and gain may play central roles in design decisions.

1.3.2 Matrix Iteration Using Optimization

This section aims to demonstrate that optimization methodologies can play an important role in iteration procedures and hence lay a foundation for the use of optimization in Iterative Learning Control. The approach is an embryonic form of the topic of Optimal Iterative Learning Control that forms the central theme of this text. It is based on the matrix model (1.1) and will lead to choices of K_0 and K_1 (Eq. (1.2)) and well-defined, and useful, algorithm properties.

1.3.2.1 The Performance Index and Optimization Problem

The purpose of the iteration is, sequentially, to reduce the error e to zero whenever possible. The approach chosen here is that of reducing a quadratic form $e^T Qe$ on every iteration, where Q is a symmetric, positive definite matrix. In the process, it is necessary to control the magnitude of the change $u - u_k$. This can be approached by choosing a symmetric, positive definite matrix R and forming the *performance index* (or *objective function*)

$$\underbrace{J(u) \;=\; e^T Q e + \epsilon^2 (u - u_k)^T R (u - u_k)\,.}_{\textit{Quadratic Objective Function}} \tag{1.9}$$

Here ϵ^2 is a positive scalar parameter introduced to provide a simple mechanism for varying the relative contributions of the two terms to J.

Consider the choice of u_{k+1} by minimizing $J(u)$ subject to the the "input–output" relation in (1.1). That is

$$\underbrace{u_{k+1} \;=\; argmin\{\, J(u) \;:\; e = r - y,\; y = Gu + d\,\}\,.}_{\textit{The Optimal Iteration}} \tag{1.10}$$

Note 1.6 As ϵ^2 increases the optimal change $u_{k+1} - u_k$ (as measured by the quadratic form) in the input will be reduced and the change in the error will be small. In contrast, as ϵ^2 decreases the optimal error will be significantly reduced (as measured by its quadratic form) and the input change may be substantial.

1.3.2.2 Some Properties of Optimal Iteration

The optimization problem has a very important property, namely that of *monotonic reduction in the error* from iteration to iteration. The word monotonic is used to emphasize that the reduction is achieved on *every* iteration. To demonstrate the property, note that, as u_{k+1} minimizes $J(u)$, it follows that every other u produces a larger value. Choosing $u = u_k$ means that $e = e_k$ and $J(u_{k+1}) \le J(u_k)$, i.e.,

$$e_{k+1}^T Q e_{k+1} + \epsilon^2 (u_{k+1} - u_k)^T R (u_{k+1} - u_k) \le e_k^T Q e_k, \quad k \ge 0. \tag{1.11}$$

As R is positive definite this leads to the conclusion that

$$\underbrace{e_{k+1}^T Q e_{k+1} \le e_k^T Q e_k,\;\; k \ge 0}_{\textit{Monotonic Error Reduction}} \tag{1.12}$$

equality holding if, and only if, $u_{k+1} = u_k$ when the iterative process stops.

Note 1.7 The error reduction is defined by reductions in a measure of error magnitude, namely the quadratic form $e^T Q e$. The detail of the nature of the reduction therefore depends on Q. As Q it is chosen by the user, it, together with R, represents an object that could, in principle, be used to provide/control the form of convergence preferred—it is an algorithm design parameter.

The monotonicity of the error leads to the conclusion that

$$\lim_{k\to\infty} e_k^T Q e_k = E_\infty \geq 0, \quad (Convergence) \tag{1.13}$$

with $E_\infty = 0$ if, and only if, $e_k \to 0$ as $k \to \infty$.

1.3.2.3 Input Computation and Error Evolution

The conditions for the minimization of $J(u)$ can be expressed as a formula. The basis of this is (Frechet) differentiation but, for this section, a constructive approach is used. Let u be a choice of input and perturb it to produce a nearby input $u + \delta u$. This means that e will be perturbed by δe and y by δy with $\delta e = -\delta y = -G\delta u$. It follows that

$$J(u+\delta u) - J(u) = \underbrace{2(\delta e^T Qe + \epsilon^2 \delta u^T R(u-u_k))}_{linear\ in\ perturbations} + \delta e^T Q \delta e + \epsilon^2 \delta u^T R \delta u. \tag{1.14}$$

The last term is positive for all δu. It follows that, when J is minimized, the first, linear, term must be zero for all δu and all δe such that $\delta e = -G\delta u$. That is, writing this term as $\delta u^T z$, it is necessary that $z = 0$. More precisely, substituting for δe gives

$$\underbrace{u_{k+1} = u_k + \epsilon^{-2} G^* e_{k+1}}_{Feedback\ Input\ Update}, \quad where \ \ G^* = R^{-1} G^T Q. \tag{1.15}$$

Note 1.8 The matrix G^* is an example of an adjoint operator and satisfies

$$a^T QGb = (G^*a)^T Rb, \quad (the\ Adjoint\ Property) \tag{1.16}$$

for all column vectors a and b of appropriate dimensions. Other examples of such operators will play an important role in the rest of this text.

Equation (1.15) defines the next input u_{k+1} in the iteration sequence. It has a "feedback" form but this can be converted to a feedforward form by noting that (1.15) can be written as $e_{k+1} = e_k - \epsilon^{-2} GG^* e_{k+1}$ which, rearranging, reveals the *Error Update Equation* (also called the *Error Evolution Equation*)

$$\underbrace{e_{k+1} = Le_k, \quad where \ \ L = (I + \epsilon^{-2} GG^*)^{-1}}_{Error\ Evolution}. \tag{1.17}$$

Note 1.9 The inverse exists as GG^* has real and positive eigenvalues. This follows as, if $Q^{\frac{1}{2}}$ is the unique, symmetric, positive definite square root of Q then the eigenvalue equation $GG^*v = \lambda v$ can be written as $Q^{\frac{1}{2}} GR^{-1} G^T Q^{\frac{1}{2}} w = \lambda w$ with $w = Q^{\frac{1}{2}} v$. λ is hence an eigenvalue of the positive, symmetric matrix $Q^{\frac{1}{2}} GR^{-1} G^T Q^{\frac{1}{2}}$ and must be real and positive.

Using (1.17) in (1.15) gives the feedforward formula

$$\underbrace{u_{k+1} = u_k + \epsilon^{-2} G^* L e_k}_{Feedforward\ Input\ Update}, \tag{1.18}$$

which generates a new input directly from previous iteration data (e_k, u_k).

1.3.2.4 Convergence and the Properties of L

The optimization approach guarantees a type of convergence in the form of the existence of E_∞. The details of the convergence properties depend on the eigenstructure of GG^*. More precisely, if $\{\sigma_j^2\}$ are the eigenvalues of GG^*, they are real and positive. The eigenvalues of L are precisely $\{(1 + \epsilon^{-2}\sigma_j^2)^{-1}\}$ which, if the $\{\sigma_j^2\}$ are ordered in decreasing order of magnitude $\sigma_1^2 \geq \sigma_2^2 \geq \cdots$, then they must lie in the interval between $(1 + \epsilon^{-2}\sigma_1^2)^{-1}$ and one. As a consequence:

1. If no eigenvalue of GG^* takes the value of zero (i.e., GG^* is nonsingular), then all eigenvalues of L are strictly less than unity. It follows that $\lim_{k\to\infty} e_k = \lim_{k\to\infty} L^k e_0 = 0$, i.e., the error converges to zero as desired.
2. Alternatively, GG^* has some eigenvalues equal to zero which implies that there exist non-zero initial errors satisfying $GG^* e_0 = 0$. This implies that $e_1 = Le_0$ and the error is unchanged. It will continue to remain unchanged as iterations progress, i.e., $E_\infty = e_0^T Q e_0 > 0$.

Note 1.10 The set of initial errors that have the property $GG^* e_0 = 0$ are precisely those $e_0 \in ker[G^*] = \{\ e_0\ :\ G^* e_0 = 0\ \}$. This is true as, for example, $GG^* e_0 = 0$ gives $0 = e_0^T QGG^* e_0 = (G^* e_0)^T RG^* e_0$ which implies that $G^* e_0 = 0$ as R is positive definite.

1.4 Approaches to Algorithm Design

Mathematical and modelling techniques form the toolbox proposed to develop the ideas and algorithms for optimization-based Iterative Learning Control presented in this text. The topics contain a number of threads and discussion items including the following:

1. Mathematical models of Iterative Learning Control take many forms as the plant, although linear, could take many forms including the following:

   a. continuous/analogue state-space models of either single-input–single-output (SISO) or multi-input–multi-output (MIMO or multivariable) form, and
   b. discrete-time/sampled-data state-space models of either SISO or MIMO form.

The model description will take several, equivalent, forms including the state-space model forms, transfer function matrix (TFM) forms, convolution, and super-vector descriptions. Many of these descriptions will be familiar and to bring them all together the text will use operator-based notations to simplify the analysis. The simplicity of the presentation is enhanced by this choice of approach but interpretation of the equations needs to be carefully considered. Readers who are inexperienced in the area should find the material accessible by interpreting the operators and vectors as matrices and the systems as being discrete-time.

2. Formal definitions of the problem of control algorithm design are used throughout. The primary objective is always error convergence but other issues include the following:

   a. identifying the influence of user-defined parameters such as gains and the weights used in optimization criteria,
   b. characterizing the robustness of the convergence to errors in the plant model, and
   c. controlling the nature of the error convergence in terms of ensuring reductions of defined error measures (norms) from iteration to iteration.

3. For linear systems, there is a wide range of different convergent algorithms including the following:

   a. The *Inverse Model Algorithm*, Chap. 3, for feedforward ILC plays a useful role in defining simple approaches and typical properties of many other algorithms.
   b. *Gradient Algorithms*, Chap. 4, take the familiar concept of gradient methods for function optimization into an ILC framework and form a simple mechanism to control convergence and reduce errors on each iteration.
   c. *Norm Optimal Iterative Learning Control (NOILC)*, Chap. 5, represents the basic algorithm that forms the core of the work and is strongly related to the ideas of optimal control.
   d. *Error Corrected NOILC Algorithms*, Chap. 7, that improve the convergence rate of NOILC algorithms using different reference signals on each iteration.
   e. *Predictive Norm Optimal Iterative Learning Control*, Chap. 8, which can accelerate convergence of NOILC by taking the effect of current input decisions on future iterations into account in input selection.
   f. The *inclusion of convex constraints, Chap. 6, on inputs and output signals* (to allow for applications where such limits are present). In these approaches, the system dynamics and the control objectives can be regarded as defining separate convex sets in suitable vector spaces (called Hilbert spaces). This interpretation forms the basis of algorithm construction using sequential projection of signals onto these sets. The many different ways of choosing the Hilbert space topology and choosing the order of the sequence of projections provide a wide range of different convergent algorithms. This also allows incorporating additional performance requirements as an *auxiliary optimization* problem.
   g. *Trajectory Tracking* is the most natural interpretation of ILC applications but there is an important place for *algorithms whose objective is limited* to, for

example, ensuring that an output signal $y(t)$ takes specified values at times $t_1, t_2, \cdots, t_M$—the so-called *Intermediate Point Control Problem*. In the case of $M = 1$ and t_M equal to the final time, the problem is termed an *End-point Control Problem*.

h. *Multi-task Algorithms* that unify some of these ideas under one algorithmic structure suitable for problems where the task has defined time periods where signal tracking is needed with other periods where the output must only satisfy intermediate value requirements.

All such algorithms ensure error convergence but the nature of each convergence differs in substantive, but useful, ways that give flexibility to the control design process. In general terms, the computational basis of each iteration is optimal control related and, for state-space systems, provides an implementation either in terms of a state feedback and an iteration to iteration feedforward structure or as a pure feedforward controller.

4. Convergence rates depend in a simple way upon the choice of different topologies in the input and output spaces but also upon plant dynamical characteristics. The biggest problems are met when controlling systems which are *non-minimum phase* in the sense that the system transfer function has zeros in the unstable region of the complex plane. The presence of such zeros leads to rapid error reduction in the first few iterations but, from that point onwards, the convergence rate is infinitesimally slow. An analysis of this phenomenon is possible and leads, in practical terms, to design guidelines aimed at improving the situation.
5. Finally, some engineers might balk at the complexities of optimal control implementations of the algorithms. Fortunately, the beneficial properties of optimization can be retained using either

a. state estimation using Kalman Filters or
b. simpler implementations such as a pure feedforward controller where all computations are undertaken off-line or
c. the use of the idea of *Parameter Optimal Iterative Learning Control*.

Parameter Optimal ILC (POILC) uses simple optimization but typically pays a price in algorithm performance. Rather than using optimal control, the approach uses control laws based on function or time series optimization by control laws expressed, in a linear way, in terms of known data and unknown parameters. That is a low-order optimization problem. The results of function optimization and parameter optimization can be shown to be identical under certain well-defined, but rarely satisfied conditions. They can be regarded as realistic alternative approaches but introduce a number of new dynamical phenomena in the iterative process, including convergence but to non-zero errors. The reasons for this are related to ideas of positivity of certain matrices. These problems can be avoided or alleviated using suitable parameterizations and/or empirical switching strategies.

1.5 A Note on Implementation

The preceding discussions have concentrated on the structure of the Iterative Learning Process as decision making based on the results of previous iterations and data obtained from plant measurements during those iterations. The remainder of this text will concentrate on the development, description, and details of effective ways of doing this using algorithms that produce desired properties from the iterative process. In practice, of course, there will be many other issues to address, most of which will be applications-specific and beyond the scope of this text.

The issue discussed in this section is just one of those that will, almost inevitably, be met in practice. That is, coping with disturbances after the initial stage of Iterative Learning stage has been completed and the system goes into its future repetitive mode. More precisely, if, after a number of iterations N_c , the input signal u_{N_c} is producing/replicating the desired reference r to an acceptable accuracy, then the iterative process will be terminated to save time and resources. The second stage of a design is to use this input signal, unchanged, for subsequent operations which will, in theory, simply replicate the tracking accuracy observed at the end of the iterative process. However, in reality, it is likely that all future operations will be subject to disturbances, including the following:

1. variations in the resetting of the initial condition,
2. variations in payload in, for example, robotics applications when the items being manipulated differ in weight, and
3. other systematic or unpredicted disturbances to the plant dynamics.

If these variations are large, then performance could be seriously degraded. If, however, they are small then attempts can be made to compensate by a second-stage implementation that incorporates feedback loops. More precisely, an implementation could take the form shown in Fig. 1.6. The block diagram has the structure of a combined feedforward and feedback systems but rather than the feedforward block operating in real time it feeds the input u_{N_c} forward from past ILC activity. The feedforward component ensures that the system will be in the right ballpark for future operations with the feedback loop(s) providing some compensation for errors introduced by the disturbances.

1.6 Discussion and Further Reading

1.6.1 General Discussion

The chapter has attempted to set the scene for Iterative Learning Control with a motivation and some examples of application. The ideas have both a familiar and novel form that describes the development of highly accurate control signals for systems that repeat an identical operation. The algorithmic approach uses past data

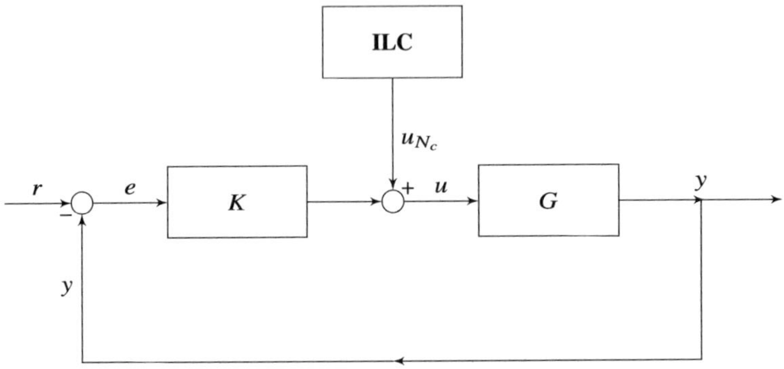

Fig. 1.6 Feedback implementation for the (feedforward) computed signal u_{N_c} from the ILC phase

to sequentially improve the input and, in the end, "learn" the input that produces a target reference signal "precisely". Practical application of the ideas will throw up many issues, most of which are applications-specific and beyond the scope of this text.

A matrix iterative problem was used to provide an introduction to the ideas of feedforward and feedback in iteration and demonstrated the benefits of optimization as an iterative mechanism. This problem has the general structure of most of the problems that appear in this text and the reader is encouraged to spend some time understanding the process and issues described.

The subject is algorithmic and hence requires some knowledge of relevant mathematical methods and notation. The text centres on the use of *linear* models of plant dynamics/behaviour and relies heavily on the use of ordinary differential and difference equations and their associated state-space models and optimal control problems. Because of this the techniques of Laplace and Z-transforms, matrix theory, and optimal control theory play central roles. Following the approach of the companion text Owens [37] the notation is simplified by the use of operator notation and some simple methodologies from vector space theory and Functional Analysis. This has the effect of making the theoretical development and algorithm structure take on a familiar transfer function or matrix structure, which, properly interpreted, always reduces to a sequence of familiar control computations.

1.6.2 Background Reading

Control Engineering is a topic of considerable depth and scope with many texts and published papers. The idea of iteration in control probably emerged from unpublished empirical techniques used in industry. There was little theoretical support or analysis. The first academic paper on Iterative Learning Control written in English was published by Arimoto and colleagues [7] under the title of a *"betterment process"* and has received considerable attention since that time. Readers wishing to read alternative and more recent motivations for Iterative Control problems can find additional material in survey papers such as Longman [29], Ahn et al. [1], Owens and Hatonen [39], Owens and Daley [38], Bristow et al. [12], Rogers et al. [42], Wang et al. [44], Xu [47], and Li et al. [27] and published texts including the first in the area Moore [32] together with Bien and Xu [10], Chen and Wen [14], Norrlof [33], Ahn et al. [2], Xu et al. [48], Rogers et al. [43], and the companion book Owens [37]. The range of applications is growing with examples to be found in Gunnarsson and Norrlof [24], Freeman et al. [19], Freeman et al. [20], Daley et al. [16], Ratcliffe et al. [40], Rogers et al. [42], Barton and Alleyne [9], Johansen et al. [25], Wu et al. [46], Blanken et al. [11], Wang et al. [45], and elsewhere.

Readers may also find it useful to refresh their knowledge of classical control techniques using texts such as Goodwin et al. [22], Gopal [23], Landau and Zito [26], and Franklin [18]. The extensions to state-space and MIMO (multivariable) control systems can be found in a number of texts including Albertos and Sala [3], Friedland [21], Owens [35], and Maciejowski [31]. The well-established ideas of optimal control seen in texts such as Athans and Falb [8], Anderson and Moore [4], Anderson and Moore [5], and Chen and Francis [13] are central to what follows. An undergraduate-level introduction can be found in Owens [36]. In particular, the geometric approach used in Luenberger [30] is particularly suited to the optimization content of this text.

The ideas are related to those used in multipass systems theory which, historically, arose out of the pioneering work of Edwards and Owens and can be explored in some detail in Edwards and Owens [17] with related work on the mathematical analysis of repetitive dynamical systems in Rogers et al. [41]. The main preoccupation of this text is linear systems but many of the ideas apply to nonlinear systems with little conceptual change but with considerable changes needed in detail and methodology. The computational load increases greatly and problems such as convergence to local minima cannot be excluded. The overall nature of the iterative paradigm remains unchanged however.

Finally, readers with a mathematical interest might find it useful to relate the material to the linear and nonlinear iterative techniques found in the mathematical texts such as Collatz [15] and Ortega and Rheinboldt [34] where the general ideas of fixed point or Newton-like iteration have potential application to Iterative Control (as illustrated by many nonlinear Iterative Control methods in Xu [47] and Lin et al. [28]).

References

1. Ahn HS, Chen Y, Moore KL (2007) Iterative learning control: brief survey and categorization. IEEE Trans Syst Man Cybern Part C Appl Rev 37(6):1099–1121
2. Ahn HS, Moore KL, Chen Y (2007) Iterative learning control: robustness and monotonic convergence for interval systems. Series on communications and control engineering. Springer
3. Albertos P, Sala A (2004) Multivariable control systems: an engineering approach. Springer
4. Anderson BDO, Moore JB (1971) Linear optimal control. Prentice-Hall, New Jersey
5. Anderson BDO, Moore JB (1989) Optimal control: linear quadratic methods. Prentice-Hall, New Jersey
6. Anthonis J, Vaes D, Engelen K, Ramon H, Swevers J (2007) Feedback approach for reproduction of field measurements on a hydraulic four poster. Biosys Eng 96(4):435–445
7. Arimoto S, Kawamura S, Miyazaki F (1984) Bettering operation of robots by learning. J Robot Syst 1(2):123–140
8. Athans M, Falb PL (1966) Optimal control. McGraw Hill
9. Barton KL, Alleyne AG (2011) A norm optimal approach to time-varying ILC with application to a multi-axis robotic testbed. IEEE Trans Control Syst Technol 19(1):166–180
10. Bien Z, Xu JX (1998) Iterative learning control: analysis, design, integration and applications. Springer, US
11. Blanken L, Koekebakker S, Oomen T (2020) Data-driven feedforward tuning using non-causal rational basis functions: with application to an industrial flatbed printer. Mechatronics 71(102):424
12. Bristow D, Tharayil M, Alleyne AG (2006) A survey of iterative learning control: a learning-based method for high-performance tracking control. IEEE Control Syst Mag 26(3):96–114
13. Chen T, Francis B (1995) Optimal sampled-data control systems. Springer, London
14. Chen Y, Wen C (1999) Iterative Learning Control: Convergence, Robustness and Applications. Lecture Notes in Control and Information Sciences. Springer, London
15. Collatz L (1966) Functional analyis and numerical mathematics. Academic Press, New York and London
16. Daley S, Owens DH, Hatonen J (2007) Application of optimal iterative learning control to the dynamic testing of mechanical structures. Proc Inst Mech Eng Part I J Syst Control Eng 221(2):211–222
17. Edwards JB, Owens D (1982) Analysis and control of multipass processes. Research Studies Press, Wiley
18. Franklin G (1997) Digital control of dynamic systems. Prentice-Hall
19. Freeman CT, Hughes AM, Burridge JH, Chappell PH, Lewin PL, Rogers E (2009) Iterative learning control of FES applied to the upper extremity for rehabilitation. Control Eng Pract 17(3):368–381
20. Freeman CT, Rogers E, Hughes AM, Burridge JH, Meadmore KL (2012) Iterative learning control in healthcare: electrical stimulation and robotic-assisted upper-limb stroke rehabilitation. IEEE Control Syst Mag 32(1):18–43
21. Friedland B (1986) Control systems design: an introduction to state space methods. Dover, New York
22. Goodwin G, Graebe S, Salgado M (2000) Control systems design. Prentice-Hall
23. Gopal M (2002) Control systems: principles and design. Tata McGraw-Hill Education
24. Gunnarsson S, Norrlof M (2001) On the design of ILC algorithms using optimization. Automatica 37(12):2011–2016
25. Johansen SV, Jensen MR, Chu B, Bendtsen JD, Mogensen J, Rogers E (2019) Broiler FCR optimization using norm optimal terminal iterative learning control. IEEE Trans Control Syst Technol 29(2):580–592
26. Landau I, Zito G (2006) Digital control systems: design, identification and implementation, Springer, London
27. Li Y, Chen Y, Ahn HS, Tian G (2013) A survey on fractional order iterative learning. J Optim Theory Appl 156:127–140

28. Lin T, Owens DH, Hatonen J (2006) Newton method based iterative learning control for discrete nonlinear systems. Int J Control 79(10):1263–1276
29. Longman R (2000) Iterative learning control and repetitive control for engineering practice. Int J Control 73(10):930–954
30. Luenberger DG (1969) Optimization by vector space methods. Wiley, New York, Chichester
31. Maciejowski JM (1989) Multivariable feedback design. Addison-Wesley
32. Moore KL (1993) Iterative learning control for deterministic systems. Springer, London
33. Norrlof M (2000) Iterative learning control: analysis, design and experiments. Linkoping studies in science and technology. Dissertation No. 653
34. Ortega J, Rheinboldt W (1970) Functional analyis and numerical mathematics. Academic Press, London
35. Owens DH (1978) Feedback and multivariable systems. Peter Peregrinus, Stevenage
36. Owens DH (1981) Multivariable and optimal systems. Academic Press, London
37. Owens DH (2015) Iterative learning control—an optimization paradigm. Springer
38. Owens DH, Daley S (2008) Iterative learning control-monotonicity and optimization. Int J Appl Math Comput Sci 18(3):279–293
39. Owens DH, Hatonen J (2005) Iterative learning control-an optimization paradigm. Annu Rev Control 29(1):57–70
40. Ratcliffe JD, Lewin PL, Rogers E, Hatonen JJ, Owens DH (2006) Norm-optimal iterative learning control applied to gantry robots for automation applications. IEEE Trans Rob 22(6):1303–1307
41. Rogers E, Galkowski K, Owens DH (2007) Control systems theory and applications for linear repetitive processes. Lecture notes in control and information sciences. Springer No, p 349
42. Rogers E, Owens D, Werner H, Freeman C, Lewin P, Kirchhoff S, Schmidt C, Lichtenberg G (2010) Norm-optimal iterative learning control with application to problems in acceleration-based free electron lasers and rehabilitation robotics. Eur J Control 5:496–521
43. Rogers E, Chu B, Freeman C, Lewin P (2023) Iterative learning control algorithms and experimental benchmarking. John Wiley & Sons
44. Wang Y, Gao F, Doyle FJ III (2009) Survey on iterative learning control repetitive control and run-to run control. J Process Control 19:1589–1600
45. Wang Z, Pannier CP, Barton K, Hoelzle DJ (2018) Application of robust monotonically convergent spatial iterative learning control to microscale additive manufacturing. Mechatronics 56:157–165
46. Wu RB, Chu B, Owens DH, Rabitz H (2018) Data-driven gradient algorithm for high-precision quantum control. Phys Rev A 97(4):042122
47. Xu JX (2011) A survey on iterative learning control for nonlinear systems. Int J Control 84(7):1275–1294
48. Xu JX, Panda S, Lee TH (2008) Real time iterative learning control: design and applications. Series on advances in industrial control. Springer

Chapter 2
State-Space Models and Optimization

In applications of any form of control, there is a system to be controlled, a means of influencing its behaviour (system *input signals*), measurements of performance outcomes (system *output signals*) plus, in some cases, auxiliary measurements made for purposes of monitoring, diagnostics, or enhancement of control signals. The *system* to be controlled is sometimes called the *plant* to be controlled. The construction of control systems for a given plant inevitably required the addition, via feedback or some other form of interconnection, of designed systems called *control systems* or, more simply, *controllers*. Controllers process current and available past data as a means of constructing current input values. The computational process is designed to achieve the desired performance and is often called the *control algorithm*. The general aim of control systems analysis and design is to ensure that the composite, interconnected system consisting of the combined plant plus controller has a behaviour that is acceptable to the application and achieves defined performance objectives to a satisfactory standard.

To find systematic approaches to successful controller design, *mathematical models* of plant behaviour have played a central role. Such models not only provide understanding, but also permit the investigation of behaviours that financial or safety considerations might preclude in experiments. Models can take many forms, the two most significant being the possession of the properties of *linearity* or *nonlinearity*. Linear systems have dominated the research and development of control design techniques and insights into control possibilities and difficulties. The mathematics of linear systems is more fully understood and has a structure that gives engineers both computational and conceptual links between plant behaviour and the structure and parameters appearing in the plant model, e.g., the links between eigenvalues/poles and stability, frequency domain analysis and parameter choice using root-locus methodologies, and transfer function matrix/frequency domain analysis.

The systems considered in this text will be almost exclusively *linear dynamical systems* with behaviours described by differential or difference equations but the presentation does not preclude the application to more complex systems containing,

B. Chu and D. H. Owens, *Optimal Iterative Learning Control*, Advances in Industrial Control, https://doi.org/10.1007/978-3-031-80236-2_2

for example, transport delays. The text makes use of matrix and operator notation in the description and construction of models and controllers. The construction of the actual relationships needed to build the hardware and software used in the physical control system will be based on the challenge of converting operator relationships into feasible, model and measurement-based computations.

This text assumes an underlying representation of plant dynamics by *linear state-space models* and, from time to time, the related *transfer function matrix* representations. These assumptions make possible a level of detailed analysis that links the work closely to engineering practice and reveals much of the detail needed for computational studies and the properties that add confidence that successful applications are, indeed, possible.

In general terms, state-space models describe system dynamics by mathematical relationships between inputs signals u, output signals y, and a *system state* x representing the internal (possibly unmeasurable) variables needed to fully describe the dynamical behaviour. The two most familiar forms of relationship are those describing continuous-time state-space models and those describing discrete-time (sampled-data) state-space models.

Consider a dynamical system with m measured outputs $y_1(t), y_2(t), \cdots, y_m(t)$ influenced by ℓ input signals $u_1(t), u_2(t), \cdots, u_\ell(t)$. The symbol t is used to represent a measure of time. Assume also that a full description of the dynamics can be obtained using n *state variables* $x_1(t), x_2(t), \cdots, x_n(t)$. If one or both of m and ℓ are greater than one, then the system is said to be *multivariable, multi-input–multi-output, multi-loop or multi-channel*, or MIMO for short. If $m = \ell = 1$, the system is said to be *single-input–single-output* (or SISO for short).

One concept of note is that of *causality*. That is, the state of a system at a time t is a consequence of its past history and current influences only (defined as relevant data and variables on the interval $t' \leq t$) and is independent of what may happen or be done to it at future times $t' > t$. This concept is important in Iterative Learning Control where formal mathematical descriptions, which appear to be non-causal must be converted into a causal form for control implementation purposes.

2.1 Models of Continuous State-Space Systems

A general *nonlinear* continuous-time, state-space model can be written in the form

$$\begin{aligned} \dot{x}(t) &= f(x(t), u(t), t), \quad x(0) = x_0 \\ y(t) &= h(x(t), u(t), t), \quad 0 \leq t \leq T \end{aligned} \tag{2.1}$$

where

1. T is the (finite or infinite) time interval of interest.
2. The column matrices

$$y(t) = \begin{bmatrix} y_1(t) \\ y_2(t) \\ \vdots \\ y_m(t) \end{bmatrix}, \quad u(t) = \begin{bmatrix} u_1(t) \\ u_2(t) \\ \vdots \\ u_\ell(t) \end{bmatrix} \quad and \quad x(t) = \begin{bmatrix} x_1(t) \\ x_2(t) \\ \vdots \\ x_n(t) \end{bmatrix} \tag{2.2}$$

are the $m \times 1$ *output vector*, $\ell \times 1$ *input vector*, and $n \times 1$ *state vector*, respectively. The shorthand for these statements is as follows:

$$y(t) \in \mathcal{R}^m\,, \quad u(t) \in \mathcal{R}^\ell \;\; \& \;\; x(t) \in \mathcal{R}^n \quad (resp.). \tag{2.3}$$

3. If H is a $p \times q$ matrix H with elements that depend on time and letting $H_{ij}(t)$ denote the element of H on row i and column j, then the symbols "$\dot{H}(t)$" and $\frac{dH(t)}{dt}$ both denote the $p \times q$ matrix with (i, j) element equal to $\frac{dH_{ij}(t)}{dt}$, i.e., the matrix of derivatives of the elements of H. The jth element of $\dot{x}(t)$ is hence $\frac{dx_j(t)}{dt}$.
4. The $n \times 1$ vector x_0 is the column matrix of *initial conditions* $x_1(0), x_2(0), \cdots, x_n(0)$ for the state variables.
5. The vector valued function $f(x, u, t)$ (respectively, $h(x, u, t)$) describes the functional dependency of the state derivatives (respectively, outputs) at time t on the values of the state vector $x(t)$ at time t, the input vector $u(t)$ at time t, and the value of t itself.
6. If both $f(x, u, t)$ and $h(x, u, t)$ have no explicit dependence on t, the system is said to be *time-invariant*. Otherwise it is said to be *time-varying*.

If both f and g are linear in x and u, the state-space model is said to be linear. The linear model used to describe this situation takes the general form

$$\begin{aligned} \dot{x}(t) &= A(t)x(t) + B(t)u(t), \quad x(0) = x_0 \\ y(t) &= C(t)x(t) + D(t)u(t), \quad 0 \leq t \leq T \end{aligned} \tag{2.4}$$

where the matrices A, B, C, D are of dimension $n \times n, n \times \ell, m \times n$, and $m \times \ell$, respectively. The following terminology and notation will be adopted:

1. If $D(t)$ is identically zero (written $D(t) \equiv 0$), the system is said to be *strictly proper*. Otherwise it is *proper*.
2. If any of the matrices A, B, C, D vary with time, then the system is *time-varying*. Otherwise it is said to be *time-invariant*.
3. As the linear state-space model can be defined by the four matrices A, B, C, D, the system will be referred to as the system $S(A, B, C, D)$. If it is strictly proper, then the notation $S(A, B, C, 0)$ is simplified to $S(A, B, C)$.

2.1.1 Solution of the State Equations

For a given initial state x_0 and input signal $u(t)$, the solutions ($x(t)$ and $y(t)$) of the state equations can be obtained using computational simulation software. For the purposes of theory and algorithm development, however, a theoretical description of the solution is extremely useful. This is only possible for linear state-space models and takes the form

$$\begin{aligned} x(t) &= \Phi(t,0)x_0 + \int_0^t \Phi(t,t')B(t')u(t')dt', \quad 0 \le t \le T, \\ with \; y(t) &= C(t)x(t) + D(t)u(t), \quad 0 \le t \le T. \end{aligned} \tag{2.5}$$

Here

1. If $H(t)$ is a $p \times q$ time-varying matrix, then $\int_a^b H(t)dt$ denotes the $p \times q$ matrix with (i,j)th element $\int_a^b H_{ij}(t)dt$.
2. The $n \times n$ *state transition matrix* $\Phi(t,t')$ is the solution of the family of differential equations on the range $0 \le t \le T$ and $0 \le t' \le T$,

$$\frac{\partial \Phi(t,t')}{\partial t} = A(t)\Phi(t,t'), \quad with \; \Phi(t',t') = I_n, \tag{2.6}$$

where I_n denotes the $n \times n$ unit matrix. A simple formula for $\Phi(t,t')$ is obtained by solving the equation for $t' = 0$ and then setting

$$\Phi(t,t') = \Phi(t,0)(\Phi(t',0))^{-1}. \tag{2.7}$$

3. The state transition matrix hence satisfies the relation

$$\Phi(t',t)\Phi(t,t') = I_n \quad so \; that \quad \Phi(t',t) = (\Phi(t,t'))^{-1}. \tag{2.8}$$

In particular,

$$\frac{\partial \Phi(t',t)}{\partial t}\Phi(t,t') + \Phi(t',t)\frac{\partial \Phi(t,t')}{\partial t} = 0 \tag{2.9}$$

so that

$$\frac{\partial \Phi(t',t)}{\partial t} = -\Phi(t',t)A(t) \quad and \; hence \quad \frac{\partial \Phi^T(t',t)}{\partial t} = -A^T(t)\Phi^T(t',t). \tag{2.10}$$

4. If $A(t)$ is a constant matrix A, then

$$\Phi(t,t') = e^{A(t-t')} \tag{2.11}$$

where e^M is the matrix exponential of a square matrix M, defined formally as the absolutely convergent power series,

$$e^M = \sum_{j=0}^{\infty} \frac{1}{j!} M^j. \tag{2.12}$$

5. If $S(A, B, C, D)$ is linear and time-invariant, then it is asymptotically stable if, and only if, all eigenvalues of A have strictly negative real parts. That is, all eigenvalues take values in the open left half of the complex plane.

2.1.2 *The Convolution Operator and Impulse Response*

Using the above solution in the form, for $0 \le t \le T$,

$$\begin{aligned} y(t) &= C(t)\Phi(t,0)x_0 + \textstyle\int_0^t C(t)\Phi(t,t')B(t')u(t')dt' + D(t)u(t), \\ i.e., \quad y(t) &= C(t)\Phi(t,0)x_0 + \underbrace{\int_0^t H(t,t')u(t')dt'} + D(t)u(t), \end{aligned} \tag{2.13}$$

with the obvious identification of $H(t, t')$. In the linear time-invariant, strictly proper case

$$H(t, t') = Ce^{A(t-t')}B = H(t - t') \tag{2.14}$$

so that the underbraced term is a convolution integral. For linear, time-invariant, strictly proper SISO systems, $H(t)$ is the familiar impulse response function. For linear, time-invariant, strictly proper MIMO systems, the $m \times \ell$ matrix $H(t) = Ce^{At}B$ is termed the *impulse response matrix* of $S(A, B, C)$.

2.1.3 *The System as an Operator Between Function Spaces*

Analysis and algorithm developments will benefit from the use of a simplified notation. To this end, it will be useful to consider the solution $y(t)$ on the interval $0 \le t \le T$ as the response to the signal $u(t)$ on the interval $0 \le t \le T$ when the initial condition $x_0 = 0$ plus the response from the initial condition x_0 with $u(t) \equiv 0$ on $0 \le t \le T$. Using notation reminiscent of matrix algebra or transfer function methodology, the solution can be written as an operational relationship

$$y = Gu + d = z + d \tag{2.15}$$

between signals $u \in \mathcal{U}$ and $y \in \mathcal{Y}$ in suitable function spaces $\mathcal{U}$ and $\mathcal{Y}$. Here, $d \in \mathcal{Y}$ represents the function $d(t) = C(t)\Phi(t, 0)x_0,\ 0 \le t \le T$, and $G : \mathcal{U} \mapsto \mathcal{Y}$ is a linear operator mapping $u(t),\ 0 \le t \le T$, into $z(t)$ where

$$z(t) = \int_0^t H(t,t')u(t')dt' + D(t)u(t), \quad 0 \le t \le T. \tag{2.16}$$

The choice of function space is non-unique. Consider the case of T finite and the choices of Cartesian product Hilbert spaces

$$\begin{aligned} \mathcal{U} &= L_2^{\ell}[0,T] = \underbrace{L_2[0,T] \times L_2[0,T] \times \cdots \times L_2[0,T]}_{\ell - times} \\ \mathcal{Y} &= L_2^{m}[0,T] = \underbrace{L_2[0,T] \times L_2[0,T] \times \cdots \times L_2[0,T]}_{m - times} \end{aligned} \tag{2.17}$$

where $L_2[0,T]$ denotes the linear vector space of functions g on $[0,T]$ such that $\int_0^T g^2(t)dt$ is well-defined and finite. $\mathcal{U}$ and $\mathcal{Y}$ are *Hilbert spaces* with *inner products* defined by symmetric, positive definite, time-varying matrices $R(t)$ and $Q(t)$ as follows:

$$\langle u, v\rangle_{\mathcal{U}} = \int_0^T u^T(t)R(t)v(t)dt \quad \& \quad \langle y, w\rangle_{\mathcal{Y}} = \int_0^T y^T(t)Q(t)w(t)dt. \tag{2.18}$$

Note 2.1 For notational simplicity, the inner products $\langle u, v\rangle_{\mathcal{U}}$ and $\langle y, w\rangle_{\mathcal{Y}}$ may be denoted by $\langle u, v\rangle_R$ and $\langle y, w\rangle_Q$ to identify the associated matrices.

Note 2.2 The Cartesian product $L_2^m[0,T]$ can be seen as a vector space of column vectors whose elements are functions in $L_2[0,T]$.

2.1.4 *The Adjoint Operator, Inner Products, and Norms*

Using the ideas outlined in Sect. 2.15, if G is a linear operator mapping a Hilbert space $\mathcal{U}$ into a Hilbert space $\mathcal{Y}$, the associated *adjoint operator* G^* maps $\mathcal{Y}$ into $\mathcal{U}$ and satisfies the inner product identity for all $w \in \mathcal{Y}$ and $u \in \mathcal{U}$

$$\langle w, Gu\rangle_{\mathcal{Y}} = \langle G^* w, u\rangle_{\mathcal{U}}. \tag{2.19}$$

This operator plays an important role in the theoretical analysis of Optimal Iterative Learning Control analysis and algorithm development. For linear, time-invariant, state-space systems defined above, and inner products,

$$\langle u, v\rangle_{\mathcal{U}} = \int_0^T u^T(t)R(t)v(t)dt \quad \& \quad \langle y, w\rangle_{\mathcal{Y}} = \int_0^T y^T(t)Q(t)w(t)dt, \tag{2.20}$$

the adjoint G^* is defined by the relationship $v = G^* w$, which can be written in more detail as the computation

$$\begin{aligned}\dot{p}(t) &= -A^T(t)p(t) - C^T(t)Q(t)w(t), \qquad p(T) = 0, \\ v(t) &= R^{-1}(t)B^T(t)p(t) + R^{-1}(t)D^T(t)Q(t)w(t).\end{aligned} \tag{2.21}$$

This can be derived using manipulation of the integrals defining the inner product $\langle w, Gu\rangle_{\mathcal{Y}}$ and is left as an exercise for the reader. Note that the differential equations have no initial condition. Instead, the value of the *costate* $p(t)$ is specified at the final time.

Remark 2.1 $R(t)$ must be positive definite and symmetric on $[0, T]$. The same normally applies to $Q(t)$ although situations where it is symmetric and positive semi-definite can be tolerated if the space $\mathcal{Y}$ (and the control problem considered) consists of L_2 functions defined on a number of subintervals of $[0, T]$ and undefined elsewhere (and hence ignored).

Finally, associated with each inner product is a *norm*. More precisely, the norms in $\mathcal{U}$ and $\mathcal{Y}$ are obtained using the definitions

$$\|u\|_{\mathcal{U}} = \langle u, u\rangle_{\mathcal{U}}^{\frac{1}{2}} \;\; \& \;\; \|y\|_{\mathcal{U}} = \langle y, y\rangle_{\mathcal{Y}}^{\frac{1}{2}}. \tag{2.22}$$

For the continuous systems considered,

$$\|u\|_{\mathcal{U}} = \sqrt{\int_0^T u^T(t)R(t)u(t)dt} \;\; \& \;\; \|y\|_{\mathcal{Y}} = \sqrt{\int_0^T y^T(t)Q(t)y(t)dt}. \tag{2.23}$$

An early insight into the link between these ideas and Optimal Control can be found in the expression

$$J = \|y\|_{\mathcal{Y}}^2 + \|u\|_{\mathcal{U}}^2 = \int_0^T \left(y^T(t)Q(t)y(t) + u^T(t)R(t)u(t)\right) dt \tag{2.24}$$

which is the familiar performance index/objective function for the Linear Quadratic Optimal Regulator Problem.

2.2 Laplace Transforms

Laplace Transform methodologies have been used widely in control systems analysis and design. The definition of the Laplace Transform of a scalar function $f(t)$ is most easily presented in the case of an infinite time interval $T = \infty$. More precisely, the Laplace Transform of a function $f(t)$, $0 \le t < \infty$ is a function of the complex variable s defined by

$$\mathcal{L}[f(t)] = \int_0^\infty e^{-st} f(t)dt, \tag{2.25}$$

for all complex number s such that the integral exists. A sufficient condition is that f is *exponentially bounded*, i.e., there exist real numbers $M \geq 0$ and λ such that

$$|f(t)| \leq Me^{\lambda t} \quad for \;\; all \;\; t \geq 0. \tag{2.26}$$

The integral then has a finite value for all s with real part strictly greater than λ. Although only valid for values of s such that the integral converges, the functions can be extended by analytic continuation to all points of the complex plane other than singularities. For this reason the issues of integral convergence play only an occasional role in practical use of the ideas.

Useful properties of the Laplace Transform used in the text are summarized below for any function $f(t)$ defined on $0 \leq t < \infty$ with $f(0)$ written as f_0:

$$\begin{array}{rclcrcl} \mathcal{L}[\frac{df(t)}{dt}] &=& s\mathcal{L}[f(t)] - f_0, & & \mathcal{L}[t^k f(t)] &=& (-1)^k \frac{d^k}{ds^k}\left(\mathcal{L}[f(t)]\right), \\ \mathcal{L}[e^{\alpha t}] &=& \frac{1}{s-\alpha} & and & \mathcal{L}[\frac{t^k}{k!}e^{\alpha t}] &=& \frac{1}{(s-\alpha)^{k+1}}. \end{array} \tag{2.27}$$

As a simplification to the notation, the Laplace Transform $\mathcal{L}[f(t)]$ will often be denoted $f(s)$. This clearly could lead to some confusion. However, despite this possibility, the notation will be used as the differences between $f(t)$ and $f(s)$ will be clear to the reader from the context. With this convention, the *Inverse Laplace Transform* is simply the inverse process and is denoted by $\mathcal{L}^{-1}$, for example,

$$f(t) = \mathcal{L}^{-1}[f(s)]. \tag{2.28}$$

The extension of the ideas to matrix-valued functions of time is easily achieved. For a $p \times q$ time-dependent matrix $H(t)$ the Laplace Transform $\mathcal{L}[H(t)]$ is denoted by $H(s)$ which has (i, j) element equal to the Laplace Transform of $H_{ij}(t)$.

2.3 Transfer Function Matrices, Poles, Zeros and Relative Degree

For linear, time-invariant, state-space models, taking Laplace Transforms of the state equations gives

$$sx(s) - x_0 = Ax(s) + Bu(s), \quad y(s) = Cx(s) + Du(s) \tag{2.29}$$

from which simple matrix algebra to eliminate $x(s)$ gives

$$\begin{array}{ll} x(s) = (sI - A)^{-1}\left(Bu(s) + x_0\right) & and \;\; hence \\ y(s) = G(s)u(s) + C(sI_n - A)^{-1}x_0 & \end{array} \tag{2.30}$$

where the system $m \times \ell$ *Transfer Function Matrix* $G(s)$ is defined by

$$\underbrace{G(s) = C(sI_n - A)^{-1}B + D}_{Trnasfer\ Function\ Matrix}. \tag{2.31}$$

It has elements that are rational polynomials in the complex variable s. If $m = \ell = 1$, $G(s)$ is the familiar transfer function used to great effect in the analysis and design of SISO feedback controllers.

Poles and zeros play an important role in classifying system's dynamics in design. These two concepts are defined as follows:

1. The denominator polynomial of $G(s)$ is the *characteristic polynomial* $det[sI_n - A]$ of A. As a consequence, the *poles* of $G(s)$ and their multiplicities are precisely the set of eigenvalues of the matrix A and their multiplicities.
2. The *zeros* of $G(s)$ are defined by rank conditions on a partitioned matrix dependent on (A, B, C, D). More precisely, the complex number s is a zero of $S(A, B, C, D)$ if, and only if,

$$rank \begin{bmatrix} sI_n - A & -B \\ C & D \end{bmatrix} < n + \min\{m, \ell\}. \tag{2.32}$$

These zeros are the multivariable analogue of familiar transfer function zeros. If all zeros lie in the open left half complex plane then the system is said to be *minimum phase*. Otherwise it is said to be *non-minimum phase*. Non-minimum phase systems are known to lead to control problems and performance limitations in the design of feedback control systems. Similar problems will be seen to be the case for inverse model-based, gradient, and Norm Optimal Iterative Control methodologies described in this text.

Note that if $m = \ell$, the zero definition reduces to the computation of the roots of a determinant, namely, the solutions of

$$\begin{vmatrix} sI_n - A & -B \\ C & D \end{vmatrix} = 0. \tag{2.33}$$

Note also that, when $m \neq \ell$, the generic situation is that the system has no zeros. If $m = \ell$ then the use of *Schur's Formula* yields

$$\begin{vmatrix} sI_n - A & -B \\ C & D \end{vmatrix} \equiv |sI_n - A| \ \det[G(s)]. \tag{2.34}$$

When $m = \ell = 1$, the zeros are precisely those of the system transfer function.

Other useful properties and parameters associated with $G(s)$ can be constructed as follows:

1. By using the power series expansion of $(sI_n - A)^{-1}$ to obtain the expression

$$G(s) = D + s^{-1}CB + s^{-2}CAB + \cdots = D + \sum_{k=1}^{\infty} s^{-k}CA^{k-1}B \tag{2.35}$$

which is convergent for $|s| > r(A)$, where $r(A)$ denotes the *spectral radius* of A, i.e., the maximum modulus of the eigenvalues of A. The matrices $CB, CAB, CA^2B, \cdots$ are called the system *Markov Parameter Matrices* or, for simplicity, the system Markov parameters.
2. The *relative degree* k^* *of* $G(s)$ is defined as the unique integer such that $G_\infty = \lim_{s\to\infty} s^k G(s)$ exists and is non-zero. From the series expansion, it is clear that either $k^* = 0$ when $D \neq 0$ or, when $D = 0$, k^* is the smallest integer k for which $CA^{k-1}B$ is non-zero. Clearly the system is strictly proper if, and only if, $k^* > 0$. In the SISO case of $m = \ell = 1$, the relative degree is precisely equal to the number of poles of $G(s)$ minus the number of zeros of $G(s)$.
3. Clearly $G_\infty \neq 0$ from the definition but it does not necessarily have full rank as a matrix. This complication requires the development of useful practical refinements of the idea of relative degree.

2.4 The System Frequency Response

The frequency response function of the linear, time-invariant, system $S(A, B, C, D)$ is defined to be the (matrix-valued) function $G(i\omega)$ where i is the familiar "square root of -1" and ω is a real frequency with units radians/unit time. This function is a foundation of classical feedback control systems design.

It can also be used to describe the response of an asymptotically stable system from zero initial conditions $x_0 = 0$ to sinusoidal input signals. More generally, if the data (A, B, C, D) contains only real numbers and α is a constant, possibly complex, $\ell \times 1$ vector, define inputs $u_c(t)$ and $u_s(t)$ to be the real and imaginary parts of the complex input $u(t) = \alpha e^{st}$ where s is any complex number. The resultant output responses from zero initial conditions are, respectively, equal to the real and imaginary parts of the response $y(t)$ from zero initial conditions to the complex input $u(t)$. It is left as an exercise for the reader to verify that

$$y(t) = G(s)u(t) - C(sI_n - A)^{-1}e^{At}B\alpha \tag{2.36}$$

which provides a direct link between the transfer function matrix $G(s)$ and well-defined time domain behaviours. In particular, if $s = i\omega$ (with ω a real frequency), the stability assumption indicates that, for all large enough values of time t, the steady-state oscillation $y(t) \approx G(s)u(t) = G(i\omega)\alpha e^{i\omega t}$ is an accurate description of steady-state oscillatory output behaviour. This is because e^{At} becomes infinitesimally small as t becomes large.

2.5 Discrete-Time, Sampled-Data State-Space Models

In many applications of control systems engineering, the basis of measurement and control is *sampling* of continuous signals with simultaneous changing of control input values at sample times $0,\ h,\ 2h,\ 3h,\ \cdots$. This situation is commonly described as uniform synchronous sampling with *sample interval* h and *sampling rate* h^{-1}. In such circumstances, the sampled output signal is a *time series* $y(0),\ y(h),\ y(2h), \cdots$ and the input signal takes the form of a *piece-wise constant* input

$$u(t) = u(kh),\quad for\ \ kh \le t < (k+1)h,\quad and\ \ 0 \le k \le N-1 \tag{2.37}$$

where $T = Nh$ is the time interval of interest to the control study. The corresponding input time series is hence $u(0), u(h), u(2h), \cdots$. For notational simplicity, the dependence on h will be suppressed and the input and output time series will be denoted $u(0), u(1), u(2), u(3), \cdots$ and $y(0), y(1), y(2), y(3), \cdots$. That is, they will be indexed by the sample number only, the existence of the invisible h being understood.

2.5.1 *State-Space Models as Difference Equations*

A general *nonlinear* discrete-time, state-space model can be written in the following form indexed by the integer sample time index $t = 0, 1, 2, 3, \cdots$

$$\begin{aligned} x(t+1) &= f_d(x(t), u(t), t),\ \ 0 \le t \le N-1,\ \ x(0) = x_0 \\ y(t) &= h_d(x(t), u(t), t),\ \ 0 \le t \le N \end{aligned} \tag{2.38}$$

where

1. N describes the (finite or infinite) time interval of interest.
2. The column matrices $y(t), u(t), x(t)$ define output vectors, input vectors, and state vectors as for the continuous-time case.
3. The $n \times 1$ vector x_0 is the column matrix of *initial conditions* $x_1(0), x_2(0), \cdots, x_n(0)$ for the state variables $x_1(t), x_2(t), \cdots, x_n(t)$.
4. If both $f_d(x, u, t)$ and $h_d(x, u, t)$ have no explicit dependence on t, the system is said to be *time-invariant*. Otherwise it is said to be *time-varying*.

If both f_d and g_d are linear in x and u, the discrete state-space model is said to be linear. A linear model takes the general form

$$\begin{aligned} x(t+1) &= \Phi_d(t)x(t) + \Delta_d(t)u(t), \quad 0 \le t \le N-1, \quad x(0) = x_0 \\ y(t) &= C(t)x(t) + D(t)u(t), \quad 0 \le t \le N, \end{aligned} \tag{2.39}$$

where the possibly sample-dependent, matrices Φ_d, Δ_d, C, D are of dimension $n \times n$, $n \times \ell$, $m \times n$, and $m \times \ell$, respectively. The following terminology is adopted from the continuous-time case:

1. If $D(t)$ is identically zero (written $D(t) \equiv 0$), the system is said to be *strictly proper*. Otherwise it is said to be *proper*.
2. If any of the matrices Φ_d, Δ_d, C, D vary with sample number, then the system is said to be time-varying. Otherwise it is said to be time-invariant.
3. As the linear state-space model can be defined by the four matrices Φ_d, Δ_d, C, D, the system will be referred to as the discrete system $S(\Phi_d, \Delta_d, C, D)$. If it is strictly proper, then the notation is simplified to $S(\Phi_d, \Delta_d, C)$.

2.5.2 Discrete Models for Linear, Time-Invariant Continuous Systems

The discrete system is always linear and time-invariant if it is constructed from an underlying linear, time-invariant system $S(A, B, C, D)$. The constant matrices Φ_d and Δ_d can be computed, in this case, from the equations

$$\Phi_d = e^{Ah} \quad and \quad \Delta_d = \int_0^h e^{A't} B dt'. \tag{2.40}$$

If A is nonsingular, Δ_d can be expressed directly as

$$\Delta_d = A^{-1}\left(e^{Ah} - I_n\right) B. \tag{2.41}$$

Note 2.3 (Important Note) In the remainder of the text the notation used for discrete systems will vary to suite the circumstance and to provide simplicity of notation in the presentation. For example, the notation $S(A, B, C, D)$ will be used, on occasion, to denote a continuous system but also a discrete system with Φ_d equal to some matrix A and Δ_d equal to a matrix B. This should not cause any confusion in cases where any underlying continuous systems dynamics play no role in the discussions.

2.5.3 Solution of Linear, Discrete-Time State Equations

For a given initial state x_0 and input time series $u(t)$, the solution can be obtained using computational simulation software. For the purposes of theory and algorithm development, however, a theoretical description of the solution is extremely useful. This is only possible for linear state-space models and takes the form

$$
\begin{aligned}
x(t) &= \Phi(t-1,0)\Phi_d(0)x_0 + \textstyle\sum_{t'=0}^{t-1} \Phi(t-1,t')\Delta_d(t')u(t'), \quad t \geq 1, \\
y(t) &= C(t)x(t) + D(t)u(t), \quad t \geq 0, \\
\textit{so that} & \\
y(0) &= C(0)x_0 + D(0)u(0) \qquad \textit{and} \\
y(t) &= \underbrace{D(t)u(t) + \sum_{t'=0}^{t-1} C(t)\Phi(t-1,t')\Delta_d(t')u(t')}_{\text{“input dependent term”}} \\
&\quad + C(t)\Phi(t-1,0)\Phi_d(0)x_0, \quad t \geq 1.
\end{aligned}
\tag{2.42}
$$

Here

1. The *discrete state transition matrix* $\Phi(t, t')$ is the solution of the family of difference equations

$$\Phi(t,t') = \Phi_d(t)\Phi(t-1,t'), \quad \textit{with } \Phi(t',t') = I_n, \;\; 0 \leq t' < t. \tag{2.43}$$

2. If $\Phi_d(t)$ is a constant matrix Φ_d, then

$$\Phi(t,t') = \Phi_d^{t-t'}. \tag{2.44}$$

3. If $S(\Phi_d, \Delta_d, C, D)$ is linear and time-invariant, then

   a. the solution of the state equations is computed from the following *discrete convolution* equations, $t \geq 0$,

$$
\begin{aligned}
x(t) &= \Phi_d^t x_0 + \textstyle\sum_{t'=0}^{t-1} \Phi_d^{t-t'-1}\Delta_d u(t'), \\
y(t) &= Du(t) + \textstyle\sum_{t'=0}^{t-1} C\Phi_d^{t-t'}\Delta_d u(t') + C\Phi_d^t x_0,
\end{aligned}
\tag{2.45}
$$

   b. a linear time-invariant, discrete system is asymptotically stable if, and only if, all eigenvalues of the matrix Φ_d have modulus strictly less than unity. Equivalent statements are that all eigenvalues have values in the open unit circle in the complex plane and, as a consequence, for some $M \geq 1$ and $0 \leq \lambda < 1$,

$$\|\Phi_d^t\| \leq M\lambda^t, \quad t = 0, 1, 2, 3, \cdots, \tag{2.46}$$

   where $\|\cdot\|$ denotes any desired matrix norm.

2.5.4 The Discrete Convolution Operator and the Discrete Impulse Response Sequence

In a similar manner to the continuous-time case, the mapping between the input time series u and the output time series y on the time interval $0 \leq t \leq N$ has the operational form

$$y = Gu + d \tag{2.47}$$

where d represents the time series $C(0)x_0$ followed by $C(t)\Phi(t-1,0)\Phi_d(0)x_0,\ 1 \le t \le N$, and G represents the linear mapping that maps u into the times series $z(0) = D(0)u(0)$ followed by

$$z(t) = D(t)u(t) + \sum_{t'=0}^{t-1} C(t)\Phi(t-1,t')\Delta_d(t')u(t'), \quad 1 \le t \le N. \tag{2.48}$$

In the linear time-invariant case, the sequence $\{D, C\Delta_d, C\Phi_d\Delta_d, C\Phi_d^2\Delta_d, \cdots\}$ is termed the *discrete systems impulse response sequence*.

To formalize this representation, regard the time series $u \in \mathcal{U}$ and $y \in \mathcal{Y}$ as signals in suitable sequence spaces $\mathcal{U}$ and $\mathcal{Y}$. Here, $G : \mathcal{U} \mapsto \mathcal{Y}$ is a linear operator mapping $u(t),\ 0 \le t \le N$, into $z(t),\ 0 \le t \le N$. The choice of sequence space is non-unique. For the purposes of the current discussion, consider the case of N finite and the choices of Cartesian product Hilbert spaces

$$\mathcal{U} = \underbrace{\mathcal{R}^\ell \times \mathcal{R}^\ell \times \cdots \times \mathcal{R}^\ell}_{(N+1)-times} \quad \& \quad \mathcal{Y} = \underbrace{\mathcal{R}^m \times \mathcal{R}^m \times \cdots \times \mathcal{R}^m}_{(N+1)-times}. \tag{2.49}$$

Define inner products by symmetric positive definite matrices R_t and $Q_t,\ 0 \le t \le N$, using

$$\langle u, v\rangle_{\mathcal{U}} = \sum_{t=0}^{N} u^T(t)R_t v(t) \quad \& \quad \langle y, w\rangle_{\mathcal{Y}} = \sum_{t=0}^{N} y^T(t)Q_t w(t). \tag{2.50}$$

Note the following:

1. For notational simplicity, these inner products $\langle u, v\rangle_{\mathcal{U}}$ and $\langle y, w\rangle_{\mathcal{Y}}$ may be denoted by $\langle u, v\rangle_R$ and $\langle y, w\rangle_Q$ to identify the associated weighting matrices.
2. Time variation of the matrices puts different weights on the contributions of the signals at different times.
3. The simplest case is the case where the matrices are t-independent. In this case they take values R and Q at all times.

2.6 Discrete-Time Supervectors and the Matrix Model

Consider a linear, time-invariant, discrete-time state-space recursion on a discrete, finite, time interval $0 \le t \le N$ generating an output time series $\{y(t)\}_{t\ge0}$ from an input time series $\{u(t)\}_{t\ge0}$. The relationship between these times series can be represented by a single matrix equation using the idea of "stacking" or "lifting" data into output (respectively, input) *supervectors*. These are defined as follows:

$$y = \begin{bmatrix} y(0) \\ y(1) \\ y(2) \\ \vdots \\ y(N) \end{bmatrix} \in \mathcal{R}^{m(N+1)}, \quad u = \begin{bmatrix} u(0) \\ u(1) \\ u(2) \\ \vdots \\ u(N) \end{bmatrix} \in \mathcal{R}^{\ell(N+1)} \tag{2.51}$$

of dimension $m(N+1) \times 1$ (respectively, $\ell(N+1) \times 1$). Defining the block, lower triangular $m(N+1) \times \ell(N+1)$ matrix

$$G(\Phi_d, \Delta_d, C, D) = \begin{bmatrix} D & 0 & \cdots & 0 & 0 & 0 \\ C\Delta_d & D & \cdots & 0 & 0 & 0 \\ C\Phi_d\Delta_d & C\Delta_d & \cdots & 0 & 0 & 0 \\ \vdots & \vdots & & & \vdots & \vdots \\ C\Phi_d^{N-2}\Delta_d & C\Phi_d^{N-3}\Delta_d & \cdots & C\Delta_d & D & 0 \\ C\Phi_d^{N-1}\Delta_d & C\Phi_d^{N-2}\Delta_d & \cdots & C\Phi_d\Delta_d & C\Delta_d & D \end{bmatrix}, \tag{2.52}$$

and the vector

$$d(C, \Phi_d, x_0) = \begin{bmatrix} Cx_0 \\ C\Phi_d x_0 \\ C\Phi_d^2 x_0 \\ \vdots \\ C\Phi_d^N x_0 \end{bmatrix} \in \mathcal{R}^{m(N+1)} \tag{2.53}$$

then $d(C, \Phi_d, x_0)$ represents the effect of initial conditions whilst the $m(N+1) \times \ell(N+1)$ matrix $G(\Phi_d, \Delta_d, C, D)$ represents the state-space model as a linear (matrix) mapping between the supervectors of the input times series and output time series. The equations of motion now have the simple, linear matrix form, notationally identical to that of the continuous-time case (2.15),

$$y = Gu + d. \tag{2.54}$$

Note 2.4 For notational simplicity, the arguments in the matrix symbols will be suppressed if this leads to no confusion. Any possible confusion should be resolved by the context.

In principle, this matrix form releases the full power of matrix algebra for application to analysis and design of discrete dynamical systems on finite time intervals. It also places the model clearly into the framework of operator descriptions of system dynamics used later in the text. However, the matrices involved typically have very large dimension as the time series length $N+1$ is almost always long. As a consequence, the direct use of matrix algebra in computation may be limited in practice. Note however that the equation is exactly equivalent to the original state-space recursion equations so an output time series y described by $y = Gu + d$ is best computed using $S(\Phi_d, \Delta_d, C, D)$ and simulation. This matrix description does, however, have great theoretical value and possesses a number of other properties, some of which are summarized in the following sections.

2.7 Adjoints of Supervector Models

The matrix G in the supervector description $y = Gu$ of the discrete-time sampled dynamics of $S(\Phi_d, \Delta_d, C, D)$ can also be associated with an adjoint operator/matrix. The ideas use the same definitions as those introduced in Sect. 2.1.4. In the discrete system case let $\mathcal{U}$ be the space of $\ell(N+1) \times 1$ column matrices of inputs and $\mathcal{Y}$ be the space of $m(N+1) \times 1$ column matrices of outputs. The inner products in these spaces are taken to be, respectively,

$$\langle u, v \rangle_{\mathcal{U}} = u^T R_d v \quad \& \quad \langle y, w \rangle_{\mathcal{Y}} = y^T Q_d w, \tag{2.55}$$

where R_d and Q_d are symmetric and positive definite matrices of appropriate dimension. The identity

$$\langle v, Gu \rangle_{\mathcal{Y}} = v^T Q_d G u = (R_d^{-1} G^T Q_d v)^T R_d u = \langle G^* v, u \rangle_{\mathcal{U}} \tag{2.56}$$

leads to the conclusion that the adjoint is defined by the matrix equation

$$\underbrace{G^* = R_d^{-1} G^T Q_d}_{Matrix\ Adjoint} . \tag{2.57}$$

Note 2.5 The adjoint reduces to the familiar transpose of G if Q and R are identity/unit matrices. The use of non-unit Q and R changes the adjoint in a substantial but structured way that fits the needs of optimization theory very well.

The structures used most often in iterative learning control (ILC) are block diagonal choices for R_d and Q_d of the form,

$$Q_d = \begin{bmatrix} Q_0 & 0 & \cdots & 0 \\ 0 & Q_1 & \cdots & 0 \\ \vdots & & & \vdots \\ 0 & \cdots & 0 & Q_N \end{bmatrix}, \quad R_d = \begin{bmatrix} R_0 & 0 & \cdots & 0 \\ 0 & R_1 & \cdots & 0 \\ \vdots & & & \vdots \\ 0 & \cdots & 0 & R_N \end{bmatrix} \tag{2.58}$$

which give the following alternative expressions for the inner products:

$$\langle u, v \rangle_{\mathcal{U}} = u^T R_d v = \sum_{t=0}^{N} u(t) R_t v(t)$$

$$\langle y, w\rangle_{\mathcal{Y}} = y^T Q_d w \;=\; \sum_{t=0}^{N} y(t) Q_t w(t). \tag{2.59}$$

In a similar way to the continuous-time case, introducing the *norms*,

$$\underbrace{\|u\| = \langle u, u\rangle_{\mathcal{U}}^{\frac{1}{2}} \;\; \& \;\; \|y\| = \langle y, y\rangle_{\mathcal{Y}}^{\frac{1}{2}},}_{\textit{Input and Output Norms}} \tag{2.60}$$

links the above to Optimal Control via the expression

$$J = \|y\|_{\mathcal{Y}}^2 + \|u\|_{\mathcal{U}}^2 = \sum_{t=0}^{N} \left(y^T(t) Q_t y(t) + u(t) R_t u(t) \right) \tag{2.61}$$

which is the performance index/objective function for the discrete Linear Quadratic Regulator Problem.

2.8 The Transpose System and Time Reversal

This section links the response characteristics of the two systems $S(\Phi_d, \Delta_d, C, D)$ and $S(\Phi_d^T, C^T, \Delta_d^T, D)$ to the idea of time reversal. *Time reversal* for a time series $f = \{f(0), f(1), \cdots, f(N)\}$ of $p \times 1$ matrices is the replacement of f by the time series $\{f(N), f(N-1), \cdots, f(0)\}$ where the order of the terms is reversed. Writing f in supervector form, the reverse time supervector can be written as $\mathcal{T}(p, N) f$ where $\mathcal{T}(p, N)$ is the square, $p(N+1) \times p(N+1)$ *matrix time reversal operator*

$$\mathcal{T}(p, N) = \begin{bmatrix} 0 & 0 & \cdots & 0 & 0 & I_p \\ 0 & 0 & \cdots & 0 & I_p & 0 \\ 0 & 0 & \cdots & I_p & 0 & 0 \\ \vdots & & & & & \vdots \\ 0 & I_p & \cdots & 0 & 0 & 0 \\ I_p & 0 & \cdots & 0 & 0 & 0 \end{bmatrix}. \tag{2.62}$$

The proof of the following statements is left as an exercise in matrix manipulation for the reader. The time reversal matrix has the following properties:

1. Invariance under transposition and inversion,

$$\begin{aligned} \mathcal{T}(p, N) = \mathcal{T}^T(p, N), \;\; (\mathcal{T}(p, N))^2 = I_{p(N+1)} \\ \textit{and} \;\; (\mathcal{T}(p, N))^{-1} = \mathcal{T}(p, N). \end{aligned} \tag{2.63}$$

2. The following identity relates the transpose of $G(\Phi_d, \Delta_d, C, D)$ to time reversal operators and the "transpose system" $G(\Phi_d^T, C^T, \Delta_d^T, D^T)$

$$G^T(\Phi_d, \Delta_d, C, D) = \mathcal{T}(\ell, N)G(\Phi_d^T, C^T, \Delta_d^T, D^T)\mathcal{T}(m, N). \tag{2.64}$$

3. If $S(\Phi_d, \Delta_d, C, D)$ has a transfer function matrix $G(z)$, then $S(\Phi_d^T, C^T, \Delta_d^T, D^T)$ has transfer function matrix $G^T(z)$.
4. If v is the supervector corresponding to a time series $\{v(0), v(1), \cdots, v(N)\}$ in $\mathcal{R}^m$, then the computation of the time series in $\mathcal{R}^\ell$ generated by the supervector $G^T(\Phi_d, \Delta_d, C, D)v$ can be expressed as

$$G^T(\Phi_d, \Delta_d, C, D)v = \mathcal{T}(\ell, N)\left(G(\Phi_d^T, C^T, \Delta_d^T, D^T)(\mathcal{T}(m, N)v)\right). \tag{2.65}$$

   That is, the time series in $\mathcal{R}^\ell$ represented by $G^T(\Phi_d, \Delta_d, C, D)v$ is the time reversal of the response of the system $S(\Phi_d^T, C^T, \Delta_d^T, D^T)$ from zero initial conditions to the time reversal of the time series v.

Finally, note that the time reversal operators will often be simply denoted $\mathcal{T}$ with the dimensions understood by the context of the discussion.

2.9 $\mathcal{Z}$-Transforms and the Discrete Transfer Function Matrix

The $\mathcal{Z}$-transform is a methodology of analysing *infinite* time series in the form of a sequence of scalar- or matrix-valued data points $\{f(0), f(1), f(2), \cdots\}$ using functions of the complex variable.

Note 2.6 The case of a finite time series $\{f(t)\}_{0 \le t \le N}$ can often be regarded as infinite in length with $f(t) = 0,\ t > N$.

It has been used widely in control systems analysis and design. The basic definition of the $\mathcal{Z}$-Transform of a scalar time series $f(t),\ 0 \le t \le N$ is most easily presented in the case on an infinite interval $N = \infty$. More precisely, the $\mathcal{Z}$-Transform of the (real or complex) series $f = \{f(0), f(1), f(2), \cdots\}$ is defined as follows:

$$\mathcal{Z}[f] = \sum_{j=0}^{\infty} z^{-j} f(t), \tag{2.66}$$

for all complex number z such that the series converges. It is noted that a sufficient condition for the existence of such values of z is that f is *exponentially (or geometrically) bounded*. That is, there exist real numbers $M \ge 0$ and λ such that

$$|f(t)| \le M\lambda^t \quad for\ all\ t \ge 0. \tag{2.67}$$

For such a time series, the infinite sum converges for all z satisfying $|z| > \lambda$. Although only valid for values of z such that the series converges, the functions can be extended by analytic continuation to all points of the complex plane other than singularities. For this reason the issue of convergence plays only an occasional role in practical use of the ideas.

For notational convenience, the $\mathcal{Z}$-transform of the sequence f will be denoted by $f(z)$. In the discrete-time case this does not lead to any confusion as f denotes a times series whilst $f(z)$ is a function of the complex variable. With this convention, the *Inverse $\mathcal{Z}$-Transform* is simply the inverse process of constructing the time series f from a given transform $f(z)$ and is denoted by $\mathcal{Z}^{-1}$, for example,

$$f = \mathcal{Z}^{-1}[f(z)]. \tag{2.68}$$

The extension of the ideas to matrix-valued time series is easily achieved. For a $p \times q$ time series $H(t),\ t \geq 0$, its $\mathcal{Z}$-Transform $\mathcal{Z}[H]$ is denoted by the $p \times q$ matrix function $H(z)$ which has (i, j) element equal to the $\mathcal{Z}$-Transform of the time series $H_{ij}(t),\ t \geq 0$.

Finally, a number of useful properties of the $\mathcal{Z}$-Transform are used in the text and are summarized below for any sequence $f(t)$ defined on $0 \leq t < \infty$ with $f(0)$ written as f_0,

$$\begin{aligned} \mathcal{Z}[\{0, f(0), f(1), \cdots\}] = z^{-1} f(z)\,, \quad & \mathcal{Z}[\{1, \alpha, \alpha^2, \cdots\}] = \tfrac{z}{z-\alpha}, \\ \mathcal{Z}[\{1, , 2\alpha, 3\alpha^2, \cdots\}] = z\tfrac{\partial}{\partial\alpha}\left(\tfrac{z}{(z-\alpha)}\right), \quad & \mathcal{Z}[\{f(1), f(2), \cdots\}] = zf(z) - zf_0. \end{aligned} \tag{2.69}$$

2.9.1 Discrete Transfer Function Matrices, Poles, Zeros, and the Relative Degree

A linear, time-invariant, discrete, state-space model has a frequency domain description obtained simply by taking $\mathcal{Z}$-Transforms of the state equations to obtain

$$zx(z) - zx_0 = \Phi_d x(z) + \Delta_d u(z), \quad y(z) = Cx(z) + Du(z) \tag{2.70}$$

from which simple matrix algebra to eliminate $x(z)$ gives

$$y(z) = G(z)u(z) + zC(zI_n - \Phi_d)^{-1} x_0. \tag{2.71}$$

The system $m \times \ell$ *Discrete Transfer Function Matrix* $G(z)$ is defined by

$$G(z) = C(zI_n - \Phi_d)^{-1}\Delta_d + D. \tag{2.72}$$

Poles and zeros are defined as follows:

1. The *poles* of $G(z)$ and their multiplicities are precisely the set of eigenvalues of the matrix Φ_d and their multiplicities.
2. The *zeros* of $G(z)$ are defined by rank conditions on a partitioned matrix dependent on (Φ_d, Δ_d, C, D). More precisely, the complex number z is a zero of $S(\Phi_d, \Delta_d, C, D)$ if and only if

$$rank \begin{bmatrix} zI_n - \Phi_d & -\Delta_d \\ C & D \end{bmatrix} < n + \min\{m, \ell\}. \tag{2.73}$$

If all zeros lie in the open unit circle in the complex plane, the system is said to be *minimum phase*. Otherwise, the system is said to be *non-minimum phase*.

A comparison with the equivalent formula for the transfer function matrix of a continuous-time system shows the algebraic similarity, the discrete-time case being obtained from the continuous-time formula by the mapping $s \mapsto z$ and $(A, B, C, D) \mapsto (\Phi_d, \Delta_d, C, D)$. It is natural therefore that the discrete-time case inherits the properties and nomenclature described previously for continuous-time case. In particular, its relative degree is the uniquely defined integer k^* such that $\lim_{z\to\infty} z^k G(z) = G_\infty$ exists and is non-zero.

2.9.2 The Discrete System Frequency Response

The linear time-invariant discrete system $S(\Phi_d, \Delta_d, C, D)$ has frequency response function defined to be the (matrix-valued) function $G(z)$ where $|z| = 1$ or, equivalently, $z = e^{i\theta}, \ 0 \le \theta \le 2\pi$. This function has found considerable application in multivariable feedback control systems design. It also describes the response of an asymptotically stable system from zero initial conditions $x_0 = 0$ to sinusoidal input time series u_c, u_s defined to be the real, (respectively) imaginary parts of the complex sequence $u = \{\alpha, z\alpha, z^2\alpha, \cdots\}$ (that is, $u(t) = \alpha z^t, \ t \ge 0$). More generally, if the data (Φ_d, Δ_d, C, D) contains only real numbers and z is an arbitrary complex number, then the resultant output responses from zero initial conditions are, respectively, equal to the real and imaginary parts of the response $y(t)$ from zero initial conditions to the complex input $u(t)$. It is left as an exercise for the reader to verify that

$$y(t) = G(z)u(t) - C(zI_n - \Phi_d)^{-1}\Phi_d^t \Delta_d \alpha, \quad t = 0, 1, 2, \cdots . \tag{2.74}$$

In particular, if $z = e^{i\theta}$, the stability assumption indicates that, for all large enough values of sample index t, the steady-state oscillation $y(t) \approx G(z)\alpha z^t$ is an accurate description of steady-state oscillatory output behaviour as asymptotic stability ensures that Φ_d^t becomes infinitesimally small if t is large.

2.10 Controllability, Observability, Minimal Realizations, and Pole Allocation

This section considers continuous systems only but, being algebraic in nature, the results apply with no change to uni-rate discrete systems simply by replacing the complex variable s by z and the quadruple of matrices (A, B, C, D) by (Φ_d, Δ_d, C, D).

The analysis of linear, time-invariant, continuous state-space models relies heavily on properties of (A, B, C, D). From time to time, the ideas of *complete state controllability* and *complete state observability* are crucial. More precisely:

1. The linear system $S(A, B, C, D)$ is completely state controllable if, and only if, for any initial state $x(0)$ and any desired final state x_f, there exist a control input $u(t)$ and a time t_f such that $u(t)$ drives the state $x(t)$ from x_0 at $t = 0$ to x_f at time $t = t_f$.
2. The linear system $S(A, B, C, D)$ is completely state observable if, and only if, for any initial state $x(0)$ and any control input $u(t),\ t \geq 0$, there exists a time $t_f > 0$ such that the initial state x_0 can be computed uniquely from input and output records only on some time interval $0 \leq t \leq t_f$.

For linear time-invariant linear systems the relevant algebraic tests are as follows:

1. The system $S(A, B, C, D)$ is completely state controllable if, and only if, one of the following two (equivalent) conditions is true:

$$\begin{gathered} rank[B, AB, A^2B, \cdots, A^{n-1}B] = n, \\ rank[sI - A, -B] = n \quad for\ all\ complex\ numbers\ s. \end{gathered} \tag{2.75}$$

2. The system $S(A, B, C, D)$ is completely state observable if, and only if, one of the following two (equivalent) conditions is true:

$$\begin{gathered} rank \begin{bmatrix} C \\ CA \\ \vdots \\ CA^{n-1} \end{bmatrix} = n, \\ rank \begin{bmatrix} sI - A \\ C \end{bmatrix} = n \quad for\ all\ complex\ numbers\ s. \end{gathered} \tag{2.76}$$

3. As a consequence:
   a. The system $S(A, B, C, D)$ is completely state controllable (respectively, observable) if, and only if, $S(A^T, C^T, B^T, D^T)$ is completely state observable (respectively, controllable).
   b. For all state feedback (respectively, output injection) matrices F (respectively, H), the completely state controllability (respectively, observability) of

the system $S(A, B, C, D)$ implies the complete state controllability (respectively, observability) of the system $S(A - BF, B, C, D)$ (respectively, $S(A - HC, B, C, D)$).

A *realization* of a proper or strictly proper transfer function matrix $G(s)$ is any linear, time-invariant model $S(A, B, C, D)$ whose transfer function matrix coincides with $G(s)$. A *minimal realization* is any realization where the state dimension n takes the lowest possible value. Any realization is minimal if, and only if, it is controllable and observable.

A useful property of controllable systems is the potential for *pole allocation*. More precisely, a *state feedback control law* takes the form

$$u(t) = Kv(t) - Fx(t) \tag{2.77}$$

where F is an $\ell \times n$ real matrix of state feedback gains, K is an $\ell \times \ell$ matrix of gains, and $v(t)$ is the new input signal. The closed-loop state-space model relating the input v to the output y is based on the same state vector model but with matrices $S(A - BF, BK, C - DF, DK)$. Note the following:

1. If $S(A, B, C, D)$ is completely state controllable and K is nonsingular, then the closed-loop system $S(A - BF, BK, C - DF, DK)$ is completely state controllable.
2. The stability of the new feedback system is described by the characteristic polynomial $\rho_F(s) = \det[sI - A + BF]$. The important property of completely state controllable systems is the pole allocation property. That is, for any set of complex numbers $\{\mu_j\}_{1\le j\le n}$ (invariant under complex conjugation), there exists a real state feedback matrix F such that the eigenvalues of $A - BF$ are precisely the points $\{\mu_j\}_{1\le j\le n}$.

2.11 Inverse Systems

Dynamical systems G are normally regarded as mapping from inputs to outputs. The general concept of an *Inverse System* is a system that reconstructs the input from the observed output. As the output depends also on the initial state condition, it is usual to assume zero initial conditions. For square systems (where $m = \ell$) the operation parallels the simple observation that $y(s) = G(s)u(s)$ can be reconstructed using $u(s) = G^{-1}(s)y(s)$ provided that the inverse transfer function matrix exists.

Note 2.7 The analysis that follows applies equally to both continuous-time and discrete-time systems as the definitions and manipulations used are entirely algebraic.

2.11.1 The Case of $m = \ell$ and Zeros

If $m = \ell$ and $S(A, B, C, D)$ has transfer function matrix satisfying $\det[G(s)] \neq 0$ at all but a finite number of points s, then the system is said to be *invertible*. The *inverse transfer function matrix* takes the form

$$G^{-1}(s) \equiv P(s) + H(s) \tag{2.78}$$

where P and H are uniquely defined by the following statements:

1. $H(s)$ is strictly proper matrix of rational polynomials.
2. $P(s)$ is either a constant matrix or has elements that are polynomials in s, i.e., the inverse reconstructs the input from computations including derivatives of the output.

If $S(A, B, C, D)$ has a transfer function matrix with inverse of the form of (2.78) it is left as an exercise for the reader to show the following:

1. The matrix $P(s)$ is invertible for almost all values of s with

$$\begin{aligned} \lim_{s\to\infty} G(s)P(s) = \lim_{s\to\infty} P(s)G(s) = I, \quad \textit{so that} \\ \lim_{s\to\infty} P^{-1}(s)H(s) = \lim_{s\to\infty} H(s)P^{-1}(s) = 0. \end{aligned} \tag{2.79}$$

2. Moreover, $\lim_{s\to\infty} P^{-1}(s) = D$ and hence $P^{-1}(s)$ has a minimal realization in terms of a proper state-space model. This model is strictly proper if $G(s)$ is strictly proper.

Assuming zero initial conditions, the input–output relations take the form $y(s) = G(s)u(s)$ and hence $u(s) = G^{-1}(s)y(s)$. That is, the inverse system reconstructs the input vector uniquely from the output vector. The transfer function matrix $G^{-1}(s)$, typically, is neither proper nor strictly proper. Its construction is computationally complex but has a simple state-space form if D is nonsingular when a simple manipulation of the state equations shows that $G^{-1}(s)$ has a state-space realization $S(A - BD^{-1}C, BD^{-1}, -D^{-1}C, D^{-1})$. Moreover, the poles of this system are the zeros of $S(A, B, C, D)$ as using Schur's Formula

$$\begin{vmatrix} sI - A & -B \\ C & D \end{vmatrix} \equiv \det[D]\det[sI - A + BD^{-1}C]. \tag{2.80}$$

More generally, for a controllable and observable system with TFM $G(s)$, the poles of $H(s)$ are zeros of $S(A, B, C, D)$ and hence $H(s) \equiv 0$ if, and only if, the system has no zeros. Conversely, any $m \times m$ matrix $P(s)$ of polynomials in s that is invertible and such that the limit $\lim_{s\to\infty} P^{-1}(s) = P_\infty$ exists has a minimal (controllable and observable) state-space realization that has no zeros.

2.11.2 Left and Right Inverses when $m \neq \ell$

When $m \neq \ell$, the transfer function matrix is non-square and matrix inversion techniques do not apply. The relevant ideas are those of left inverse and right inverse.

Let $S(A, B, C, D)$ have transfer function matrix $G(s)$ and suppose that $m > \ell$. Then a *Left Inverse* of $G(s)$ can be regarded as *any* $\ell \times m$ matrix $L(s)$ of rational polynomials in s such that

$$L(s)G(s) = I_\ell. \tag{2.81}$$

In conceptual terms, assuming zero initial conditions, an input $u(s)$ produces an output $y(s) = G(s)u(s)$ from G. From the definition of L, the input is reconstructed using $u(s) = L(s)y(s)$. The simplest way of approaching the problem is to choose a constant $\ell \times m$ matrix K such that the *square* $\ell \times \ell$ system $KG(s)$ system is invertible. With such a choice, a left inverse of $G(s)$ can be computed as

$$L(s) = (KG(s))^{-1}K. \tag{2.82}$$

The zeros of KG will depend on both $G(s)$ and K so attempts should be made, if possible, to ensure that $KG(s)$ is minimum phase or at least that no additional right-half-plane zeros are introduced.

The left inverse hence, in general, does not have a state-space realization of an ℓ-output, m-input system $S(A_L, B_L, C_L, D_L)$. The only case when such a state-space realization exists is defined as follows: suppose that $m > \ell$ and that $S(A, B, C, D)$ satisfies the condition $rank[D] = \ell$. Suppose also that K is any $\ell \times m$ matrix such that KD is nonsingular. Then the state-space system

$$S(A - B(KD)^{-1}KC, B(KD)^{-1}K, -(KD)^{-1}KC, (KD)^{-1}K)$$

maps $y \mapsto u$ and hence is a left inverse of $S(A, B, C, D)$.

The case of $m < \ell$ is analysed in a similar manner with a *Right Inverse* of $G(s)$ being regarded as any $m \times \ell$ matrix $R(s)$ of rational polynomials in s such that

$$G(s)R(s) = I_m. \tag{2.83}$$

It is a simple matter to see that $R^T(s)$ is a left inverse of $G^T(s)$. Any algorithm that constructs a left inverse of the transpose of a system has a transpose that is a right inverse of that system.

2.12 Quadratic Optimal Control of Linear Continuous Systems

This section describes an Optimal Control problem that forms the basis of many Optimal Iterative Learning Control problems described in the literature. The presentation can be approached using the operator formulation described in previous sections.

For the purpose of this text, however, a simplified derivation is used directly from the state-space model that is constructive in form and uses nothing more complicated than integration by parts.

2.12.1 Problem Formulation

Consider a linear, time-varying, continuous, state-space system (2.4) (with $D(t) \equiv 0$) and the Optimal Control problem of minimizing a quadratic performance index/ objective function $J(u)$,

$$J(u) = J_f(u) + J_x(u) + J_u(u). \tag{2.84}$$

More precisely, the objective function is defined by

$$\begin{aligned} J_f(u) &= \tfrac{1}{2}\textstyle\sum_{j=1}^{M}(e^e(t_j))^T Q_j e^e(t_j), \quad e^e(t_j) = r^e(t_j) - F_j y(t_j), \\ J_x(u) &= \tfrac{1}{2}\textstyle\int_0^T e^T(t)Q(t)e(t)dt, \quad e(t) = r(t) - y(t), \\ J_u(u) &= \tfrac{1}{2}\textstyle\int_0^T (u(t) - v(t))^T R(t)(u(t) - v(t))dt \end{aligned} \tag{2.85}$$

where

1. $r(t)$ and $v(t)$ are $m \times 1$ and $\ell \times 1$ vector functions,
2. $\{t_j\}_{1 \le j \le M}$ are distinct times ordered $t_1 < t_2 < \cdots < t_M$ on the time interval of interest,
3. $r^e(t_j),\ 1 \le j \le M$ are $f_j \times 1$ vectors and $F_j,\ 1 \le j \le M$ are $f_j \times m$ matrices,
4. the $Q_j,\ 1 \le j \le M$, are symmetric, positive definite $f_j \times f_j$ matrices, $Q(t)$ is $m \times m$, symmetric and positive semi-definite, and $R(t)$ is $\ell \times \ell$, symmetric and positive definite.

Note this objective function is an extension of the more familiar form seen in Optimal Control texts as it includes the general set of *intermediate times points* $\{t_j\}_{1 \le j \le M}$ and the associated extra terms in $J_f(u)$.

The minimization of J is control based on a balance between the objective of reducing tracking errors at intermediate points and improving the tracking of the reference signal $r(t)$ along the control interval by using control signals that do not deviate from $v(t)$ too much. The choice of matrices F_j, $Q(t)$ and $R(t)$ have a serious influence on the outcome in changing that balance in any attempt to achieve the control characteristics needed. In particular, special cases include the following:

1. All such problems are forms of *Tracking*.
2. Setting $M = 0$ removes the intermediate points from the control objective by setting $J_f(u) = 0$.
3. The time variation of $Q(t)$ reflects the importance of achieving good tracking accuracy of $r(t)$ along the different parts of the solution. If $Q(t)$ is zero on some subinterval, then the problem does not require tracking on that interval.

4. If $M = 0$ and $r(t) \equiv 0$ then the Optimal Control Problem is a *Regulator Problem.*
5. If $Q(t) \equiv 0$, the problem is an *Intermediate Point Problem* whilst if, in addition, $M = 1$ and t_1 is the final time, then the problem is an *End-Point Problem.*

2.12.2 Optimization Conditions

Let u_1 and u_2 be two input signals with state and output responses $x_1(t)$, $x_2(t)$ and $y_1(t)$, $y_2(t)$, respectively. Consider the difference $J(u_2) - J(u_1)$ written in the form

$$J(u_2) - J(u_1) + E(u_1, u_2) \tag{2.86}$$

where $E(u_1, u_2)$ is defined by the model equations and a "costate" vector $p(t)$ that is piece-wise continuous with $p(T+) = 0$ and possessing possible discontinuities only at the intermediate points. More precisely, using the state equations,

$$\begin{aligned} 0 &\equiv E(u_1, u_2) \\ &= \int_0^T p^T(t)\,(A(x_2(t) - x_1(t)) + B(u_2(t) - u_1(t)) - (\dot{x}_2(t) - \dot{x}_1(t)))\,dt \\ &= \int_0^T p^T(t)\,(A(x_2(t) - x_1(t)) + B(u_2(t) - u_1(t)))\,dt \\ &\quad -\textstyle\sum_{j=1}^M \left[p^T(t)(x_2(t) - x_1(t))\right]_{t=t_{j-1}+}^{t=t_j-} \\ &\quad +\textstyle\sum_{j=1}^M \int_{t_{j-1}+}^{t_j-} \dot{p}^T(t)(x_2(t) - x_1(t))dt \end{aligned} \tag{2.87}$$

where $f(t_j-)$ (resp. $f(t_j+)$) denotes the left and right limits for $f(t)$ at $t = t_j$.

Writing

$$\begin{aligned} J_x(u_2) - J_x(u_1) = &- \int_0^T (C^T(t)Q(t)e_1(t))^T(x_2(t) - x_1(t))dt \\ &+\tfrac{1}{2}\int_0^T (e_2(t) - e_1(t))^T Q(t)(e_2(t) - e_1(t))dt \end{aligned} \tag{2.88}$$

and

$$\begin{aligned} J_f(u_2) - J_f(u_1) = &-\textstyle\sum_{j=1}^M (C^T(t_j)F_j^T Q_j e_1^e(t_j))^T(x_2(t_j) - x_1(t_j)) \\ &+\tfrac{1}{2}\textstyle\sum_{j=1}^M (e_2^e(t_j) - e_1^e(t_j))^T Q_j (e_2^e(t_j) - e_1^e(t_j)). \end{aligned} \tag{2.89}$$

The next step is to define the vector $p(t)$ more precisely using the constructions described above and the continuity of the state vectors. Set

$$\begin{aligned} \dot{p}(t) = -A^T(t)p(t) + C^T(t)Q(t)e_1(t), &\quad t \neq t_j, \quad j = 1, 2, \cdots, M \\ \textit{with jump conditions}& \\ p(t_j+) - p(t_j-) = C^T(t_j)F_f^T Q_j e_1^e(t_j), &\quad 1 \leq j \leq M \\ \textit{where} \qquad & p(t_M+) = 0 \end{aligned} \tag{2.90}$$

to obtain, from (2.84), (2.85), (2.86), (2.87), (2.88), and (2.89), the simplified expression

$$
\begin{aligned}
J(u_2) - J(u_1) = & \int_0^T (H_r(u_2(t)) - H_r(u_1(t)))\,dt \\
& + \tfrac{1}{2}\sum_{j=1}^{M} (e_2^e(t_j) - e_1^e(t_j))^T Q_j (e_2^e(t_j) - e_1^e(t_j))dt \\
& + \tfrac{1}{2}\int_0^T (e_2(t) - e_1(t))^T Q(t)(e_2(t) - e_1(t))dt
\end{aligned} \tag{2.91}
$$

where the "Hamiltonian" is defined by

$$
H_r(u(t)) = \frac{1}{2}(u(t) - v(t))^T R(t)(u(t) - v(t)) + p^T(t)B(t)u(t). \tag{2.92}
$$

If u_1 is the optimal input minimizing J, then $J(u_2) - J(u_1) > 0$ whenever $u_2 \neq u_1$. As the quadratic terms in (2.91) are positive, a sufficient condition for u_1 to be optimal is that $H_r(u_2(t)) - H_r(u_1(t)) > 0$ at every point of the time interval, i.e., H is minimized by u_1. If the input is not subject to constraints, then the relevant formula

$$
u_1(t) = v(t) - R^{-1}(t)B^T(t)p(t) \tag{2.93}
$$

for the minimizing input can be derived by observing that

$$
\begin{aligned}
\tfrac{1}{2}(u - v)^T R(u - v) + p^T Bu = & \tfrac{1}{2}(u - v + R^{-1}B^T p)^T R(u - v + R^{-1}B^T p) \\
& + p^T Bv - \tfrac{1}{2}(B^T p)^T R^{-1}(B^T p)
\end{aligned} \tag{2.94}
$$

with R positive definite.

2.12.3 Optimal Control TPBVP

From the results of the previous section, the minimizing input $u^*(t)$ of $J(u)$ is the solution of the *Two-Point Boundary Value problem* (TPBVP), consisting of the state equations

$$
\begin{aligned}
\dot{x}(t) &= A(t)x(t) + B(t)u^*(t), \quad x(0) = x_0 \\
y(t) &= C(t)x(t),
\end{aligned} \tag{2.95}
$$

with

$$
u^*(t) = v(t) - R^{-1}(t)B^T(t)p(t) \tag{2.96}
$$

together with *costate vector* equations

$$
\dot{p}(t) = -A^T(t)p(t) + C^T(t)Q(t)e(t), \qquad t \neq t_j, \;\; j = 1, 2, \cdots, M \tag{2.97}
$$

satisfying the *jump conditions*

$$
\begin{aligned}
p(t_j+) - p(t_j-) = C^T(t_j)F_f^T Q_j e^e(t_j), \;\; 1 \leq j \leq M, \\
\& \;\; p(t_M+) = 0.
\end{aligned} \tag{2.98}
$$

Note the boundary conditions at the initial, $t = 0$, and final, $t = t_M$, times.

2.12.4 Solving the TPBVP: The Riccati Equation

The solution of the TPBVP in Sect. 2.12.3 can be obtained by setting

$$p(t) = K(t)x(t) - \xi(t) \tag{2.99}$$

where $K(t)$ is the $n \times n$, time-varying Riccati state *feedback* matrix and $\xi(t)$ is an $n \times 1$ *feedforward* or *predictive* term representing the effects of $\upsilon(t)$ and $r(t)$. The control input is then given by

$$u(t) = \upsilon(t) - \underbrace{R^{-1}(t)B^T(t)K(t)x(t)}_{state\ feedback} + \underbrace{R^{-1}(t)B^T(t)\xi(t)}_{feedforward}. \tag{2.100}$$

Calculation of the necessary properties of $K(t)$ and $\xi(t)$ is approached by differentiation at any point on the intervals $t_{j-1} < t < t_j$. This gives

$$\dot{p}(t) = \dot{K}(t)x(t) + K(t)\dot{x}(t) - \dot{\xi}(t). \tag{2.101}$$

Substituting for $\dot{p}(t)$ and $\dot{x}(t)$ suggests the following choices after a little algebra to collect all terms containing the state $x(t)$ together, namely,

$$\begin{aligned} \dot{K}(t) + K(t)A(t) + A^T(t)K(t) - K(t)B(t)R^{-1}(t)B^T(t)K(t) \\ +C^T(t)Q(t)C(t) = 0 \\ \& \\ \dot{\xi}(t) = -\left(A^T(t) - K(t)B(t)R^{-1}(t)B^T(t)K(t)\right)\xi(t) \\ -C^T(t)Q(t)r(t) + K(t)B(t)\upsilon(t). \end{aligned} \tag{2.102}$$

These equations must also satisfy the costate jump conditions (2.98), namely

$$\begin{aligned} \left(K(t_j+) - K(t_j-)\right)x(t_j) - \left(\xi(t_j+) - \xi(t_j-)\right) \\ = C^T(t_j)F_f^T Q_j\left(r^e(t_j) - F_jC(t_j)x(t_j)\right), \ 1 \le j \le M, \\ with \quad K(t_M+) = 0 \ \& \ \xi(t_M+) = 0. \end{aligned} \tag{2.103}$$

These equations suggest the *jump conditions* for $K(t)$ and $\xi(t)$,

$$\begin{aligned} K(t_j+) - K(t_j-) = -C^T(t_j)F_f^T Q_j F_j C(t_j), \ 1 \le j \le M, \\ with \quad K(t_M+) = 0, \\ \& \\ \xi(t_j+) - \xi(t_j-) = C^T(t_j)F_f^T Q_j r^e(t_j), \ 1 \le j \le M, \\ with \quad \xi(t_M+) = 0. \end{aligned} \tag{2.104}$$

Both $K(t)$ and $\xi(t)$ can be computed using simulation in reverse time from the final time to $t = 0$.

2.13 Quadratic Optimal Control of Linear Discrete Systems

The Optimal Control problem for the ℓ-input, m-output, linear, time-invariant, discrete-time system (using the notational Note 2.3)

$$\begin{aligned} x(t+1) &= Ax(t) + Bu(t), \quad t = 0, 1, 2, \cdots, N-1, \quad x(0) = x_0 \\ and \quad y(t) &= Cx(t) + Du(t) \quad for \quad t = 0, 1, 2, 3, \cdots, N, \end{aligned} \tag{2.105}$$

uses a reference output time series r together with a nominal input time series u_0 in a quadratic objective function

$$J(u) = \frac{1}{2} \sum_{j=0}^{N} \left((r(t) - y(t))^T Q(t)(r(t) - y(t)) + (u(t) - u_0(t))^T R(t)(u(t) - u_0(t)) \right), \tag{2.106}$$

where the "time" index t denotes the sample index/number. The time-varying weight matrices $R(t) = R^T(t)$ are positive definite and the $Q(t) = Q^T(t)$ are positive semi-definite. The aim of the optimization is to reduce the "error" $r - y$ along the solution whilst not allowing the input to deviate too much from the nominal input u_0.

Note the following:

1. The notation Φ_d, Δ_d, C, D has been replaced by A, B, C, D for notational simplicity and should cause no confusion with the continuous-time case as the context of any discussion will easily resolve the identification problem.
2. In addition, the time variation of the $Q(t)$ makes it possible to adapt the optimization to suit different objectives. For example:
   a. If the $Q(t)$ are sample independent, then the optimization is aiming to reduce the "error" $r(t) - y(t)$ along the whole interval with no priority being given to any particular subinterval.
   b. If any $Q(t)$ is symmetric and positive semi-definite then it can be written in the form $Q(t) = \Gamma^T(t)\Gamma(t)$ where $\Gamma(t)$ has full row rank. In this circumstance the expression $\gamma^T Q(t)\gamma = (\Gamma(t)\gamma)^T(\Gamma(t)\gamma)$. Minimizing J therefore only considers the signal component $\Gamma(t)\gamma$ ignoring any component of γ in the kernel of $\Gamma(t)$ and hence $Q(t)$.
   c. For example, if all $Q(t)$ are zero except at intermediate sample points $t = t_1, t_2, \cdots, t_M$, then the aim of control is to make the errors small at these times without using too large a control action (represented by $R(t)$, $0 \le t \le N$).
   d. If $Q(t) = 0$ on some subinterval $t_1 \le t \le t_2$, then reducing the error in this interval is of no importance in the optimization. Indeed, the values of $r(t)$ in this interval play no role in the problem and can be ignored.

The supervector description provides a simple way to characterize the solution procedure. The relevant system model is simply

$$y = G(A, B, C, D)u + d, \tag{2.107}$$

where the supervectors $y \in \mathcal{Y} = \mathcal{R}^{m(N+1)}$ and $u \in \mathcal{U} = \mathcal{R}^{\ell(N+1)}$. Defining the block matrices

$$\begin{aligned} R &= blockdiag\,[R(0), R(1), \cdots, R(N)], \\ Q &= blockdiag\,[Q(0), Q(1), \cdots, Q(N)], \end{aligned} \tag{2.108}$$

the relevant inner products in $\mathcal{U}$ (respectively, $\mathcal{Y}$) are taken to be

$$\langle u, v\rangle_{\mathcal{U}} = u^T R v \quad (\,respectively \quad \langle y, w\rangle_{\mathcal{Y}} = y^T Q w\,). \tag{2.109}$$

Denoting the supervectors corresponding to the time series $\{r(0), r(1), \cdots, r(N)\}$ (respectively, $\{u_0(0), u_0(1), \cdots, u_0(N)\}$) as r (respectively, u_0), the objective function takes the form

$$J(u) = \frac{1}{2}\|r - y\|^2_{\mathcal{Y}} + \frac{1}{2}\|u - u_0\|^2_{\mathcal{U}} \tag{2.110}$$

where a "norm" $\|\cdot\|$ is defined by the formula $\|a\| = \langle a, a\rangle^{\frac{1}{2}}$.

The optimal solution is then obtained by setting the Jacobian (derivative) of $J(u)$ to zero to obtain the supervector equation

$$u = u_0 + G^*(A, B, C, D)(r - y) \quad where \quad G^* = R^{-1}G^T Q. \tag{2.111}$$

This matrix characterization of the solution is precise but will normally lead to computational problems if N is large. A characterization of solutions using two-point boundary value problems and Riccati matrices is given in the following sections.

2.13.1 The Discrete Two-Point Boundary Value Problem

The relationship defining the optimal solution is now converted into a two-point boundary value problem. First note that, as $S(A, B, C, D)$ is assumed to be time-invariant, the time variation in the problem is contained solely in the matrices R and Q. The transpose $G^T(A, B, C, D)$ is associated with the matrix representation of the linear, time-invariant system $G(A^T, C^T, B^T, D^T)$. The precise form of this relationship is expressed using time reversal matrices $\mathcal{T}$,

$$G^T(A, B, C, D) = \mathcal{T}(\ell, N)G(A^T, C^T, B^T, D^T)\mathcal{T}(m, N) \tag{2.112}$$

so that

$$u - u_0 = R^{-1}\mathcal{T}(\ell, N)G(A^T, C^T, B^T, D^T)\mathcal{T}(m, N)Q(r - y). \tag{2.113}$$

This calculation is now done in a number of simple steps. For simplicity, introduce the supervector $z = \mathcal{T}(m, N)Q(r - y)$ and note that the components of the supervector $w = G(A^T, C^T, B^T, D^T)z$ can be computed from the state-space simulation

$$\begin{array}{rrcl} & \psi(t'+1) & = & A^T\psi(t') + C^T z(t'), \quad t' = 0, 1, 2, \cdots, N-1, \\ & \psi(0) & = & 0, \\ \textit{with} & w(t') & = & B^T\psi(t') + D^T z(t') \\ \textit{and} & z(t') & = & Q(N-t')\big(r(N-t') - y(N-t')\big) \;\; \textit{for} \;\; t' = 0, 1, 2, \cdots, N. \end{array} \tag{2.114}$$

The equations defining the time reversed supervector $\eta = \mathcal{T}(\ell, N)w$ of w are obtained by the substitution $t = N - t'$ and the introduction of the *costate vector* $p(t) = \psi(N - t)$ and notation $\eta(t) = w(N - t)$ to give

$$\begin{array}{rcl} p(t-1) & = & A^T p(t) + C^T Q(t)\,(r(t) - y(t))\,, \quad t = 1, 2, \cdots, N, \\ p(N) & = & 0 \quad \textit{and} \\ \eta(t) & = & B^T p(t) + D^T Q(t)\,(r(t) - y(t))\,, \quad t = 0, 1, 2, \cdots, N. \end{array} \tag{2.115}$$

The final computation is that of $u - u_0 = R^{-1}\eta$ to give the optimal input as the solution of the two-point boundary value problem

$$\begin{array}{rrcl} & x(t+1) & = & Ax(t) + Bu(t), \quad \textit{and} \\ & p(t) & = & A^T p(t+1) + C^T Q(t+1)\,(r(t+1) - y(t+1)) \\ & & & \textit{for} \quad t = 0, 1, 2, \cdots, N-1, \\ \textit{with} & u(t) & = & u_0(t) + R^{-1}(t)B^T p(t) + R^{-1}(t)D^T Q(t)\,(r(t) - y(t)) \\ & & & \textit{for} \quad t = 0, 1, 2, \cdots, N\,. \\ & & & \textit{The boundary conditions are} \\ & & & \qquad x(0) = x_0 \quad \textit{and} \quad p(N) = 0. \end{array} \tag{2.116}$$

This representation is a discrete equivalent to the two-point boundary value problem for continuous systems derived in Sect. 2.12.3. There are clear structural similarities between the two solutions and, in fact, the reader might like to demonstrate that they become equivalent if T is fixed and the sampling interval $h = T/N$ tends to zero as $N \to \infty$.

2.13.2 A State Feedback/Feedforward Solution

The two-point boundary value representation can be converted into a state feedback plus feedforward solution in a similar manner to that described for continuous-time systems in Sect. 2.12.4. The first step in such a conversion is to propose a linear relationship between $p(t)$ and $x(t)$. The simplest such representation takes the form

$$p(t) = -K(t)x(t) + \xi(t), \quad t = 0, 1, 2, \cdots, N \tag{2.117}$$

where, as in the continuous-time case, the $n \times n$ matrix $K(t)$ and the term $\xi(t) \in \mathcal{R}^n$ are, as yet, unspecified. The optimal input is then given by

$$u(t) = u_0(t) + R^{-1} D^T Q(t) (r(t) - y(t)) + R^{-1}(t) B^T (-K(t)x(t) + \xi(t)), \quad for \quad t = 0, 1, 2, \cdots, N. \tag{2.118}$$

The characterization of a suitable $K(t)$ and $\xi(t)$ is now described when $D(t) \equiv 0$.
The Case of $D(t) \equiv 0$: Substituting for the costate in the equations for the two-point boundary problem gives

$$\begin{aligned} x(t+1) &= \left(A - BR^{-1}(t)B^T K(t)\right) x(t) + Bu_0(t) + BR^{-1}(t)B^T \xi(t) \\ &and \\ -K(t)x(t) + \xi(t) &= -\left(A^T K(t+1) + C^T Q(t+1)C\right) x(t+1) + A^T \xi(t+1) \\ &\quad + C^T Q(t+1) r(t+1) . \end{aligned} \tag{2.119}$$

Using the first equation to substitute for $x(t+1)$ into the second of these equations, collecting terms, and setting the coefficient matrix of $x(t)$ to zero then imply that these equations are satisfied if the feedback matrix satisfies the nonlinear recursion relations

$$K(t) = \left(A^T K(t+1) + C^T Q(t+1)C\right) \left(A - BR^{-1}(t)B^T K(t)\right), \quad for \quad t = 0, 1, 2, \cdots, N-1 \tag{2.120}$$

and the sequence $\{\xi(t)\}$ satisfies the equations

$$\begin{aligned} \xi(t) = &-\left(A^T K(t+1) + C^T Q(t+1)C\right) \left(Bu_0(t) + BR^{-1}(t)B^T \xi(t)\right) \\ &+ A^T \xi(t+1) + C^T Q(t+1) r(t+1), \\ &for \quad t = 0, 1, 2, \cdots, N-1. \end{aligned} \tag{2.121}$$

The boundary conditions are obtained by requiring that $p(N) = 0$ which is satisfied if

$$K(N) = 0 \quad and \quad \xi(N) = 0. \tag{2.122}$$

The equations must hence be solved by reverse time recursion.

More explicit representations for the solutions can be obtained by a simple rearrangement of the equations into the form, for $t = 0, 1, \cdots, N-1$,

$$\begin{aligned} \tilde{K}(t+1) &= A^T K(t+1) + C^T Q(t+1)C, \\ K(t) &= \left(I + \tilde{K}(t+1) BR^{-1}(t) B^T\right)^{-1} \tilde{K}(t+1) A \end{aligned} \tag{2.123}$$

and

$$\begin{aligned} \xi(t) = &\left(I + \tilde{K}(t+1) BR^{-1}(t) B^T\right)^{-1} (A^T \xi(t+1) \\ &- \tilde{K}(t+1) Bu_0(t) + C^T Q(t+1) r(t+1)). \end{aligned} \tag{2.124}$$

This representation is a basis for the construction of the time-varying state feedback gain matrix $-R^{-1}(t)B^T K(t)$ from the model data A, B, C and objective function weighting matrices $Q(t)$ and $R(t)$, $t = 0, 1, 2, \cdots, N$. The feedforward or predictive term $\xi(t)$, $t = 0, 1, 2, \cdots, N$ can then be constructed by backwards recursion from this data plus the signals $r(t)$ and $u_0(t)$, $t = 0, 1, 2, \cdots, N$.

2.14 Frequency Domain Relationships

The supervector description of plant dynamics is matrix-based and is equivalent, on any finite time interval, with the frequency domain, transfer function matrix description. The problems in using TFM descriptions in ILC are as follows:

1. they more naturally apply to infinite time intervals whereas ILC is, inevitably, considered only on finite intervals and
2. the inequalities must relate the properties of real matrices to those of complex matrices.

This section provides some links between supervector relationships and frequency domain properties. It is of particular value when considering the robustness of ILC schemes and throws some light on some aspects of performance.

2.14.1 Notation for Complex Matrices

The transpose notation M^T is a familiar concept in matrix theory. For complex matrices, additional operations are needed. In what follows $M^\dagger$ will be used to denote the matrix $\overline{M}^T$, i.e., the transpose of the complex conjugate of M. That is

$$M^\dagger = \overline{M}^T. \tag{2.125}$$

Traditionally, if $M^\dagger = M$, the matrix is said to be Hermitian, a concept that reduces to symmetric for real matrices.

A $p \times p$ matrix of real (resp. complex) numbers can be regarded as a linear operator mapping the space $\mathcal{U}$ (resp. $\mathcal{U}_c$) of $p \times 1$ columns of real (resp. complex) numbers. The inner products in $\mathcal{U}$ and $\mathcal{U}_c$ are defined by

$$\langle u, v\rangle_{\mathcal{U}} = u^T R v = \langle v, u\rangle_{\mathcal{U}}, \quad \langle u, v\rangle_{\mathcal{U}_c} = u^\dagger R v = \overline{\langle v, u\rangle_{\mathcal{U}_c}}, \tag{2.126}$$

respectively, where R is a real, symmetric, positive definite matrix. In both cases, the adjoint of M is the matrix

$$M^* = R^{-1} M^\dagger R \quad satisfying \quad \langle u, Mv\rangle = \langle M^* u, v\rangle. \tag{2.127}$$

Note the following:

1. $(M^*)^* = M$.
2. M is self-adjoint if and only if $M = M^*$.
3. The use of $R = I$ reduces these definitions to more familiar Euclidean concepts in complex matrix spaces.
4. In particular $\langle u, u\rangle_{\mathcal{U}_c}$ is a real number.
5. Equation (2.126) reveals the identities in the real and complex cases, respectively,

$$\begin{aligned} \langle u, Mu\rangle_{\mathcal{U}} = \langle u, \tfrac{1}{2}(M + M^*)u\rangle_{\mathcal{U}} = \tfrac{1}{2}u^\dagger(RM + M^\dagger R)u \quad \& \\ 2Re\left[\langle u, Mu\rangle_{\mathcal{U}_c}\right] = \langle u, (M + M^*)u\rangle_{\mathcal{U}_c} = u^\dagger(RM + M^\dagger R)u, \end{aligned} \tag{2.128}$$

all of which are real numbers.

2.14.2 Notions of Matrix Inequality

Inequality is an established concept for real numbers. Central to the results here is the notion of inequalities between real matrices and inequalities between complex matrices and the use of inequalities based on inner product descriptions. More precisely, using the constructions of the previous section yields the definition of matrix inequality as follows:

1. **Real Matrices:** The notation $M_1 \geq M_2$ (resp. $M_1 > M_2$) between *real* matrices is used to represent the situations when $M_1 - M_2$ is positive semi-definite (resp. positive definite), i.e., for all non-zero real vectors u,

$$\langle u, (M_2 - M_1)u\rangle_{\mathcal{U}} \geq 0 \quad (resp.\ \langle u, (M_2 - M_1)u\rangle_{\mathcal{U}} > 0). \tag{2.129}$$

The property of positivity is dependent on the matrix R. It can be reduced to the case of $R = I$ using the concept of the *symmetric part (or component)* of a real matrix M, namely $\frac{1}{2}\left(M + M^T\right)$. With this terminology, M is positive semi-definite (resp. positive definite) in $\mathcal{U}$ if the symmetric part of RM is positive definite in the more familiar case. A necessary and sufficient condition for this is that all eigenvalues of $\frac{1}{2}(RM + M^T R)$ are real and ≥ 0 (resp > 0).
2. **Complex Matrices:** The notation $M_1 \geq M_2$ (resp. $M_1 > M_2$) between *complex* matrices is used to represent the situations when $M_1 - M_2$ is positive semi-definite (resp. positive definite), i.e., for all non-zero complex vectors u,

$$Re\left[\langle u, (M_2 - M_1)u\rangle_{\mathcal{U}_c}\right] \geq 0 \quad (resp.\ Re\left[\langle u, (M_2 - M_1)u\rangle_{\mathcal{U}_c}\right] > 0). \tag{2.130}$$

The property of positivity is dependent on the matrix R. It can be reduced to the case of $R = I$ using the concept of the *Hermitian part (or component)* of a complex matrix M, namely $\frac{1}{2}\left(M + M^\dagger\right)$. More generally, M is positive semi-definite (resp. positive definite) in $\mathcal{U}_c$ if the Hermitian part of RM is positive definite in the more familiar case. A necessary and sufficient condition for this is that all eigenvalues of $\frac{1}{2}(RM + M^{\dagger}R)$ are real and ≥ 0 (resp > 0).

2.14.3 Useful Equalities

The fundamental link between a time series $\{u(0, u(1), \cdots, u(N)\}$ and the frequency domain is its (finite) $\mathcal{Z}$-transform

$$u(z) = \sum_{t=0}^{N} z^{-t} u(t), \tag{2.131}$$

which is an example of a more general class of time series that are infinite in length (i.e., $N = \infty$).

Note 2.8 Indeed, any finite times series becomes infinite if it is continued with zero entries, e.g., $\{u(0, \cdots, u(N), 0, 0, 0, \cdots \; to \; \infty\}$.

Note 2.9 The use of the notation $u(z)$ should not be confusing as the context of the analysis will easily indicate whether a frequency domain value is an element of the time series or a value of $u(z)$ at some particular value of z.

The basic results from complex variable theory that underpins what follows are as follows: suppose that $y(z)$ and $u(z)$ are $p \times 1$ vector functions of the complex variable z with infinite power series in z^{-1} that have the following properties:

1. Of being absolutely convergent for $|z| \geq 1 - \gamma$ for some $0 < \gamma < 1$.
2. Of being square summable in the sense that

$$\sum_{t=0}^{\infty} u^T(t)u(t) < \infty \quad \& \quad \sum_{t=0}^{\infty} y^T(t)y(t) < \infty. \tag{2.132}$$

   These conditions are always satisfied, for example, for signals of finite length, e.g., those generated by linear, asymptotically stable discrete state-space models with an input sequence that is zero for all t greater than some value N.
3. If R is any $p \times p$ symmetric, positive definite matrix, then the contour integral (evaluated by setting $z = e^{i\theta}$ in the interval $0 \leq \theta \leq 2\pi$)

$$\begin{aligned} \underbrace{\frac{1}{2\pi i}\oint_{|z|=1} u^T(z^{-1})Ry(z)\frac{dz}{z} = \sum_{t=0}^{\infty} u^T(t)Ry(t)}_{frequency\ to\ time\ domain\ equality} \\ = \underbrace{\sum_{t=0}^{N} u^T(t)Ry(t)}_{finite\ interval\ contribution} + \underbrace{\sum_{t=N+1}^{\infty} u^T(t)Ry(t)}_{rest\ of\ series}, \end{aligned} \tag{2.133}$$

which, from the Cauchy–Schwarz inequality, is absolutely convergent. Note, in particular, the following:

a. If, for example, $u(t) = 0,\ t > N$, then the infinite summation converges (becomes finite) for any signal values $y(t),\ t > N$.
b. Otherwise, the absolute convergence of the series shows that the contour integral is a good approximation to the finite interval component if N is large enough.
c. In particular, if $y = u$, then

$$\frac{1}{2\pi i}\oint_{|z|=1} y^T(z^{-1})Ry(z)\frac{dz}{z} = \sum_{t=0}^{N} y^T(t)Ry(t) + \sum_{t=N+1}^{\infty} y^T(t)Ry(t) \tag{2.134}$$

showing that the integral is always an upper bound on the finite interval component, the difference converging to zero as $N \to \infty$.

2.14.4 Positive Real Systems

Consider a discrete, linear, time-invariant, asymptotically stable, state-space system with $m = \ell$ and a supervector description $y = G(A, B, C, D)u$ (assuming zero initial conditions) on a finite interval $0 \le t \le N$. Let the input and output signal spaces be identical with inner product

$$\langle u, y\rangle_{\mathcal{U}} = \sum_{t=0}^{N} u^T(t)Ry(t). \tag{2.135}$$

Consider conditions that ensure that

$$G + G^* \ge \underline{g}_1 I \tag{2.136}$$

with respect to the specified inner product. Here, G^* is the adjoint of G and $\underline{g}_1$ is a real scalar. Extend u to the infinite time interval by adding zeros to the time series and write $y(z) = G(z)u(z)$. In general, the time series for y will be infinite in length and contain non-zero entries along its length. Despite this, the following equality will hold:

$$\langle u, y\rangle_{\mathcal{U}} = \langle u, Gu\rangle_{\mathcal{U}} = \frac{1}{2\pi i}\oint_{|z|=1} u^T(z^{-1})RG(z)u(z)\frac{dz}{z}. \tag{2.137}$$

As the left-hand side is real and $\langle u, G^*u\rangle_{\mathcal{U}} = \langle Gu, u\rangle_{\mathcal{U}}$, it follows that

$$\langle u, (G + G^*)u \rangle_{\mathcal{U}} = \frac{1}{2\pi i} \oint_{|z|=1} u^T(z^{-1})(RG(z) + G^T(z^{-1})R)u(z)\frac{dz}{z}. \tag{2.138}$$

If the complex matrix

$$RG(z) + G^T(z^{-1})R \geq \underline{g}_1 R, \quad |z| = 1, \tag{2.139}$$

then

$$\langle u, (G + G^*)u \rangle_{\mathcal{U}} \geq \frac{\underline{g}_1}{2\pi i} \oint_{|z|=1} u^T(z^{-1})u(z)\frac{dz}{z} = \underline{g}_1 \langle u, u \rangle_{\mathcal{U}}. \tag{2.140}$$

It follows that the frequency domain condition (2.139) guarantees the validity of the operator/matrix inequality (2.136).

Example 2.1 If the system is SISO, then (2.139) takes the simplified form

$$2Re\,[G(z)] \geq \underline{g}_1, \quad |z| = 1, \tag{2.141}$$

i.e., the real part of the transfer function is greater than $\frac{1}{2}\underline{g}_1$ at every point on the unit circle of the complex plane. This is a version of the concept of *positive real* transfer functions.

2.14.5 Frequency Attenuation Properties

This section outlines a useful approximation valid for discrete systems on long time intervals.

Consider a ℓ-input, m-output, linear, time-invariant, asymptotically stable, discrete-time state-space system (with TFM $G(z)$) in its supervector description $G(A, B, C, D)$. "Sinusoidal" inputs are represented by time series denoted by

$$u = \psi(z, \alpha) = \{\alpha, \alpha z, \alpha z^2, \cdots, \alpha z^N\}. \tag{2.142}$$

The response of G to this is, approximately, the oscillatory time series

$$y = \psi(z, G(z)\alpha), \tag{2.143}$$

plus a decaying exponential from $t = 0$ to satisfy the zero initial condition requirement. Suppose, now, that the adjoint system G^* is represented by

$$G^* = \mathcal{T} G(A^T, C^T Q, R^{-1}B^T, R^{-1}D^T Q)\mathcal{T}, \tag{2.144}$$

where $\mathcal{T}$ denotes the appropriate time reversal operator. Then the time series resulting from the input u, using the time reversal interpretation of adjoint system computations, can be approximated by a time series with terms

$$y = \psi(z, R^{-1}G^T(z^{-1})Q\alpha), \tag{2.145}$$

with errors at the end of the interval $0 \le t \le N$. The error being an exponentially decaying function in reverse time from $t = N$. It follows that

$$\psi(z, \underbrace{R^{-1}G^T(z^{-1})QG(z)}_{frequency\ attenuation}\alpha) \quad (resp.\ \psi(z, \underbrace{G(z)R^{-1}G^T(z^{-1})Q}_{frequency\ attenuation}\alpha)) \tag{2.146}$$

are the approximate responses of G^*G (resp. GG^*) to u. The error in the representation consists of exponentially bounded signals from $t = 0$ and (in reverse time) $t = N$, arising from the need to ensure that the initial and terminal conditions are satisfied.

Note 2.10 A more detailed treatment can be found in the companion text Owens [18]. For long enough time intervals, the "end effects" make an infinitesimal contribution to the response characteristics on the majority of the interval $0 \le t \le N$. This leads to the intuitive expectation that the operator GG^* has frequency domain performance issues affected in several ways including the following:

1. the frequency response characteristics of the TFM dominate the response of GG^* and G^*G;
2. resonances in the system G will appear as resonances in both GG^* and G^*G;
3. high frequency characteristics of GG^* and G^*G are inherited from G including low-pass properties if $D = 0$; and
4. if $m = \ell = 1$, attenuation is described by the (squared) gain characteristics of $G(z)$.

Clearly, the classical interpretations of frequency domain behaviour can play an important role in interpretation of the system time response. The interpretations are only approximate as end effects will change the details of what would be observed in measured data.

Remark 2.2 (Response to Frequency Domain Eigenvectors) If, for example, α is an eigenvector of $G(z)R^{-1}G^T(z^{-1})Q$ (or $R^{-1}G^T(z^{-1})QG(z)$) with eigenvalue $\sigma^2(z)$, then, if N is large, the output response is approximately equal to $\sigma^2(z)u(z, \alpha)$, i.e., a scaled version of the input.

2.14.6 Estimating the Norm of $G(A, B, C, D)$

Suppose that $G(A, B, C, D)$ represents a linear, time-invariant, asymptotically stable, discrete-time state-space system and let G be the matrix operator in the supervector description. The norm of G, denoted $\|G\|$, mapping $\mathcal{U} = \mathcal{R}^{\ell(N+1)}$ into $\mathcal{Y} = \mathcal{R}^{m(N+1)}$ can be deduced from this supervector description for the inner products,

$$\langle u, v \rangle_{\mathcal{U}} = u^T R_d v = \sum_{t=0}^{N} u^T(t) R v(t), \quad \langle y, w \rangle_{\mathcal{Y}} = y^T Q_d w = \sum_{t=0}^{N} y^T(t) Q w(t), \tag{2.147}$$

by solving the constrained maximization problem

$$maximize \sqrt{(Gu)^T Q_d G u}, \quad subject\ to \quad u^T R_d u = 1. \tag{2.148}$$

Note 2.11 Here, R_d (resp. Q_d) is the $\ell(N+1) \times \ell(N+1)$ (resp.$m(N+1) \times m(N+1)$) matrix of zeros with the symmetric, positive definite matrix R_d (resp. Q_d) copied $N+1$ times along the diagonal.

Note 2.12 If u is allowed to be any *complex* time series, the norm of G is computed from

$$maximize \sqrt{(Gu)^\dagger Q_d G u}, \quad subject\ to \quad u^\dagger R_d u = 1, \tag{2.149}$$

where the notation $\phi^\dagger$ denotes the complex conjugate transpose of ϕ.

The solutions yield the norm as being equal to the maximum eigenvalue of $G^* G = R_d^{-1} G^T Q_d G$. The important point to note is that this can be estimated using frequency domain calculations. Write the supervector equation $y = Gu$ on the interval $0 \leq t \leq N$, and construct the input transform $u(z)$ by extending u to the infinite interval with zeros and computing the resultant output sequence (with $\mathcal{Z}$-transform $y(z)$) on $0 \leq t < \infty$. It follows, using complex integration techniques, that

$$\begin{aligned} y^T Q_d y &= \tfrac{1}{2\pi i} \oint_{|z|=1} y^T(z^{-1}) Q y(z) \tfrac{dz}{z} - \underbrace{\sum_{t>N} y^T(t) Q y^T(t)}_{\to 0 \text{ as } N \to \infty} \\ &= \tfrac{1}{2\pi i} \oint_{|z|=1} u^T(z^{-1}) G^T(z^{-1}) Q G(z) u(z) \tfrac{dz}{z} - \textstyle\sum_{t>N} y^T(t) Q y^T(t) \\ &\leq \sup_{|z|=1} r\left(R^{-1} G^T(z^{-1}) Q G(z)\right) \tfrac{1}{2\pi i} \oint_{|z|=1} u^T(z^{-1}) R u(z) \tfrac{dz}{z} \\ &= \sup_{|z|=1} r\left(R^{-1} G^T(z^{-1}) Q G(z)\right) u^T R_d u, \end{aligned} \tag{2.150}$$

where $r(\cdot)$ denotes the *spectral radius* of its argument, i.e., the largest modulus of the eigenvalues. This shows that

$$\underbrace{\|G\|^2 \leq \sup_{|z|=1} r\left(R^{-1} G^T(z^{-1}) Q G(z)\right).}_{Bound\ for\ Norm\ of\ G} \tag{2.151}$$

As N increases, the inequality approaches equality,

$$\underbrace{\lim_{N\to\infty} \|G\|^2 = \sup_{|z|=1} r\left(R^{-1}G^T(z^{-1})QG(z)\right)}_{Asymptotic\ Equality}, \tag{2.152}$$

suggesting that the use of the frequency domain characterization is often a good approximation to the norm.

Note 2.13 The statement (2.152) can be proven by noting the following:

1. The norm of $\|G\|$ is the same whether or not $\mathcal{U}$ and $\mathcal{Y}$ are real or complex. It is therefore useful to consider *complex* u generating *complex* y.
2. With this assumption, the use of the oscillating input u generated by $u(t) = \frac{1}{\sqrt{N+1}} v_1 z_1^t$, $0 \le t \le N$, where z_1, satisfying $|z_1| = 1$, generates the maximum eigenvalue η_1 of $r\left(R^{-1}G^T(z_1^{-1})QG(z_1)\right)$ and v_1 is the corresponding eigenvector. If $v_1^\dagger R v_1 = 1$, then $u^\dagger R_d u = 1$. Using the methods of Sect. 2.14.5 it follows that $y(t) \approx R^{-1}G^T(z_1^{-1})QG(z_1)v_1 = \eta_1 v_1$ with error going to zero as $N \to \infty$. In this case it follows that $\lim_{N\to\infty} y^\dagger Q_d y = \eta_1$ which is what was asserted.

2.14.7 Another Useful Inequality

The methodologies described above can be used to derive a range of time to frequency domain inequalities. For example, using the notation of the previous sections, consider the derivation of frequency domain conditions for discrete systems by providing sufficient conditions for the inequality

$$U + U^* \ge \mu U^* G^* G U \qquad \left(= \mu (GU)^* GU \right) \tag{2.153}$$

to hold, where $\mu \ge 0$ is a real number, U maps $\mathcal{U}$ into itself, and the matrix G maps $\mathcal{U}$ into $\mathcal{Y}$.

In inner product form, the operator inequality takes the form, for any real vector u,

$$\langle u, (U + U^*)u \rangle_{\mathcal{U}} \ge \mu \langle GUu, GUu \rangle_{\mathcal{Y}}. \tag{2.154}$$

Using the techniques of Sects. 2.14.4 and 2.14.6, this converts easily into the sufficient condition

$$RU(z) + U^T(z^{-1})R \ge \mu U^T(z^{-1})G^T(z^{-1})QG(z)U(z), \quad |z| = 1. \tag{2.155}$$

2.14.7.1 An Application

Suppose that $\mathcal{Y} = \mathcal{U}$ and hence that $Q = R$. Also take $G = I$. Consider the inequality

$$U + U^* \geq \mu U^* U. \tag{2.156}$$

There are two expressions that follow from the previous sections:

1. The inequality can then be written in the form,

$$(\mu^{-1}I - U)^*(\mu^{-1}I - U) \leq \mu^{-2}I \tag{2.157}$$

   which is satisfied if $\|\mu^{-1}I - U\|^2 \leq \mu^{-2}$. Using the results of Sect. 2.14.6 it is therefore sufficient that the spectral radius

$$r\left(R^{-1}(\mu^{-1}I - U(z^{-1}))^T R(\mu^{-1}I - U(z))\right) \leq \mu^{-2}, \quad |z| = 1. \tag{2.158}$$

2. An alternative condition follows from the discussion in Sect. 2.14.7, namely

$$RU(z) + U^T(z^{-1})R \geq \mu U^T(z^{-1})RU(z), \quad |z| = 1. \tag{2.159}$$

2.15 Inner Products, Norms, and Operators

The notation used in this chapter is brought together and extended in this section to form a basis for the rest of the text. Those readers familiar with the ideas can skip this material. Less experienced readers will find an overview that may form a useful point of reference.

2.15.1 Set and Related Notation

It is convenient to use set and other notations as a shorthand in the presentation of the material in the rest of this book. The main symbols used are as follows:

1. If $\mathcal{S}$ is a set, then the statement $s \in \mathcal{S}$ is a simple way of saying that the point s lies inside the set $\mathcal{S}$.
2. The statement $\mathcal{S}_1 \subset \mathcal{S}_2$ is a simple way of saying that every point in the set $\mathcal{S}_1$ lies inside the set $\mathcal{S}_2$. $\mathcal{S}_1$ is a *subset* of $\mathcal{S}_1$.
3. The *union* $\mathcal{S}_1 \cup \mathcal{S}_2$ of two sets is the set of points that lie in any one, or both, of the sets.
4. The *intersection* $\mathcal{S}_1 \cap \mathcal{S}_2$ of two sets is the set of points that lie in both of the sets.

5. The symbol $\forall$ is a simple representation of the phrase "for all", e.g., the statement $\forall s \in \mathcal{S}$ is shorthand for the statement "for all points s in the set $\mathcal{S}$".

Other notations used in this text include statements such as the following:

1. The symbol $\{ s \in \mathcal{S};\ defined\ property \}$ represents the set of points s in $\mathcal{S}$ with the property specified. For example $\{real\ number\ s\ ;\ |s-1| \le 2\}$ represents the set of numbers in the interval $-1 \le s \le 3$.
2. The statement
$$u = \arg\min\{J(u)\ ;\ u \in \mathcal{S}_0 \subset \mathcal{S}\} \tag{2.160}$$
represents a statement that $u \in \mathcal{S}$ is a point in the subset $\mathcal{S}_0$ of $\mathcal{S}$ that *minimizes* the function $J(u)$ of values in $\mathcal{S}_0$. In many cases of relevance to ILC, $\mathcal{S}_0$ is defined by equations, rather than by set inclusion statements.
3. The statement
$$J^* = \inf\{J(u)\ ;\ u \in \mathcal{S}_0 \subset \mathcal{S}\} \tag{2.161}$$
represents a statement that J^* is the smallest value (greatest lower bound) of the function $J(u)$ in $\mathcal{S}_0$. If inf is replaced by sup the statement represents the largest value (least upper bound) of $J(u)$.

2.15.2 Vector Spaces

A real vector space $\mathcal{S}$ is a set of *vectors* that is closed under addition and multiplication by real numbers. That is, if the vector s_1, s_2 are elements of $\mathcal{S}$, then so is $\lambda_1 s_1 + \lambda_2 s_2$ for any real numbers λ_1 and λ_2. Examples of such spaces are as follows:

1. The space $\mathcal{S} = \mathcal{R}^{p_1 \times p_2}$ of $p_1 \times p_2$ matrices. If $p_2 = 1$ the notation is simplified to $\mathcal{R}^{p_1}$.
2. The space $L_2[a, b]$ of real-valued functions $f(t)$ on some time interval $a \le t \le b$ (denoted $[a, b]$) and with the property that the integral $\int_a^b f^2(t)dt$ exists and is finite.
3. More generally, the space $L_2^p[a, b]$ of real-valued $p \times 1$ matrix functions $f(t)$ on $[a, b]$ with the property that the integral
$$\int_a^b f^T(t)M(t)f(t)dt \tag{2.162}$$
exists and is finite. Here $M(t)$ is a $p \times p$ piece-wise continuous, symmetric matrix function of time t satisfying the requirement that, for some scalar $\delta > 0$, $M(t) - \delta I_p$ is positive semi-definite at every point $t \in [a, b]$.
4. The space of sequences $\{s(0), s(1), \cdots, s(N)\}$ (where N is finite or infinite) with the property that each $s(t),\ t \ge 0$, lies in $\mathcal{R}^p$ and

$$\sum_{0}^{N} s^T(t)M(t)s(t) < +\infty, \tag{2.163}$$

where $\{M(t)\}_{t\geq 0}$ is bounded and, for some scalar $\delta > 0$, $M(t) - \delta I_p$ is positive semi-definite at every point t in the sequence. This is trivially satisfied if N is finite but is a convergence condition if N is infinite.

5. A set of vectors $\mathcal{B} = \{s_j\}$ are *linearly independent* if, and only if, the expression $\sum_{s_j\in\mathcal{B}} a_j s_j = 0$, where the coefficients a_j are real numbers, implies that all coefficients are equal to zero.

   a. If $\mathcal{B}$ are linearly independent and any vector in $\mathcal{S}$ can be written as a linear combination of points in $\mathcal{B}$, then $\mathcal{B}$ forms a *basis set* for $\mathcal{S}$.
   b. A basis set for $\mathcal{S}$ is an *orthogonal basis* if $\langle s_j, s_k\rangle = 0,\ j \neq k$. It is *orthonormal* if, in addition, $\langle s_j, s_k\rangle = 1$ whenever $j = k$ (inner products are defined in Sect. 2.15.4).
   c. If a basis set contains a finite number of elements only, the space $\mathcal{S}$ is said to be finite dimensional. In all other cases, the space is infinite dimensional.

2.15.3 Normed Vector Spaces

A space $\mathcal{S}$ is a *normed vector space* if it is a vector space that associates with each vector s a real number $\|s\|_{\mathcal{S}}$, its *norm*, with the properties that

$$\begin{aligned} &\|s\|_{\mathcal{S}} \geq 0, \qquad \|\lambda s\|_{\mathcal{S}} = |\lambda|\|s\|_{\mathcal{S}} \\ &\|s_1 + s_2\|_{\mathcal{S}} \leq \|s_1\|_{\mathcal{S}} + \|s_2\|_{\mathcal{S}} \quad and \\ &\|s\|_{\mathcal{S}} = 0 \quad if,\ and\ only\ if, \quad s = 0, \end{aligned} \tag{2.164}$$

for all s, s_1, and s_2 in $\mathcal{S}$ and all real numbers λ. Examples of norms include

$$\|f\|_{\mathcal{R}^p} = \sqrt{f^T f} \ \ \& \ \ \|f\|_{L_2^p[a,b]} = \sqrt{\int_a^b f^T(t)M(t)f(t)dt}, \tag{2.165}$$

if p is finite and $M(t)$ is positive definite at every point of $[a, b]$.

Norms are the extension of the concept of distance, familiar in the real world, to other areas of mathematics. They are used as a measure of distance and magnitude but also for assessing the convergence of infinite sequences $s_0, s_1, s_2, \cdots$ in a normed space. More precisely:

1. The series *converges* to a point s_∞ in $\mathcal{S}$ (the limit of the sequence) if, and only if, the maximum distance $\|s_j - s_\infty\|$ converges to zero in the familiar sense of real numbers.

2. The sequence is a *Cauchy sequence* if $\|s_j - s_k\|$ converges to zero as $min\{j, k\}$ goes to infinity. One of the quirks of mathematics is that not all normed vector spaces have the property that all Cauchy sequences converge. If all Cauchy sequences converge in a normed space, the space is said to be *complete*. This is true for all spaces relevant to this text (unless otherwise stated) and includes $\mathcal{R}^{p_1 \times p_2}$, $L_2^p[a, b]$ (a and b finite) and sequence spaces for all N (including ∞).
3. A complete normed vector space is said to be a Banach space.
4. Normally there are options available in the choice of norm for a space. Two norms $\|s\|_1$ and $\|s\|_2$ in a space $\mathcal{S}$ are *topologically equivalent* if, and only if, there exist two real numbers $0 < \alpha \leq \beta$ such that, for all s in $\mathcal{S}$,

$$\alpha\|s\|_1 \leq \|s\|_2 \leq \beta\|s\|_1 \quad (topological\ equivalence). \tag{2.166}$$

The important consequence for this property is that convergence of a sequence with respect to one norm implies the very same convergence with respect to the other.

2.15.4 Inner Products, Norms, and Cauchy–Schwarz Inequality

An *inner product* in a real vector space $\mathcal{S}$ is a bilinear function of vectors u, v in $\mathcal{S}$, denoted $\langle \cdot, \cdot \rangle_{\mathcal{S}}$ with the properties

$$\begin{array}{c} \langle u, v\rangle_{\mathcal{S}} = \langle v, u\rangle_{\mathcal{S}}, \quad \langle u, \lambda v\rangle_{\mathcal{S}} = \lambda\langle u, v\rangle_{\mathcal{S}} \\ \langle u, v + w\rangle_{\mathcal{S}} = \langle u, v\rangle_{\mathcal{S}} + \langle u, w\rangle_{\mathcal{S}} \qquad and \\ \langle u, u\rangle_{\mathcal{S}} \geq 0, \quad equality\ holding\ if,\ and\ only\ if,\ \ u = 0. \end{array} \tag{2.167}$$

A space with an inner product is said to be an *inner product space* or *pre-Hilbert space*. If $\mathcal{S}$ is also complete, it is a *Hilbert space*.

Inner products have some important properties, including the following:

1. Two vectors s_1 and s_2 are said to be *orthogonal* if $\langle s_1, s_2\rangle_{\mathcal{S}} = 0$.
2. If $\mathcal{S}_0$ is a vector subspace of $\mathcal{S}$ then the *orthogonal complement* $\mathcal{S}_0^{\perp}$ of $\mathcal{S}_0$ is the set of vectors that are orthogonal to all points in $\mathcal{S}_0$. That is,

$$\mathcal{S}_0^{\perp} = \{\, w \in \mathcal{S}\,;\ \langle s, w\rangle_{\mathcal{S}} = 0 \ \ \forall s \in \mathcal{S}_0 \,\}. \tag{2.168}$$

3. The square root of the inner product $\langle u, u\rangle_{\mathcal{S}}$ is a norm

$$\|u\|_{\mathcal{S}} \ = \ \sqrt{\langle u, u\rangle_{\mathcal{S}}}. \tag{2.169}$$

4. Norms and inner products are related by the *Cauchy–Schwarz Inequality*

$$|\langle u, v\rangle_{\mathcal{S}}| \leq \|u\|_{\mathcal{S}}\|v\|_{\mathcal{S}}, \tag{2.170}$$

equality holding if, and only if, $v = \lambda u$ for some real number λ, i.e., the vectors are colinear or aligned.

Examples of inner products are as follows:

$$\langle u, v\rangle_{\mathcal{R}^p} = u^T R v \quad \& \quad \langle u, v\rangle_{L_2^p[a,b]} = \int_a^b u^T(t) M(t) v(t) dt, \tag{2.171}$$

where the $p \times p$ matrix R is symmetric and positive definite, and $M(t)$ is symmetric and positive definite at every point t in $[a, b]$. Both of these examples are Hilbert spaces.

2.15.5 *Linear Operators*

A linear operator L in a vector space $\mathcal{S}_1$ is a mathematical operation on vectors s in a vector subspace of $\mathcal{S}_1$ called its *domain*. Normally the domain is the whole of the space $\mathcal{S}_1$ although this may not always be the case, e.g., continuous functions form a vector subspace of $L_2[a, b]$ which contains, amongst other vectors, piecewise continuous functions. The result of the operation is denoted by Ls which has the following properties:

1. Ls lies in a subset (said to be its *range* $\mathcal{R}[L]$) of a specified space $\mathcal{S}_2$.
2. For all s_1, s_2 in the domain of L and real numbers λ,

$$L(s_1 + s_2) = Ls_1 + Ls_2 \quad \& \quad L(\lambda s_1) = \lambda s_1. \tag{2.172}$$

3. The spaces are identified by writing $L : \mathcal{S}_1 \to \mathcal{S}_2$.
4. Unless otherwise stated, the domain of L will normally be the whole space $\mathcal{S}_1$. The range of L is denoted by $\mathcal{R}[L]$ and need not be the whole space $\mathcal{S}_2$.
5. The *kernel* of L is the set of vectors in the domain of L possessing the property that $Ls = 0$. It is denoted by $ker[L]$.
6. $\mathcal{S}_1$ and $\mathcal{S}_2$ could be based on the same underlying vector space but with different norms. This observation underlines the point that *normed spaces must be considered as vector spaces with an associated norm*, i.e., changing the norm changes the space in a mathematical sense even though the vectors within that space are the same.

Real $p \times q$ matrices are examples of linear operators mapping the space $\mathcal{S}_1 = \mathcal{R}^p$ of $p \times 1$ column matrices into the space $\mathcal{S}_2 = \mathcal{R}^q$. Amongst other examples of interest, this implies, in particular, that discrete, ℓ-input, m-output, state-space models (in the equivalent state-space or supervector forms) on finite intervals $0 \le t \le N$ are also linear operators $\mathcal{R}^{\ell(N+1)} \mapsto \mathcal{R}^{m(N+1)}$. In addition the convolution operator G,

$$(Gu)(t) = \int_a^t Ce^{A(t-t')} Bu(t')dt', \quad t \ge a, \tag{2.173}$$

on any finite interval, is a linear operator $L_2^\ell[a,b] \to L_2^m[a,b]$.

2.15.6 Operator Norms

The set of linear operators $L : \mathcal{S}_1 \to \mathcal{S}_2$ form a vector space. An important subclass of these operators are *bounded operators* for which the *operator norm* $\|L\|$ defined below exists and is finite. The definition implies that $\|L\|$ satisfies all the conditions required for it to be a norm in the sense of (2.164).

The norm of a linear operator L is defined by the maximization problem,

$$\underbrace{\|L\| = \sup_{s\in\mathcal{S}_1}\{\ \|Ls\|_{\mathcal{S}_2}\ ;\ \|s\|_{\mathcal{S}_1} = 1\ \},}_{Definition\ of\ the\ operator\ norm} \tag{2.174}$$

the operator being bounded if the maximum value is finite.

Operator norms have some important properties, namely the inequalities,

$$\|Ls\|_{\mathcal{S}_2} \le \|L\|\|s\|_{\mathcal{S}_1}\ \ \forall s \in \mathcal{S}_1 \ \ \& \ \ \|L_1L_2\| \le \|L_1\|\|L_2\|. \tag{2.175}$$

Example 2.2 Consider the $p_1 \times p_2$ matrix M, regarded as a linear operator from $\mathcal{R}^{p_2}$ into $\mathcal{R}^{p_1}$, i.e., a mapping from $p_2 \times 1$ column matrices into $p_1 \times 1$ column matrices. The inner products are taken to be, respectively,

$$\langle u, v\rangle_{\mathcal{R}^{p_2}} = u^T R v \quad \& \quad \langle y, w\rangle_{\mathcal{R}^{p_1}} = y^T Q w, \tag{2.176}$$

where R (resp. Q) are symmetric, positive definite $p_2 \times p_2$ (resp. $p_1 \times p_1$) matrices. The norm of M is obtained by solving the maximization problem

$$\|M\|^2 = \sup_{s\in\mathcal{R}^{p_2}}\{\ (Ms)^T QMs\ ;\ s^T Rs = 1\ \}. \tag{2.177}$$

Application of Lagrange Multiplier techniques then yields the result that the norm of M is the square root of the maximum eigenvalue of $R^{-1}M^TQM$.

2.15.7 Adjoint and Self-Adjoint Operators

Adjoint Operators play a significant role in optimization theory and, consequently, in the theory of Iterative Learning Control. Suppose that L is a bounded linear operator from an inner product space $\mathcal{S}_1$ into an inner product space $\mathcal{S}_2$. The *adjoint* of L denoted L^* is the unique, bounded, linear operator $L^* : \mathcal{S}_2 \mapsto \mathcal{S}_1$ satisfying the equality

$$\langle s_2, Ls_1 \rangle_{\mathcal{S}_2} \ = \ \langle L^* s_2, s_1 \rangle_{\mathcal{S}_1}, \ \ \forall \ s_1 \in \mathcal{S}_1 \ \ \& \ \ s_2 \in \mathcal{S}_2. \tag{2.178}$$

Note 2.14 The adjoint is a generalization of the idea of "transpose". Using the *Euclidean* inner product $\langle s_1, s_2 \rangle = s_1^T s_2$ for $p \times 1$ real matrices (vectors), then, if L is a $p \times p$ real matrix the identity $s_1^T L s_2 = (L^T s_1)^T s_2$ shows that $L^* = L^T$ in this case.

It has several properties as follows:

1. The adjoint of $L_1 + L_2$ is $L_1^* + L_2^*$ and that of λL is λL^*,
2. the adjoint of $L_1 L_2$ is $L_2^* L_1^*$,
3. the norm of the adjoint of L has the property that $\|L^*\| = \|L\|$, and
4. the adjoint of the adjoint of an operator L is the original operator, i.e., $(L^*)^* = L$.

Special cases of operators are the *self-adjoint* operators and the subset of *positive self-adjoint operators*:

1. L is *self-adjoint* whenever $\mathcal{S}_1 = \mathcal{S}_2$ and $L = L^*$.
2. Even if $\mathcal{S}_1 \neq \mathcal{S}_2$, the operators $LL^* : \mathcal{S}_2 \to \mathcal{S}_2$ and $L^* L : \mathcal{S}_1 \to \mathcal{S}_1$ are self-adjoint.
3. If $L = L^*$ and $L : \mathcal{S} \to \mathcal{S}$, the quadratic form $\langle s, Ls \rangle_{\mathcal{S}}$ is *positive semi-definite* (resp. *positive definite*), written $L \geq 0$ (resp. $L > 0$), if and only if it is ≥ 0 (resp. > 0) for all $s \neq 0$ in $\mathcal{S}$. In both cases, the operator is said to be *positive*.

Note 2.15 This is a generalization of the more familiar notions used for matrix quadratic forms $s^T Ls$.

4. The notation $L_1 \geq L_2$ (resp. $L_1 > L_2$) is equivalent to the statements $L_1 - L_2 \geq 0$ (resp. $L_1 - L_2 > 0$).
5. The norm of a self-adjoint, positive operator $L : \mathcal{S} \to \mathcal{S}$ can also be computed from the formula
$$\|L\| = \sup_{s \in \mathcal{S}} \{ \ \langle s, Ls \rangle_{\mathcal{S}} \ ; \ \|s\|_{\mathcal{S}} = 1 \ \}, \tag{2.179}$$
a formula that holds trivially with $\|L\| = \|G\|^2 = \|G^*\|^2$ if $L = GG^*$ or $L = G^* G$, even if G is not self-adjoint or positive.
6. Any positive, self-adjoint operator $L : \mathcal{S} \to \mathcal{S}$ has a unique positive, self-adjoint square root $L^{\frac{1}{2}} : \mathcal{S} \to \mathcal{S}$ with the property that $L = L^{\frac{1}{2}} L^{\frac{1}{2}}$.

2.15.7.1 Examples of Adjoint Operators

Examples of adjoint operators have been seen in the modelling sections of the text and depend on the inner products used in the spaces. Relevant examples for this text include the following:

1. Adjoint for discrete supervector models of the $p_1 \times p_2$ matrix M regarded as a linear operator from $\mathcal{R}^{p_2}$ into $\mathcal{R}^{p_1}$ of $p_2 \times 1$ column matrices into $\mathcal{R}^{p_1}$ of $p_1 \times 1$ column matrices with inner products, respectively,

$$\langle u, v\rangle_{\mathcal{R}^{p_2}} = u^T R v \quad \& \quad \langle y, w\rangle_{\mathcal{R}^{p_1}} = y^T Q w, \tag{2.180}$$

where R (resp. Q) are symmetric, positive definite $p_2 \times p_2$ (resp. $p_1 \times p_1$) matrices, is the $p_2 \times p_1$ matrix

$$M^* = R^{-1} M^T Q. \tag{2.181}$$

2. In the special case when $p_1 = p_2 = p$ and $R = Q = I_p$, then $M^* = M^T$, i.e., the adjoint of M is the transpose of M.
3. The adjoint operator of the convolution operator G given in (2.173) is $G^* : L_2^m[a, b] \mapsto L_2^\ell[a, b]$. The chosen inner products are

$$\langle u, v\rangle_{L_2^\ell} = \int_a^b u^T(t) R v(t) dt \quad \& \quad \langle y, w\rangle_{L_2^m} = \int_a^b y^T(t) Q w(t) dt \tag{2.182}$$

where R (resp. Q) are symmetric, positive definite $\ell \times \ell$ (resp. $m \times m$) matrices. G^* takes the form of the mapping, $a \le t \le b$,

$$(G^* y)(t) = \int_t^b R^{-1} B^T e^{A^T(t'-t)} C^T Q y(t') dt', \tag{2.183}$$

which is the solution of the state equations

$$\dot{p}(t) = -A^T p(t) - C^T Q y(t), \ \ (G^* y)(t) = R^{-1} B^T p(t), \quad p(b) = 0. \tag{2.184}$$

The reader is invited to consider the proof of these statements and to extend the results to the case of time-dependent R and Q. Hint: The same equations hold with R and Q replaced by time-dependent versions $R(t)$ and $Q(t)$.

2.16 Further Reading

This chapter has reviewed, briefly, the mathematical and control systems knowledge that will be used in later chapters of the book. For a more detailed introduction, please refer to the companion text Owens [18]. For more information on classical control theory, readers may refer to texts such as Goodwin et al. [9], Gopal [10], Landau and Zito [12], and Franklin [7]. The extensions to state-space and MIMO (multivariable) control systems can be found in a number of texts including Albertos and Sala [1], Friedland [8], Owens [16], and Maciejowski [14]. The well-established ideas of optimal control seen in texts such as Athans and Falb [4], Anderson and Moore

[2], Anderson and Moore [3], and Chen and Francis [5] are central to what follows. An undergraduate-level introduction can be found in Owens [17]. In particular, the geometric approach used in Luenberger [13] is particularly suited to the optimization content of this text. More information on functional analysis (including vector spaces, norms, and operators) can be found in the mathematical texts such as Collatz [6], Ortega and Rheinboldt [15], and Kreyszig [11].

References

1. Albertos P, Sala A (2004) Multivariable control systems: an engineering approach. Springer
2. Anderson BDO, Moore JB (1971) Linear optimal control. Prentice-Hall, New Jersey
3. Anderson BDO, Moore JB (1989) Optimal control: linear quadratic methods. Prentice-Hall, New Jersey
4. Athans M, Falb PL (1966) Optimal control. McGraw Hill
5. Chen T, Francis B (1995) Optimal sampled-data control systems. Springer, London
6. Collatz L (1966) Functional analyis and numerical mathematics. Academic Press, New York and London
7. Franklin G (1997) Digital control of dynamic systems. Prentice-Hall
8. Friedland B (1986) Control systems design: an introduction to state space methods. Dover, New York
9. Goodwin G, Graebe S, Salgado M (2000) Control systems design. Prentice-Hall
10. Gopal M (2002) Control systems: principles and design. Tata McGraw-Hill Education
11. Kreyszig E (1978) Introductory functional analysis with applications. John Wiley & Sons
12. Landau I, Zito G (2006) Digital control systems: design, identification and implementation. Springer, London
13. Luenberger DG (1969) Optimization by vector space methods. Wiley, New York, Chichester
14. Maciejowski JM (1989) Multivariable feedback design. Addison-Wesley
15. Ortega J, Rheinboldt W (1970) Functional analyis and numerical mathematics. Academic Press, London
16. Owens DH (1978) Feedback and multivariable systems. Peter Peregrinus, Stevenage
17. Owens DH (1981) Multivariable and optimal systems. Academic Press, London
18. Owens DH (2015) Iterative learning control—an optimization paradigm. Springer

Chapter 3
Inverse-Based Iterative Learning Control

This chapter introduces the first Iterative Learning Control (ILC) design of the book, namely, the Inverse-Based ILC Algorithm. Inverse-based ILC plays a key role in the research into ILC as the ultimate goal of ILC design can be interpreted as finding an system input u^* to generate the desired reference r by "inverting" the system dynamics. Inverse-based ILC provides a benchmark algorithm whose convergence and robustness properties give some key insights for the general properties of many other ILC algorithms to be introduced in later chapters of the book.

To illustrate the key ideas, in this chapter we use the following operator form of system representation, i.e.,

$$y = Gu + d.$$

More specifically, we use discrete-time, linear time-invariant systems as examples and consider the input and output spaces to be the familiar Euclidean spaces. The dynamical equation can therefore be most easily visualized as a matrix relationship relating input and output supervectors. The ideas in this chapter apply more generally to continuous-time systems and more general forms of underlying signal spaces—at the end of the chapter, interested readers are pointed towards further materials on these topics.

The chapter is organized as follows. To motivate the Inverse-Based ILC Algorithm, the simplest Arimoto law and its extension, the Parameter Optimal ILC (POILC) algorithm, are firstly introduced in Sect. 3.1. The Inverse-based ILC design is then presented and its convergence properties are given in Sect. 3.2. Its robustness properties against modelling mismatches are discussed in Sect. 3.3. The possible difficulties when dealing with non-minimum phase (NMP) systems are explained in Sect. 3.4, and finally application of Inverse-based ILC to address point-to-point tracking tasks is presented in Sect. 3.5.

B. Chu and D. H. Owens, *Optimal Iterative Learning Control*, Advances in Industrial Control, https://doi.org/10.1007/978-3-031-80236-2_3

3.1 From Arimoto Law to Parameter Optimal Iterative Learning Control

The following example illustrates the, so-called, *Arimoto law*-type ILC algorithm which is one of the earliest ILC design approaches [1]. It can work effectively for perfect trajectory (signal) tracking, but also has substantial limitations.

Example 3.1 (Owens and Hatonen [21]) Consider the system

$$G(s) = \frac{s+1}{s^2+5s+6}.$$

The system is sampled at intervals of $T_s = 0.1$ using zero-order hold resulting in a discrete-time plant model $S(A, B, C)$. The system is required to track a reference signal $r(t) = sin(2\pi t/6)$ defined over the interval $t \in [0, 6]$. First, the system is controlled with a Proportional–Integral–Differential (PID) controller

$$u(t) = K_P e(t) + K_I \int_0^t e(\tau)\mathrm{d}\tau + K_D \dot{e}(t) \tag{3.1}$$

where $K_p = 15$, $K_I = 8$, and $K_D = 0$, which is also sampled using zero-order hold with the same sampling time.

Figure 3.1 shows the output $y(t)$ indicating that the system is only capable of tracking the reference signal to a moderate degree of accuracy. Note that this same tracking result is obtained during each repetition, because the PID-controller has fixed parameters.

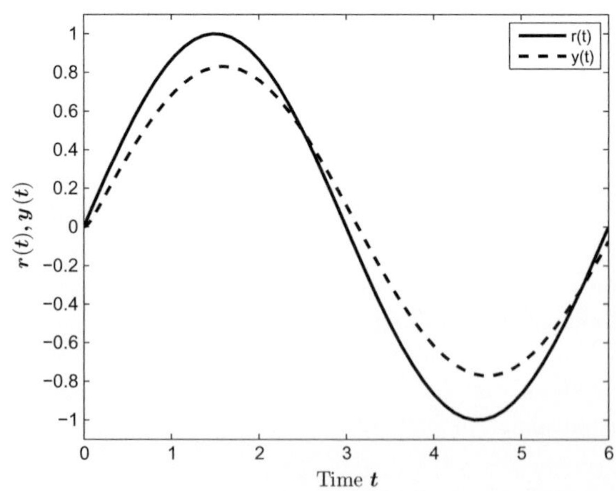

Fig. 3.1 The output response to the use of the PID-controller (3.1)

Fig. 3.2 Error norm convergence behaviour of the Arimoto algorithm (3.2) with different learning gains

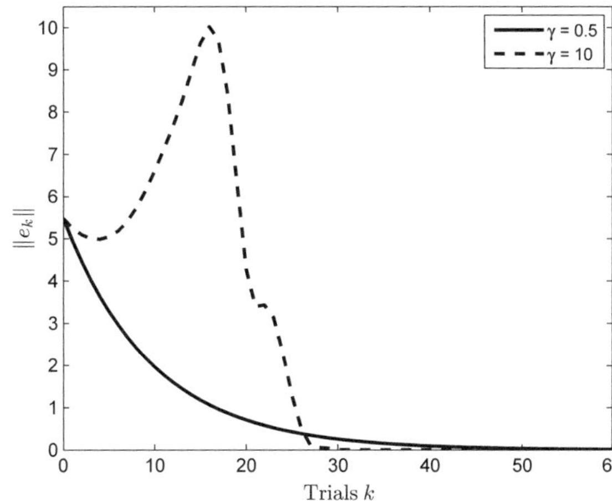

Figure 3.2, on the other hand, shows the l_2-norms (Euclidean norm) $\sqrt{e_k^T e_k}$ of the supervector corresponding to the tracking error $e_k(t) := r(t) - y_k(t)$ sequence as a function of the iteration index (number or round) k with the Arimoto ILC algorithm in its original form,

$$u_{k+1}(t) = u_k(t) + \gamma e_k(t + T_s), \quad for \quad t = 0, T_s, 2T_s, \cdots, (N-1)T_s, \tag{3.2}$$

where $\gamma = 0.5, 10$, respectively. That is, the algorithm updates the input from iteration to iteration by adding a scalar multiple of a term dependent on the error observed on the previous iteration.

Note that at first sight this algorithm seems to be non-causal, because $u_{k+1}(t)$ is a function of $e_k(t + T_s)$. However, on iteration $k + 1$, $e_k(t)$ is past data and is available for the whole range $t \in [0, T]$. In general, it is possible (and typically necessary) to make $u_{k+1}(t)$ a function of $e_k(t')$ for $t' \geq t$. Based on Fig. 3.2 it seems that the tracking error does indeed tend to zero as $k \to \infty$, for both $\gamma = 0.5$ and $\gamma = 10$, but the convergence exhibits different properties. In more detail, the convergence for $\gamma = 0.5$ is monotonic whereas the convergence for $\gamma = 10$ is not in the sense that errors do increase in early repetitions but ultimately reduce to zero.

The above example shows that the performance of the algorithm depends heavily on the choice of the parameter γ. If γ is not selected properly, the algorithm results only in asymptotic convergence and may exhibit poor transient performance. A generalization of this approach can eliminate some of these problems. Consider the following control update law in supervector form:

$$u_{k+1} = u_k + \gamma_{k+1} e_k \quad (Input\ Update\ Law), \tag{3.3}$$

where the "gain" γ_{k+1} is iteration-dependent and computed using an optimization process.

More precisely, in *Parameter Optimal ILC*, this iteration variation is constructed by finding an "optimal" γ_{k+1} for each iteration that minimizes a cost function defined by

$$\underbrace{J_{k+1}(\gamma_{k+1}) = \|e_{k+1}\|^2 + \omega\gamma_{k+1}^2}_{Parameter\ Optimal\ Objective\ Function} \tag{3.4}$$

where $\|e\|^2 = e^T e$, and $w > 0$ is a weighting parameter introduced to limit the value of γ_{k+1} that is used. Note that w can be constant or "adaptive" (change each iteration). The following form:

$$\underbrace{\omega = \omega_1 + \omega_2\|e_k\|^2, \quad \omega_1 \geq 0, \omega_2 \geq 0, \ \omega_1 + \omega_2 > 0}_{Assumed\ form\ of\ \omega} \tag{3.5}$$

was proposed to allow for varying "caution" as the algorithm progresses. It is based on the intuition that reduced values of γ_k should be used when errors are large.

To solve the above optimization problem, note that the constraint equation for the cost function (3.4) can be written as

$$\begin{aligned} e_{k+1} = r - y_{k+1} &= r - Gu_{k+1} - d \\ &= r - G(u_k + \gamma_{k+1}e_k) - d = r - Gu_k - d - \gamma_{k+1}Ge_k \\ &= (I - \gamma_{k+1}G)e_k \end{aligned}$$

where for discrete-time system in the lifted, supervector form seen in Chap. 2, G is the system matrix as defined in the preceding chapter, and $u_{k+1}, y_{k+1}, e_{k+1}$ are the lifted input, output, and tracking error supervectors. Using this description of the plant model, the reader can verify that the optimal update law for the optimal γ_{k+1} is given by the following equation:

$$\gamma_{k+1} = \frac{e_k^T G e_k}{\omega + e_k^T G^T G e_k} \qquad (Optimal\ Parameter\ Choice). \tag{3.6}$$

The next proposition shows that the optimal update γ_{k+1} generated by the algorithm results in monotonic convergence.

Theorem 3.1 (Monotonicity of the Tracking Error) The Parameter Optimal ILC algorithm (3.3), where γ_{k+1} is calculated using (3.6), results in the monotonic convergence property

$$\|e_{k+1}\| \leq \|e_k\| \quad so\ that \quad \lim_{k\to\infty} \|e_k\| = E_\infty \geq 0, \tag{3.7}$$

for some E_∞. In particular, $\|e_{k+1}\| = \|e_k\|$ if and only if $\gamma_{k+1} = 0$. Furthermore, suppose that $\omega_1 > 0$. The sequence of optimal learning gains $\{\gamma_k\}$ generated by (3.6) satisfies $\lim_{k\to\infty} \gamma_k = 0$.

Proof Note that, for the optimization problem (3.4), $\gamma_{k+1} = 0$ is suboptimal and thus

$$J_{k+1}(0) \geq J_{k+1}(\gamma_{k+1}),$$

at the optimal value. This is just

$$\|e_k\|^2 \geq \|e_{k+1}\|^2 + \omega\gamma_{k+1}^2,$$

from which $\|e_k\|^2 \geq \|e_{k+1}\|^2$ follows naturally. Note that $\|e_k\|^2 = \|e_{k+1}\|^2$ if and only if $w\gamma_{k+1}^2 = 0$ which is $\gamma_{k+1} = 0$ as required.

To show that $\gamma_k \to 0$, note that

$$\begin{aligned}\|e_0\|^2 &\geq \|e_1\|^2 + \omega\gamma_1^2 \geq \|e_2\|^2 + \omega\gamma_1^2 + \omega\gamma_2^2 \geq \cdots \\ &\geq \|e_k\|^2 + \omega\gamma_1^2 + \omega\gamma_2^2 + \ldots + \omega\gamma_k^2 \geq \omega\gamma_1^2 + \omega\gamma_2^2 + \ldots + \omega\gamma_k^2.\end{aligned}$$

According to monotone convergence theorem, we have $\omega\gamma_k^2 \to 0$, and thus $\gamma_k \to 0$ (as $\omega > 0$ since $\omega_1 > 0$). ■

Note that for the above algorithm, monotonicity in the tracking error norm arises from optimality and is independent of the plant model or weighting parameter ω. Furthermore, $\gamma_k \to 0$ shows that the algorithm becomes more cautious as k increases, possibly enhancing the robustness properties of the algorithm.

Unfortunately, the above proposition does not necessarily imply that the error $e_k \to 0$ as $k \to \infty$. It turns out that convergence to zero error is dependent on the plant dynamics, this dependence being expressed in terms of positivity conditions. More precisely, we have the following proposition.

Proposition 3.1 (Convergence to Zero Error) Consider the algorithm (3.3) with the update law (3.6) for γ_{k+1}. Suppose that $G + G^T$ is positive definite, i.e., there exists $\sigma \in R, \sigma > 0$, such that $v^T G v = v^T \left((G + G^T)/2\right) v \geq \sigma v^T v$ for all $v \in R^N$. Then the resulting error sequence satisfies $\lim_{k\to\infty} e_k = 0$.

Proof The fact that $\gamma_k \to 0$ implies $e_k^T G e_k \to 0$ or $\sigma 2 e_k^T e_k \leq e_k^T (G + G^T)$ $e_k \to 0$. ■

Remark 3.1 Note that, if $\omega_1 = 0$ in (3.5), Proposition 3.1 can be modified to show that the algorithm leads to geometric convergence to zero tracking error, i.e., the error converges whilst satisfying the bound $\|e_k\| \leq \lambda^k \|e_0\|,\ k \geq 0$ for some $\lambda < 1$.

Remark 3.2 Note that if the matrix $G + G^T$ is negative definite, a similar analysis as in Proposition 3.1 shows that the algorithm will converge monotonically to zero tracking error. However, if $G + G^T$ is negative definite, it can be always made Positive definite by multiplying G with -1. Therefore in this section (without loss of generality) the statement "$G + G^T$ is not a positive definite matrix" means that $G + G^T$ is either a semi-definite or a non-definite matrix.

Remark 3.3 In general, for a discrete state-space model with matrices (A, B, C, D), the supervector description reveals that the matrix $G + G^T$ is positive definite only if $D + D^T$ is positive definite. For strictly proper systems, therefore, a different formulation is needed. One such formulation is to let G be the matrix mapping the input sequence $\{u(t)\}_{0 \le t \le N-1}$ into $\{y(t)\}_{1 \le t \le N}$. The input update equation $u_{k+1} = u_k + \gamma_{k+1} e_k$ then becomes

$$u_{k+1}(t) = u_k(t) + \gamma_{k+1} e_k(t+1), \quad 0 \le t \le N-1 \tag{3.8}$$

which takes the form suggested by Arimoto (3.2). In this case $G + G^T$ is positive definite requires that $CB + (CB)^T$ is positive definite.
In this case, Hatonen [12] has been shown that a sufficient condition for $G + G^T$ to be positive definite is that $zG(z) = zC(zI - A)^{-1}B$ is a positive real transfer function matrix. This can be proved easily using the techniques described in Sect. 2.14.

The performance of the algorithm can be seen in the following example.

Example 3.2 Consider the same setting as in previous example, but now use the parameter optimal ILC algorithm with weighting parameter choice $\omega = 1 \times 10^{-3}, 1 \times 10^{-4}, 1 \times 10^{-5}$, respectively. It can be verified that for this example, $G + G^T > 0$ is satisfied as the system is positive real (this can be seen from its Nyquist plot in Fig. 3.3).

Figure 3.4 shows the tracking error norm for parameter optimal ILC algorithm over 60 iterations. Note that unlike the Arimoto algorithm, monotonic convergence to zero tracking error is achieved for all the weighting parameter choices, as predicted by the theory. Furthermore, the optimal learning gain γ_k converges to zero as shown in Fig. 3.5—again consistent with theoretical expectations.

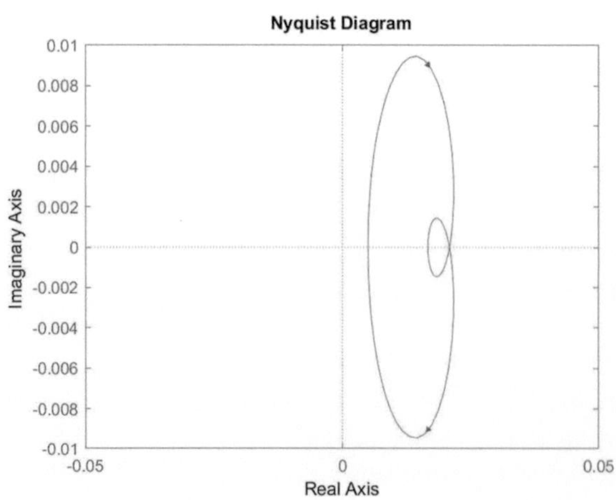

Fig. 3.3 Nyquist plot of the system in Example 3.2 showing its positive real property

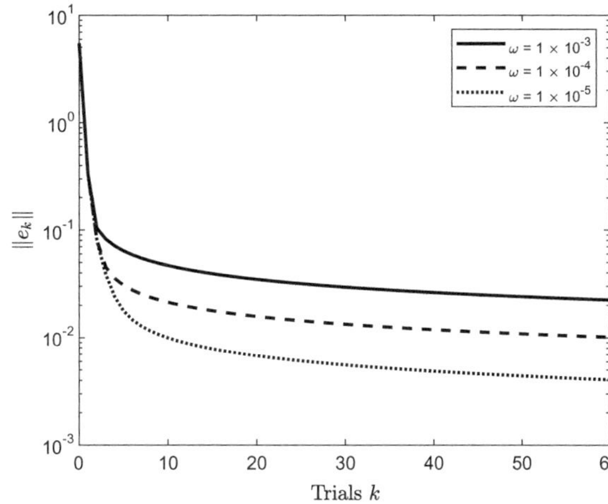

Fig. 3.4 Error convergence of the parameter optimal ILC algorithm with different weightings

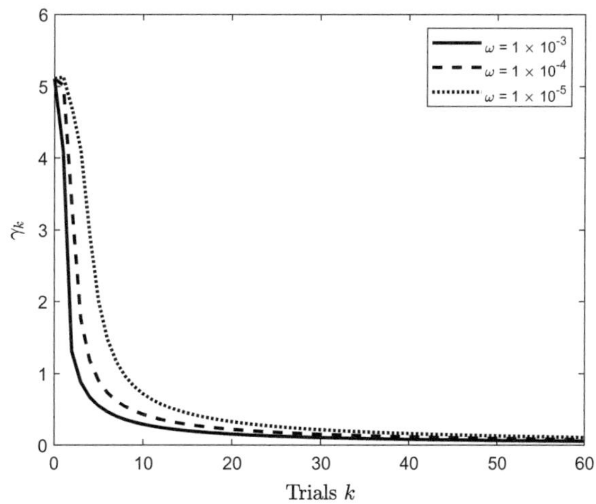

Fig. 3.5 Learning gain behaviour in the parameter optimal ILC algorithm

What happens if $G + G^T$ is not a positive definite matrix? This is a more complex question but the following proposition, based on the fact that $\gamma_{k+1} \to 0$, shows that the algorithm can indeed converge to a non-zero solution.

Proposition 3.2 Assume that $G + G^T$ is not a positive definite matrix. Then, for any non-zero point in the limit set $\{e_\infty : e_\infty^T G e_\infty = 0\}$ there exists an initial control input u_0 such that the POILC algorithm (3.3) converges to e_∞.

This can be illustrated in the following example.

Example 3.3 Consider the setting as in previous example but the system dynamics is replaced by

$$G(s) = \frac{1}{s^2 + 5s + 6}.$$

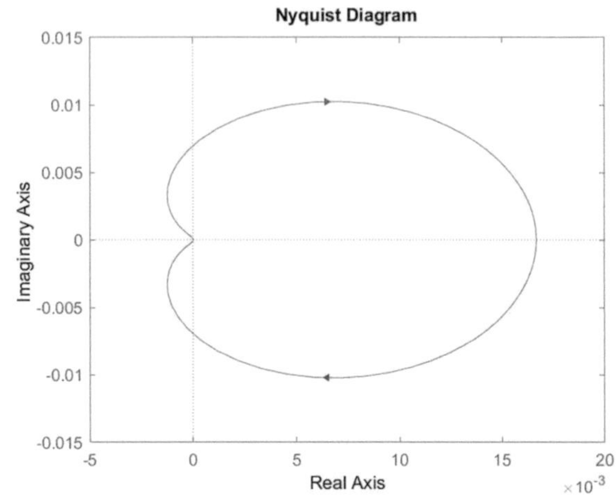

Fig. 3.6 Nyquist plot of the system in Example 3.3 showing its non-positivity

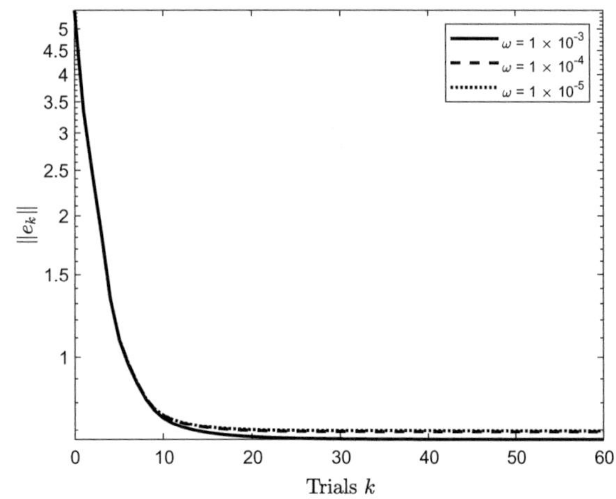

Fig. 3.7 Convergence of parameter optimal ILC for a non-positive system

It can be verified that for this example, $G + G^T$ is indefinite (as can be seen from Fig. 3.6 that the system is not positive real) and thus non-zero tracking error can be expected. Figure 3.7 shows the tracking error norm for parameter optimal ILC algorithm over 60 iterations. Although monotonic error norms are achieved for all the weighting parameter choices, the final converged value is non-zero. Also, the final converged values are different for these different weighting parameter choices. On the other hand, the optimal learning gains γ_k still converge to zero as shown in Fig. 3.8.

In summary, zero tracking error in the limit is guaranteed when $G + G^T$ is a positive definite matrix, but in other cases the limiting tracking error can be (and can be seen to be) a non-zero vector. A more detailed convergence analysis of this algorithm can be found in Owens and Feng [20].

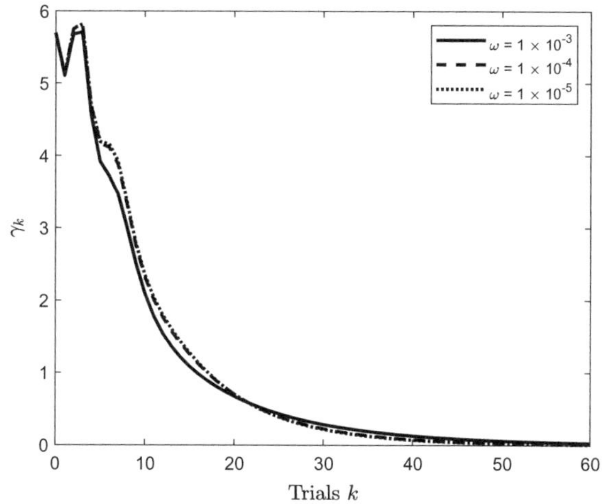

Fig. 3.8 Convergence behaviour of the Parameter Optimal ILC algorithm with different weightings

Remark 3.4 Positivity can be regarded as a property of the plant or as an objective of the control design analysis and/or implementation. In this context:

- the plant G may already contain control loops introduced to ensure that positivity conditions are satisfied, see Hatonen [12] and Owens and Feng [20], or;
- a change of norm to an exponentially weighted norm [20] may induce positivity at the expense of strict monotonicity of the Euclidean norm, see Owens and Feng [20].

Remark 3.5 The existence of non-zero limits is established above and is a reality for any non-positive system. The form and properties of these limit sets and their implications for practical POILC implementations are identified in Owens et al. [22]. The analysis is complex, if only because the limit set lies in a high dimensional space (of dimension equal to the number of samples in the time interval of interest). The analysis is needed because most systems, in practice, are non-positive. Non-zero limits will, hence, be the norm rather than the exception. The avoidance of these behaviours in practice was also investigated in Owens et al. [22] using novel switching strategies.

Remark 3.6 The discussion so far is for signal tracking where the reference is defined over the whole horizon. There are other types of ILC design problems. As an example, for point-to-point/intermediate point tracking, the reference is only defined at a, possibly small, finite number of intermediate points over the interval. For this case, the tracking error information is only available at these intermediate points. This would cause significant difficulties for the simple Arimoto-type ILC and parameter optimal ILC as there is not enough information from the tracking error to update the input signal (defined over the whole trial length). This can be easily seen for single-input–single-output discrete-time system where the dimensions of the lifted form of the input and tracking error vectors do not match (and thus cannot be added together).

3.2 Inverse-Based Iterative Learning Control

This section again considers an ℓ-input, m-output linear, time-invariant discrete state-space system

$$\begin{aligned} x(t+1) &= Ax(t) + Bu(t), \quad t = 0, 1, 2, \cdots, N-1 \\ y(t) &= Cx(t) + Du(t), \quad t = 0, 1, 2, \cdots, N \end{aligned} \tag{3.9}$$

with a transfer function

$$G(z) = C(zI - A)^{-1}B + D.$$

This representation is ideal for simulation purposes, but, for the purpose of algorithm development, it is easier to consider the model in a supervector form $y = Gu + d$.

The previous section reveals that whilst the simple structure ILC law can work well to achieve perfect reference tracking, it also has certain limitations. A natural question arises: can we design ILC algorithms with improved performance if we more fully utilize the information to be found in the underlying system model (i.e., the system operator G)?

Given the design task is to find an input u^* such that the output of the system tracks the reference r perfectly, i.e., $r = Gu^* + d$, and noting that, if G is invertible, $u^* = G^{-1}(r - d)$, then a natural approach to answering the above question is to use the system inverse (or an approximation to that inverse) in the ILC input updating law. That leads to the following Inverse-based ILC law [11]:

$$u_{k+1} = u_k + \gamma K e_k \quad (\textit{Input update law}) \tag{3.10}$$

where K represents an inverse of the system G. It is worth noting that the inverse of a system does not always exist (as introduced in the previous chapter). For notational simplicity, here we consider the case that the lifted system matrix G has a right inverse, i.e., there exists a right inverse system G_R^{-1} such that $GG_R^{-1} = I$.

Note that this covers many situations of interest. Let k^* be the unique integer defined to be $k^* = 0$ if $D \neq 0$ and, otherwise, as the integer such that $CA^jB = 0,\ 0 \leq j \leq k^* - 1$ with $CA^{k^*-1}B \neq 0$. That is $G_\infty = \lim_{|z|\to\infty} z^{k^*}G(z)$ is non-zero and finite. Let G be the operator mapping the input sequence $u = \{u(0), u(1), \cdots, u(N - k^*)\}$ into $y = \{y(k^*), y(k^* + 1), \cdots, y(N)\}$ then G has a state-space realization equal to a realization of $z^{k^*}G(z)$, e.g.,

$$\begin{aligned} \psi(t+1) &= A\psi(t) + Bu(t), \\ \phi(t) &= CA^{k^*}\psi(t) + G_\infty u(t), \quad t = 0, 1, \cdots, N - k^* \\ \textit{and} \quad y(t+k^*) &= \phi(t). \end{aligned} \tag{3.11}$$

If $m = \ell$ and G_∞ is nonsingular, then the supervector matrix of this model is invertible. In this case, $K = G_R^{-1}$ and the above algorithm becomes

$$\underbrace{u_{k+1} = u_k + \gamma G_R^{-1} e_k,}_{The\ Inverse\ ILC\ Algorithm} \tag{3.12}$$

where $\gamma > 0$ is a parameter to be chosen.

The above Inverse-Based ILC Algorithm has the following properties.

Theorem 3.2 The Inverse-Based ILC Algorithm has the following error evolution law:

$$e_{k+1} = (1 - \gamma)e_k \quad so\ that \quad e_k = (1 - \gamma)^k e_0.$$

Therefore, the tracking error norm will monotonically converge to zero for any initial input choice u_0 (with tracking error e_0) if

$$|1 - \gamma| < 1, \qquad i.e., \ \ 0 < \gamma < 2.$$

Moreover, the choice of $\gamma = 1$ will lead to zero tracking error just in one iteration.

Proof The error evolution equation can be obtained as

$$\begin{aligned} e_{k+1} &= r - y_{k+1} = r - Gu_{k+1} - d = r - G(u_k + \gamma G_R^{-1} e_k) - d \\ &= r - Gu_k - \gamma e_k - d = (r - Gu_k - d) - \gamma e_k = e_k - \gamma e_k \\ &= (1 - \gamma)e_k, \end{aligned}$$

from which $e_k = (1 - \gamma)^k e_0$ follows naturally. It can also be seen that convergence requires that $|1 - \gamma| < 1$ from which $0 < \gamma < 2$ follows. Finally setting $\gamma = 1$ immediately leads to $e_1 = 0$ for any e_0 and thus zero tracking error is achieved just in one iteration. ■

Compared to the Parameter Optimal ILC design, Inverse-based ILC results in monotonic convergence in the tracking error norm to zero, and the choice of learning gains can be easily set. Note that unlike Parameter Optimal ILC, this is achieved irrespective of the properties of the system dynamics (no requirements on the positivity of the system dynamics are needed). It is also worth commenting that the implementation of the Inverse-based ILC is no more (structurally) complex than the Parameter Optimal ILC algorithm, as, if a precise inverse is available, the system inverse can be implemented using state-space simulation (as discussed in the previous chapter) or direct matrix multiplication (in lifted supervector form for discrete-time systems).

Remark 3.7 The statement above is misleading in the sense that Parameter Optimal ILC is likely to be much more insensitive to modelling errors than the Inverse ILC Algorithm. An intuitive indicator of this is that errors in the model G are likely to transform into much larger errors in the inverse and hence larger errors in the input update. As errors are almost always present in the high frequency behaviour of the plant (where frequency response gains are very small), the errors in the high frequency component of an inverse could be extremely large.

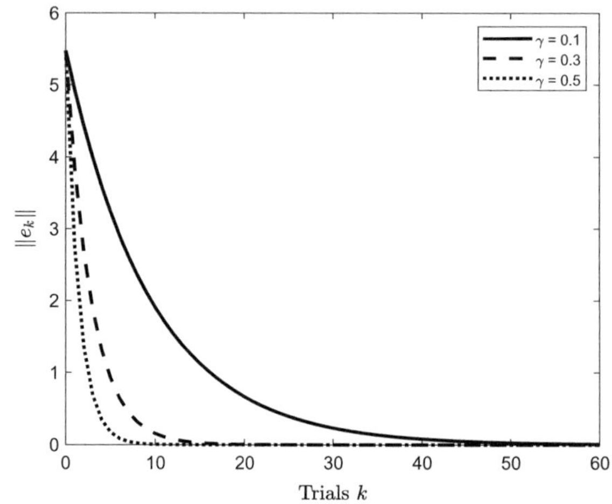

Fig. 3.9 Convergence behaviour of the inverse algorithm with different learning gains

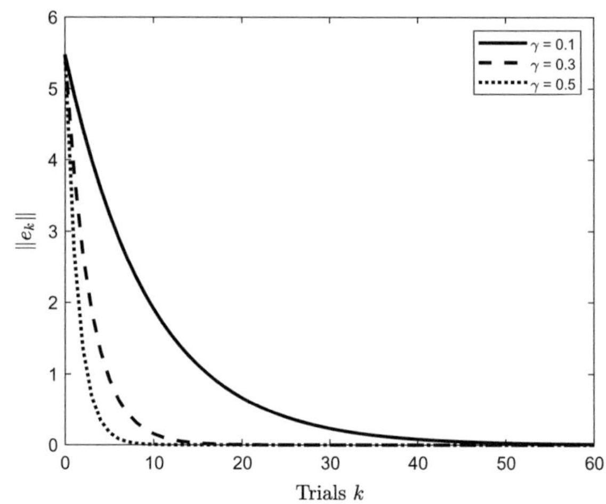

Fig. 3.10 Convergence behaviour of the inverse algorithm for a non-positive system

The performance of the Inverse-Based ILC Algorithm is illustrated in the following example.

Example 3.4 Consider the same setting as in Example 3.2. The performance of Inverse-based ILC design with learning gain choice $\gamma = 0.1, 0.3, 0.5$, respectively, is shown in Fig. 3.9.

As can be seen, the tracking error norm converges monotonically to zero tracking for all three learning gain choices as expected.

In Example 3.3, it is shown that when $G + G^T$ is indefinite then parameter optimal ILC will not be able to guarantee the achievement of zero tracking error. The Inverse-based ILC design does not suffer from this problem as can be seen for Fig. 3.10 where for all learning gain choices the tracking error norms converge monotonically to zero.

Remark 3.8 The above assumes that the system G has a right inverse. When this is not the case but a left inverse exists, a similar Inverse-based ILC design can also be developed but the convergence properties will differ. More details can be found in Owens [19].

Remark 3.9 The Inverse-based ILC law can also be implemented in the frequency domain. In fact this was some of the very earliest concepts of ILC. As an example, this idea can be found in Uchiyama [24]. However, because this publication was written in Japanese, non-Japanese researchers were not aware of it when the ILC research initially started in USA and Western Europe. Therefore, the publication Arimoto et al. [1] was referenced as being the starting point for ILC research. However, it seems that there were earlier publications than Uchiyama's on topics related to ILC. For example, in Cryer et al. [5] the authors (who worked for General Motors Truck and Bus) proposed the following discrete-time "iterative control" (as they call it) in the frequency domain

$$u_{k+1}(e^{j\omega T_s}) = u_k(e^{j\omega T_s}) + \beta G^{-1}(e^{j\omega T_s})e_k(e^{j\omega T_s})$$

where $\beta \in R, 0 < \beta < 1$ is a design parameter, and $G(e^{j\omega T_s})$ is the discrete-time Fourier transform of the impulse response $g(t)$ of the plant in question. This is clearly an ILC algorithm and the authors applied this algorithm in the context of laboratory road simulation. Another even earlier reference to ILC concepts seems to be the USA Patent No. 3, 555, 252—Learning Control of Actuators in Control Systems from 1971; see Chen and Moore [2] for details.

3.3 Robustness of Inverse-Based Iterative Learning Control

Robustness analysis aims to evaluate the possible effects of model uncertainties/errors on algorithm performance when applied to a process with an assumed model G. Errors could include mismatches in modelling process dynamics, non-repeated noise/disturbances, and imprecise initial state resetting at the beginning of each iteration. Since these factors are almost inevitable in practice, robustness analysis is vitally important. What follows focuses on the modelling mismatches as the latter two have a different nature since they can be seen as "exogenous" to the system dynamics and thus can be treated (arguably) in a relatively easier manner.

Robustness analysis is an important matter in classical control systems design. It is, however, much more complicated in ILC than traditional situations, which can be seen from the following example. Consider the following simplest Arimoto-type updating law for a strictly proper, SISO discrete-time system $S(A, B, C)$,

$$u_{k+1}(t) = u_k(t) + \gamma e_k(t+1). \tag{3.13}$$

It is a simple algebraic exercise to show that the condition for asymptotic stability (assuming system relative degree $k^* = 1$, i.e., $CB \neq 0$) is that

$$|1 - \gamma CB| < 1. \tag{3.14}$$

Suppose there exist model uncertainties. From (3.14) it can be seen that as long as the model uncertainty doesn't change the sign of the first Markov parameter (i.e., CB), the ILC system remains stable provided that γ is sufficiently small. This observation indicates that robustness analysis of the Arimoto algorithm is more complex/subtle than that in non-iterative situations and can explain, to some degree, why progress in producing useful analysis and design tools for analysing plant uncertainty has been slow.

Remark 3.10 Although a number of results have been obtained in the literature (see, for example, Chen and Wen [3], Furuta and Yamakita [8], Norrlof [18], and, for optimal ILC algorithms, [19]), it is fair to say that theoretical knowledge and understanding of these problems, for all but a few ILC methodologies, are currently limited.

To begin an analysis of the robustness of the Inverse-Based ILC Algorithms, suppose G_0 is the *nominal model* and a *left multiplicative model uncertainty* U is defined, i.e., the true plant G is given by the expression, in supervector form,

$$G = UG_0. \tag{3.15}$$

That is, the output from the nominal model will produce the plant output if it is used as an input to the error/uncertainty model U. In the frequency domain, this can be expressed as the product of transfer function matrices as follows:

$$G(z) = U(z)G_0(z).$$

Remark 3.11 Note that the case with no model mismatch is recovered by setting $U = I$ and that, in most applications, U is unknown or, at best, only partially known.

It may be expected that if an inverse model of G_0 (which could potentially be vulnerable to model mismatches) is used in the input updating, then, in the presence of model uncertainty, the Inverse-based ILC design may fail to work. However, as can be seen in the following theorem, convergence properties of the inverse-based algorithm can still be guaranteed, provided certain conditions are met.

Theorem 3.3 The Inverse-Based ILC Algorithm is *robust monotonically convergent* in the presence of a left multiplicative modelling error U for all initial errors e_0 if, and only if,

$$\|I - \gamma U\| < 1 \quad (Robustness\ Criterion\ for\ the\ Inverse\ Algorithm), \tag{3.16}$$

where $\|\cdot\|$ is the operator norm induced by the norm $\|\cdot\|_{\mathcal{Y}}$ in the space of output signals.

Proof Following a similar manner as in the convergence proof of Inverse-Based ILC Algorithm, the tracking error evolution with the presence of modelling mismatches is given by

$$\begin{aligned} e_{k+1} &= r - y_{k+1} = r - Gu_{k+1} - d = r - G(u_k + \gamma G_{0,R}^{-1} e_k) - d \\ &= r - Gu_k - \gamma G G_{0,R}^{-1} e_k - d = r - Gu_k - \gamma U G_0 G_{0,R}^{-1} e_k - d \\ &= (1 - \gamma U) e_k, \\ i.e., \quad \|e_{k+1}\|_{\mathcal{Y}} &\le \|I - \gamma U\| \|e_k\|_{\mathcal{Y}}. \end{aligned}$$

The Inverse-Based ILC Algorithm can thus tolerate a certain degree of model uncertainty by producing *monotonic convergence* in the tracking error norm if the condition $\|I - \gamma U\| < 1$ is satisfied. That completes the proof. ■

Remark 3.12 The checking of (3.16) can be undertaken if U is known. Alternatively, it describes a set of model errors that can be tolerated in Inverse ILC. The actual form of the criterion will depend upon the norm chosen in the output space $\mathcal{Y}$. For example, if U is asymptotically stable and $\mathcal{Y}$ is endowed with an inner product

$$\langle y, w \rangle_{\mathcal{Y}} = \sum y^T(t) Q w(t), \quad where \ \ Q = Q^T > 0,$$

then the robustness condition (see Sect. 2.14.7) can be replaced by the frequency domain requirement,

$$\max_{|z|=1} r\left[(I - \gamma Q^{-1} U^T(z^{-1}) Q)(I - \gamma U(z)) \right] < 1, \qquad (3.17)$$

where $r[\cdot]$ denotes the spectral radius of the argument.

The above theorem provides several key insights into the robustness properties of the Inverse Algorithm. To see this, suppose that $\mathcal{Y}$ is an inner product space so that U has an adjoint U^*. The robustness conditions can then be converted into the inequality,

$$(I - \gamma U)^*(I - \gamma U) < I.$$

Simple algebraic operations then show that this is equivalent to

$$U + U^* > \gamma U^* U > 0 \quad (Alternative\ Robustness\ Criterion), \qquad (3.18)$$

with respect to the inner product in $\mathcal{Y}$. This implies, in particular, that the modelling error needs to be positive definite in order to achieve monotonic convergence in the tracking error norm.

Remark 3.13 The above theorem also has a frequency domain inequality interpretation. More precisely, using the same assumptions as those used in Remark 3.12 and the results of Sect. 2.14.7, the alternative robustness criterion can be written as

$$U^T(z^{-1})Q + QU(z) > \gamma U^T(z^{-1})QU(z) > 0 \ \ \forall\, |z| = 1. \tag{3.19}$$

Alternatively, if $\hat{U}$ is the inverse of U, this can be written as

$$Q\hat{U}^T(z^{-1}) + \hat{U}(z)Q > \gamma Q > 0 \; (Robustness - Inverse\ Criterion) \tag{3.20}$$

for all $|z| = 1$. For SISO systems, Q is a scalar and these criteria simplify to the requirement that, for all $|z| = 1$,

$$(a)\ \left|\frac{1}{\gamma} - U(z)\right| < \frac{1}{\gamma}\ \ and\ \ (b)\ 2Re\left[\hat{U}(z)\right] > \gamma. \tag{3.21}$$

Both have graphical interpretations. More precisely, (a) requires that the frequency response plot of $U(z)$ lies inside a circle of radius γ^{-1} centred on the point $(\gamma^{-1}, 0)$ of the complex plane. In contrast, (b) requires that the frequency response plot of the inverse $\hat{U}(z)$ lies entirely to the right of the vertical line $Re[\eta] = \frac{1}{2}\gamma$ in the complex plane.

Note 3.1 A necessary condition for the robustness condition (3.19) when $U(z)$ is asymptotically stable is that

$$QU(z) + U^T(z^{-1})Q > 0, \ \ \forall |z| = 1 \ \ (Positive\ Real\ Requirement). \tag{3.22}$$

The above results can be illustrated in the following example.

Example 3.5 Consider the same setting as in the previous example where the nominal model has underlying continuous transfer function

$$G_0(s) = \frac{s+1}{s^2 + 5s + 1}.$$

Two cases are considered. For the first one, consider the case where the model mismatch has underlying transfer function

$$U(s) = \frac{s+1}{s+5}.$$

The corresponding discrete-time transfer function is positive real which implies that the Inverse Algorithm will be robust for small enough values of γ. The performance of Inverse-based ILC design with learning gain choice $\gamma = 0.1, 0.3, 0.5$, respectively, is shown in Fig. 3.11. Note that for these parameter choices, the robustness monotonic convergence condition is satisfied.

As can be seen, the tracking error norm converges monotonically to zero tracking for all three learning gain choices even with the presence of model mismatches as expected.

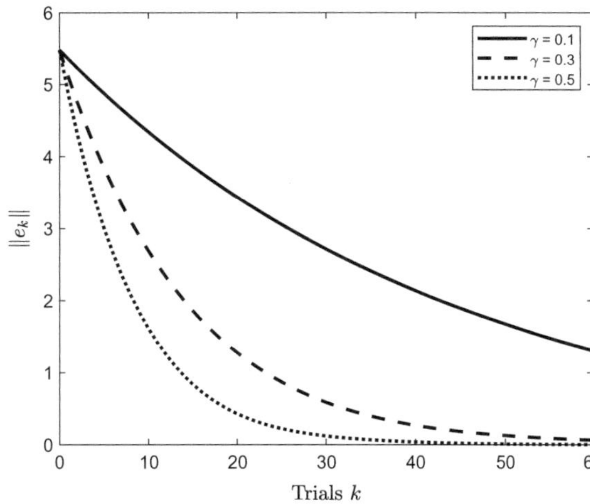

Fig. 3.11 Robust monotonic convergence of the inverse algorithm with different learning gains

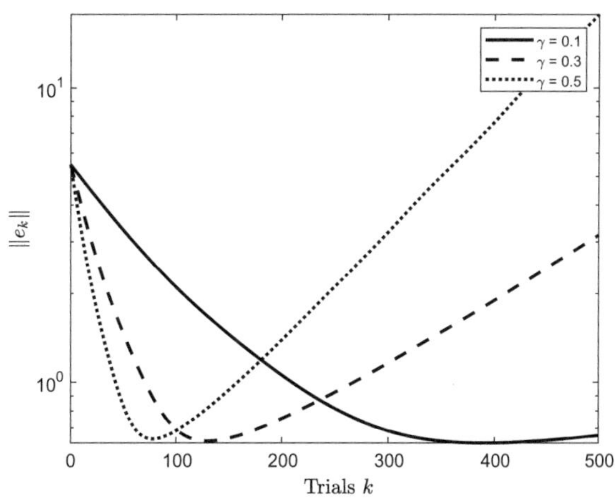

Fig. 3.12 Possible divergence if the modelling error is not positive real: variation with gains

For the second case, the model mismatch is given by

$$U(s) = \frac{1}{s^2 + 5s + 6},$$

which is not positive real. Again, the performance of Inverse-based ILC design with learning gain choice $\gamma = 0.1, 0.3, 0.5$, respectively, is shown in Fig. 3.12. Note that the robustness monotonic convergence condition is not satisfied.

As expected, the tracking error norms eventually diverge after the initial reduction, although it is intriguing that error reductions are achieved initially before divergence "begins".

3.4 Effect of Non-minimum Phase Zeros

As mentioned previously, ILC may experience practical problems when dealing with non-minimum phase systems. Whilst some progress has been made in the analysis of such situations, many of the issues that could arise in practice are only partially understood. To see this consider a discrete-time system with transfer function $G(z)$ and the inverse input updating law

$$u_{k+1} = u_k + \gamma G^{-1} e_k. \tag{3.23}$$

If $G(z)$ is non-minimum phase with a NMP zero outside of the unit circle, it can be easily seen that its inverse $G^{-1}(z)$ is unstable (as zeros of $G(z)$ become poles of $G^{-1}(z)$). When implementing the Inverse-Based ILC Algorithm using state-space simulation, the error e_k is provided as an input to the unstable system inverse. In practice, any noise or disturbances (including numerical errors) will excite the unstable pole in an unknown way, typically leading to very large errors in input updates and consequent algorithm divergence. When implementing the algorithm in lift form, the matrix representation of system G^{-1} will have very large elements and is ill conditioned (and thus will have problems too). As a result, the inverse-type ILC control law will eventually fail when applied to non-minimum phase systems.

For non-minimum phase systems, the optimal input which gives perfect tracking is, assuming zero initial conditions,

$$u^*(z) = G^{-1}(z)r(z). \tag{3.24}$$

Because $G(z)$ is non-minimum phase, its inverse, $G^{-1}(z)$, is unstable. $u^*(t)$ will contain the unstable modes and grow to a very large value with the only exception that the reference signal $r(t)$ has the unstable modes of $G^{-1}(z)$ and hence cancel the unstable poles of $G^{-1}(z)$. However, since $r(t)$ is usually selected independent of $G(z)$, this cannot be satisfied generally. Therefore, for non-minimum phase systems, we have to make a compromise between large input effort and non-zero tracking error. That is the key reason why, even for the algorithms that can be successfully applied to non-minimum phase systems, most of them will result in fast reduction in the tracking error norm initially and then slow convergence thereafter, which is undesirable in practice. This phenomenon will be analysed rigorously in later chapters.

Remark 3.14 Several methods have been proposed in the literature to solve the above problems. As examples, Wang and Chen [25] used the concept of stable inversion to construct the updating law. Jeong and Choi [14] adopted the non-causal filter to achieve perfect tracking during a subset of the trial duration. Gunnarsson and Norrlof [10] utilized input weighting to suppress the input which results in non-zero tracking error. However, it can be argued that ILC for non-minimum phase systems has not been fully understood yet and new design methods are still needed to solve this problem. For more results on non-minimum phase systems, see Ghosh and

Paden [9], Gunnarsson and Norrlof [10], Wang and Chen [25], Jeong and Choi [14], Kinosita et al. [17], Sogo [23], Freeman et al. [7], van Zunderi et al. [26], and the references therein.

3.5 Application to Point-to-Point Tracking Tasks

In this section, we will show how the inverse ILC algorithm can be applied to solve the point-to-point tracking problems. For point-to-point tracking, the system is required to track a reference defined on given M intermediate time instants $t_1, t_2, \ldots, t_M$. In vector form, the reference is defined to be a selection of values desired at a number of intermediate times,

$$r^p = \left[r^T(t_1)\ r^T(t_2)\ \cdots\ r^T(t_M)\right]^T.$$

As mentioned previously for the point-to-point tracking tasks, the Arimoto-type ILC law and the Parameter Optimal ILC design cannot be directly used as the tracking error information is only available at the above intermediate time instants (rather than the whole interval). The Inverse-based ILC design, on the other hand, can be applied to address this problem as it uses the systems dynamics in the control update and hence may be capable of creating enough data to make effective iterative control possible. The ideas are outlined below for discrete-time state-space systems.

The relationship between supervector of system outputs at the intermediate time instants, y^p, and the supervector of the input time series u can be described by a matrix relationship of the form

$$y^p = G^p u + d^p,$$

where the superscript p denotes the signal values at these intermediate points.

Remark 3.15 The description also applies to continuous-time, state-space systems, but the operator G^p can be regarded as an integral operator derived from the state-space equations. For discrete-time systems in supervector form, y^p, G^p, and d^p just consist of the $t_1^{th}, t_2^{th}, \ldots, t_M^{th}$ blocks of rows of y, G, d, respectively. For example,

$$y(t_j) = \left[CA^{t_j-1}B, CA^{t_j-1}B, \cdots, CB, D, 0, \cdots, 0\right]u + CA^{t_j}x_0.$$

For point-to-point tracking task generally speaking G^p does not have a left or right inverse. However, when $\mathcal{R}(G^p) = \mathcal{Y}$, G^p has a right inverse $(G^p)_R^{-1}$ if t_1 is not too small. This inverse satisfies

$$G^p(G^p)_R^{-1} = I.$$

Remark 3.16 For discrete-time state-space systems G^p is a matrix. It is a simple matter to show that a right inverse can be written in the form

$$(G^p)_R^{-1} = (G^p)^T (G^p (G^p)^T)^{-1} \tag{3.25}$$

if the indicated inverse exists, i.e., if G^p has full row rank.

The Inverse-Based ILC Algorithm can hence be used to solve the point-to-point tracking task using the input update relationship,

$$u_{k+1} = u_k + \gamma G_R^{p^{-1}} e_k^p$$

where $e_k^p = r^p - y_k^p$ is the point-to-point tracking error.

The above Inverse-based ILC design for point-to-point tasks has the following properties, proved using an identical argument to that used to prove the convergence properties of the original Inverse-based Algorithm:

Theorem 3.4 The Inverse-based Point-to-Point ILC algorithm has the following error evolution law:

$$e_{k+1}^p = (1 - \gamma) e_k^p.$$

As a result,

$$e_k^p = (1 - \gamma)^k e_0^p.$$

Therefore, the tracking error norm will monotonically converge to zero for any initial tracking error e_0^p if $|1 - \gamma| < 1$ or, equivalently,

$$0 < \gamma < 2.$$

A particular choice of $\gamma = 1$ will lead to zero tracking error just in one iteration.

To illustrate the performance of the above algorithm, consider the following example.

Example 3.6 Consider the same system and the setting as in previous examples. Instead of tracking the reference defined over the whole trial interval, the system is now only required to track the reference at intermediate time instants at $t = 1s, 2s, 3s, 4s, 5s, 6s$. The convergence performance of the inverse-based point-to-point ILC algorithm with learning gain choice $\gamma = 0.1, 0.3, 0.5$, respectively, is shown in Fig. 3.13.

As can be seen, the point-to-point tracking error norm converges monotonically to zero as the theory predicts. The system output on the last trial together with the point-to-point reference are shown in Fig. 3.14.

As can be seen, perfect tracking of the point-to-point reference is achieved. Note that the system outputs for different learning gain choices are on top of each other in the figure and not distinguishable.

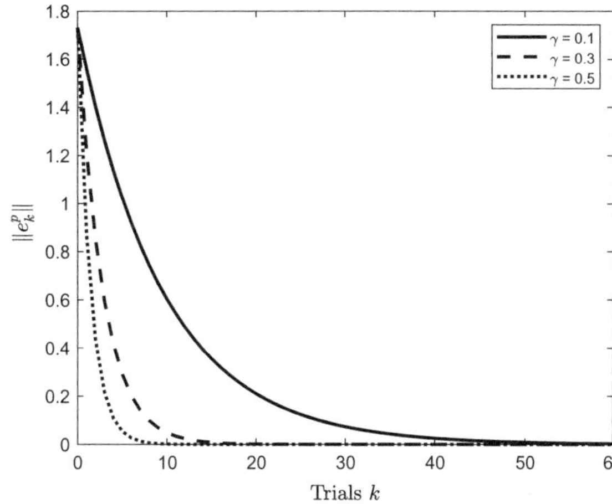

Fig. 3.13 Convergence for the point-to-point Inverse Algorithm example

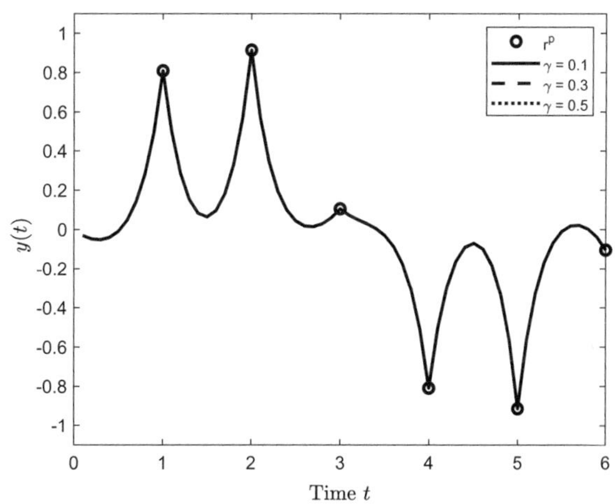

Fig. 3.14 Converged trajectory for the point-to-point example

3.6 Discussion and Further Reading

In this chapter, the feasibility of constructing convergent Iterative Algorithms for dynamical systems has been demonstrated. The basic Arimoto algorithm is sufficient to demonstrate this algorithm divergence is a real possibility. Parameter Optimal ILC removes this part by ensuring monotonic reductions of the error norm based on optimization criteria. This guarantees, theoretically, convergence of the error norm but not necessarily to a non-zero value, as the plant must satisfy a positivity condition. The Inverse-based ILC design (when an appropriate inverse exists) has been introduced as a benchmark algorithm that shows that convergence can be achieved, theoretically, in all situations. There are many issues that need to be discussed/resolved here (see below) but the method does provide some key insights on the properties of many other ILC design methods to be introduced in later chapters:

1. It demonstrates that ILC can have significant robustness properties although, using current knowledge, the modelling error is assumed to consist of multiplicative error dynamics. This constrains the form of errors considered but further constraints arise in the analysis, namely that the error model is, necessarily, constrained to satisfy defined robustness criteria that include positive real conditions. There are, of course, other forms of modelling error, not considered in this chapter such as parametric errors in the model or nonlinear modelling errors. Both present considerable theoretical difficulties and further analysis and research is needed in these areas.
2. In principle, Inverse ILC has wide application but, for discrete state-space systems, it faces considerable difficulties when applied to non-minimum phase systems. There are many reasons why this difficulty arises but, in essence, the problems arise from the choice of reference signal. More precisely, in general, tracking of a reference signal will require inputs that contain exponentially unstable components, a fact that is exacerbated by the resultant sensitivity to modelling errors and disturbances.
3. In this chapter the key ideas of Inverse-based ILC (and POILC) have been illustrated using discrete-time state-space systems where the underlying input and output spaces are familiar Euclidean spaces of times series of inputs and outputs. The ideas can, however, also be applied, in principle, to continuous-time systems as well as more general signal spaces. A more detailed discussion of this, as well as an analysis of the case when a right inverse of the system does not exist, can be found in the companion book Owens [19].
4. This chapter has assumed that a model of the system is available. When this is not case, various data-driven, data-based methods can be explored. An explicit model identification step can be included, such as that introduced in Janssens et al. [13]. Frequency domain inverse-based design is introduced in Kim and Zau [16] by implicitly estimating a system inverse from the previous trial's data. More recently, for linear systems, the so-called fundamental lemma can be explored to replace the model information [4, 15]. Furthermore, using techniques from machine learning, Gaussian process-based ILC design has been introduced in Devasia [6]. The readers are referred to the aforementioned references for more details.

These observations provide pointers to issues that arise from a preliminary analysis and consideration of Iterative Learning Control but there are many more of interest to meet the defined objectives of ILC. Essentially, any question that has been asked within Control Theory can, in some modified form, be asked of ILC. The rest of this text is concerned with more generally applicable and successful algorithms, developed using optimization methodologies. The approaches will remove many of the issues that plague the Inverse ILC Algorithm for linear systems, the remaining limitations being the issues of dealing with issues of slow convergence for non-minimum phase systems and the effect of application to severely nonlinear processes.

As with classical feedback theory, linear analysis based on linear models can be very successful but severe nonlinearities such as hysteresis offer a real and continuing challenge.

References

1. Arimoto S, Kawamura S, Miyazaki F (1984) Bettering operation of robots by learning. J Robot Syst 1(2):123–140
2. Chen Y, Moore KL (2000) Comments on United States patent No. 3,555,252 learning control of actuators in control systems. In: Proceedings of the 2000 International Conference on Automation, Robotics, and Control
3. Chen Y, Wen C (1999) Iterative learning control: robustness and applications. Springer
4. Chu B, Rapisarda P (2023) Data-driven iterative learning control for continuous-time systems. In: 2023 62nd IEEE Conference on Decision and Control (CDC). IEEE, pp 4626–4631
5. Cryer BW, Nawrocki PE, Lund RA (1976) A road simulation system for heavy duty vehicles. Technical Report No 760361 Society of Automotive Engineers
6. Devasia S (2019) Iterative machine learning for output tracking. IEEE Trans Control Syst Technol 27(2):516–526
7. Freeman CT, Lewin PL, Rogers E (2005) Experimental evaluation of iterative learning control algorithms for non-minimum phase plants. Int J Control 78(11):826–846
8. Furuta K, Yamakita M (1987) The design of learning control systems for multivariable systems. In: Proceedings of the IEEE International Symposium on Intelligent Control, pp 371–376
9. Ghosh J, Paden B (1999) Iterative learning control for nonlinear nonminimum phase plants with input disturbances. In: Proceedings of the 1999 American Control Conference, pp 2584–2589
10. Gunnarsson S, Norrlof M (2001) On the design of ILC algorithms using optimization. Automatica 37(12):2011–2016
11. Harte T, Hatonen J, Owens D (2005) Discrete-time inverse model-based iterative learning control: stability, monotonicity and robustness. Int J Control 78(8):577–586
12. Hatonen J (2004) Issues of algebra and optimality in iterative learning control. PhD thesis, University of Oulu, Finland
13. Janssens P, Pipeleers G, Swevers J (2012) A data-driven constrained norm-optimal iterative learning control framework for LTI systems. IEEE Trans Control Syst Technol 21(2):546–551
14. Jeong GM, Choi CH (2002) Iterative learning control for linear discrete time nonminimum phase systems. Automatica 38(2):287–291
15. Jiang Z, Chu B (2022) Norm optimal iterative learning control: a data-driven approach. IFAC-PapersOnLine 55(12):482–487
16. Kim KS, Zou Q (2012) A modeling-free inversion-based iterative feedforward control for precision output tracking of linear time-invariant systems. IEEE/ASME Trans Mechatron 18(6):1767–1777
17. Kinosita K, Sogo T, Adachi N (2002) Iterative learning control using adjoint systems and stable inversion. Asian J Control 4(1):60–67
18. Norrlof M (2004) Disturbance rejection using an ILC algorithm with iteration varying filters. Asian J Control 6(3):432–438
19. Owens DH (2015) Iterative learning control—an optimization paradigm. Springer
20. Owens DH, Feng K (2003) Parameter optimization in iterative learning control. Int J Control 76(11):1059–1069
21. Owens DH, Hatonen J (2005) Iterative learning control—an optimization paradigm. Annu Rev Control 29(1):57–70
22. Owens DH, Tomas-Rodriguez M, Daley S (2008) Limit sets and switching strategies in parameter-optimal iterative learning control. Int J Control 81(4):626–640

23. Sogo T (2002) Stable inversion for nonminimum phase sampled-data systems and its relation with the continuous-time counterpart. In: Proceedings of 41st IEEE Conference on Decision and Control, pp 3730–3735
24. Uchiyama M (1978) Formation of high-speed motion pattern of a mechanical arm by trial. Trans Soc Instrum Control Eng 14(2):706–712
25. Wang X, Chen D (2005) An inversion-based iterative learning control algorithm for a class of nonminimum-phase systems. IEE Proc Control Theory Appl 152(1):72–78
26. van Zundert J, Luijten F, Oomen T (2019) Exact and causal inversion of nonminimum-phase systems: A squaring-down approach for overactuated systems. IEEE/ASME Trans Mechatron 24(6):2953–2963

Chapter 4
Gradient-Based Iterative Learning Control

This chapter introduces another Iterative Learning Control (ILC) design methodology that uses system model information and associated computations. The approach is very different from the inverse ILC algorithm introduced in the last chapter where the control updating law

$$u_{k+1} = u_k + \beta K e_k \tag{4.1}$$

uses the system inverse $K = G_R^{-1}$ and a scalar parameter β chosen by the user. The aim of the algorithm is to find an input signal that ensures that a system output y tracks a desired *reference* output r exactly, or, if no such input exists, tracks that reference as closely as possible.

In this chapter, the system dynamics will be described by a general linear model

$$y = Gu + d \tag{4.2}$$

with inputs and outputs, u and y, assumed to be in Hilbert spaces $\mathcal{U}$ and $\mathcal{Y}$, respectively. The linear, bounded operator G maps $\mathcal{U}$ into $\mathcal{Y}$ and $d \in \mathcal{Y}$ represents initial condition and fixed disturbance effects. The inner products are denoted $\langle u, v\rangle_{\mathcal{U}}$ and $\langle y, w\rangle_{\mathcal{Y}}$ with associated norms $\|u\|_{\mathcal{U}} = \sqrt{\langle u, u\rangle_{\mathcal{U}}}$ and $\|y\|_{\mathcal{Y}} = \sqrt{\langle y, y\rangle_{\mathcal{Y}}}$. This description provides great generality to the ideas and possible applications and allows the user to use either continuous or discrete state-space models.

The Gradient Algorithm uses the adjoint operator G^* of the system operator G in (4.1). That is, it uses

$$K = G^* , \tag{4.3}$$

the resulting ILC algorithm iteratively reducing the tracking error norm from iteration to iteration for suitable small values of $\beta > 0$. The Gradient Algorithm, together with the Inverse Algorithm introduced in the last chapter, forms a necessary foundation and

B. Chu and D. H. Owens, *Optimal Iterative Learning Control*, Advances in Industrial Control, https://doi.org/10.1007/978-3-031-80236-2_4

provides useful insight into the key ILC design algorithm of the book to be introduced in the next chapter, i.e., Norm Optimal Iterative Learning Control (NOILC).

The chapter is organized as follows. The Algorithm 4.1 (Gradient-Based ILC Algorithm) is firstly introduced, its convergence properties are derived, and robustness properties against modelling mismatches are then discussed. The possible difficulties of Gradient-based ILC with non-minimum phase (NMP) systems are explained using the example of discrete-time state-space systems with one zero outside the unit circle and some insight is obtained into the amelioration of the problem by choice of reference and initial input. This discussion is followed by consideration of the application of Gradient-based ILC to point-to-point tracking tasks. Finally, the use of feedback design to improve convergence rates is discussed, in brief.

4.1 Description of the Algorithm

To develop the Algorithm 4.1 (Gradient-Based ILC Algorithm), consider the solution of the optimization problem

$$\underbrace{\min_u \|e\|^2_{\mathcal{Y}} \quad where \;\; e = r - y = r - Gu - d\,.}_{Ultimate\ Objective\ of\ ILC} \tag{4.4}$$

Remark 4.1 This can be seen as the ultimate design objective of ILC design, i.e., finding an input u to minimize the tracking error. If the minimum error is zero, *exact* tracking has been achieved. Otherwise, the solution is the best solution in the sense that the error norm is minimized.

4.1.1 The Euclidean Case: An Introduction

For simplicity, consider an ℓ-input m-output discrete-time, linear time-invariant system with operator (supervector/matrix) representation $y = Gu + d$ where $y \in \mathcal{R}^{m(N+1)}$, $u \in \mathcal{R}^{\ell(N+1)}$, and $G \in \mathcal{R}^{m(N+1)\times\ell(N+1)}$ in which $N+1$ is the trial length (number of sample points). Suppose also that the inner products are the normal Euclidean products $\langle u, v\rangle_{\mathcal{U}} = u^T v$ and $\langle y, w\rangle_{\mathcal{Y}} = y^T w$. Applying classical gradient optimization techniques at the data point (y_k, u_k) by using the Jacobian of $e^T e$ with respect to u leads to the following iterative update equation:

$$u_{k+1} = u_k + \beta G^T(r - Gu_k - d) = u_k + \beta G^T e_k \tag{4.5}$$

where $\beta > 0$ is a step size (here called a "gain") chosen by the user. This is equivalent to the use of $K = G^T$ in the general updating law $u_{k+1} = u_k + \beta K e_k$.

4.1.2 The General Case

If $\mathcal{Y}$ and $\mathcal{U}$ are more general Hilbert spaces, then the change in error $\delta e = e_{k+1} - e_k = -G(u_{k+1} - u_k)$ due to the change in input $\delta u = u_{k+1} - u_k$ takes the form

$$\|e_k + \delta e\|_{\mathcal{Y}}^2 - \|e_k\|^2 = 2\langle e_k, \delta e\rangle_{\mathcal{Y}} + \|\delta e\|_{\mathcal{Y}}^2 = -2\langle e_k, G\delta u\rangle_{\mathcal{Y}} + \|\delta e\|_{\mathcal{Y}}^2 \quad (4.6)$$

$$= -2\langle G^* e_k, u_{k+1} - u_k\rangle_{\mathcal{U}} + \|\delta e\|_{\mathcal{Y}}^2, \quad (4.7)$$

which shows immediately that the input update equation

$$u_{k+1} = u_k + \beta G^* e_k \quad (General\ Gradient\ Algorithm\ Update) \quad (4.8)$$

will generate a reduction in error norm if $\beta > 0$ is small enough and proves that the choice of $K = G^*$ in (4.1) is a choice that can improve algorithm performance from iteration to iteration.

The application of the Gradient Algorithm update equation requires a characterization of the adjoint operator G^* and the calculation of the signal $G^* e_k$ from the previous iteration error data. This also requires that $\mathcal{U}$ and $\mathcal{Y}$ and the associated inner products are defined. Various examples have been described earlier in this text and are not repeated here. Rather, the discussion returns to the case of a linear, discrete-time state-space system as discussed in Sect. 4.1.1 but using inner products (in the supervector description)

$$\begin{aligned} \langle u, v\rangle_{\mathcal{U}} &= u^T R_d v = \sum_{t=0}^{N} u^T R(t) v(t) \\ \langle y, w\rangle_{\mathcal{Y}} &= y^T Q_d w = \sum_{t=0}^{N} y^T Q(t) w(t) \end{aligned} \quad (4.9)$$

where $Q_d = blockdiag[Q(0), Q(1), \cdots, Q(N)]$ and $R_d = blockdiag[R(0), \cdots, R(N)]$ (and hence all $\{Q(t)\}$ and $\{R(t)\}$) are symmetric and positive definite matrices. It has been shown that G then has an adjoint,

$$G^* = R_d^{-1} G^T Q_d \quad (4.10)$$

which reduces to $G^* = G^T$ if both Q_d and R_d are unit matrices.

Remark 4.2 Note that intermediate (point-to-point) problems have an operator obtained from the above by deleting the rows of G at times when no reference signal is defined. A simple analysis shows that the above also applies in this case. The details are left for the reader to consider and to prove that the same effect is achieved by setting $Q(t) = 0$ at all sample points that are not intermediate points.

Implementing the Algorithm 4.1 (Gradient-Based ILC Algorithm) requires the computation of $G^* e_k$. This can either be done using direct matrix calculation (using the lifted/supervector matrix form of G and e_k) or via simulation using a state-space model representation of G^* as follows:

Algorithm 4.1 (Gradient-Based ILC Algorithm) Given a choice of scalar gain $\beta_{k+1} > 0$, the Gradient Algorithm input update takes the form

$$u_{k+1} = u_k + \beta_{k+1} G^* e_k$$

on the $(k+1)$th iteration. For a linear, time-invariant, discrete-time, state-space system, it can be computed as follows:

$$u_{k+1}(t) = u_k(t) + \beta_{k+1} R^{-1}(t) v_{k+1}(t), \quad 0 \le t \le N,$$

where

$$\begin{aligned} v_{k+1}(t) &= B^T p_{k+1}(t) + D^T Q(t) e_k(t) \\ p_{k+1}(t) &= A^T p_{k+1}(t+1) + C^T Q(t+1) e_k(t+1) \end{aligned} \tag{4.11}$$

for $t = 0, 1, 2, \cdots, N-1$, with $p_{k+1}(N) = 0$. These computations are done between iterations using data $e_k(t)$ and reverse time simulation of the state equations for $p_{k+1}(t)$.

Proof The validity of the computation can be demonstrated by interpreting the supervector $v = R_d^{-1} G^T Q_d e_k$ as the output of a reverse time state-space model. The proof of this uses the time reversal operator $\mathcal{T}$ to write $G^T = \mathcal{T} G \mathcal{T}$. ■

Remark 4.3 The iterative nature of these computations is clear from the computational structure given above. The algorithm is stated in a form allowing variations in gain on each iteration. It is not necessary that the gain be varied but, in principle, it does allow a degree of flexibility to the user to achieve other objectives. It remains to state the range of gain values that achieve norm reduction and, ultimately, convergence of the error norm to its minimum value.

4.2 Convergence Properties of the Algorithm

To analyse the convergence behaviour of Algorithm 4.1 (Gradient-Based ILC Algorithm), consider the case when all β_k take the same value $\beta > 0$.

First examine how the tracking error evolves from trial to trial. Note that

$$\begin{aligned} e_{k+1} &= r - y_{k+1} = r - G u_{k+1} - d \\ &= r - G(u_k + \beta G^* e_k) - d = (r - G u_k - d) - \beta G G^* e_k \\ &= e_k - \beta G G^* e_k = (I - \beta G G^*) e_k = L e_k, \end{aligned} \tag{4.12}$$

$$\textit{and hence } e_k = L^k e_0, \tag{4.13}$$

where the *error evolution operator* $L = I - \beta G G^*$. Clearly the properties of L control the evolution of the error sequence.

Remark 4.4 The operator $L : \mathcal{Y} \to \mathcal{Y}$ is self-adjoint (i.e., $L = L^*$) as GG^* is self-adjoint. Its spectral values (including all eigenvalues) are therefore real. In any application where G has a matrix representation, GG^* will also have a matrix representation and a complete set of eigenvectors that are orthonormal in the topology defined by the inner product $\langle \cdot, \cdot \rangle_{\mathcal{Y}}$. Classical matrix analysis of the expression $e_k = L^k e_0$ then shows that $e_k \to 0$ if all eigenvalues of L lie in the open interval $(-1, 1)$. In particular, this requires that $0 < \beta GG^* < 2I$, and hence that the kernel $ker[G^*] = \{0\}$, i.e., G^* has full column rank.

A more general analysis is presented in Owens [8] and is restated here in the form of the following result:

Theorem 4.1 (Convergence of Gradient-Based ILC) Consider the Gradient-based ILC law

$$u_{k+1} = u_k + \beta G^* e_k \quad with\ gain\ satisfying \quad 0 < \beta \|G\|^2 < 2. \tag{4.14}$$

Then, the tracking error norm converges monotonically, i.e.,

$$\|e_{k+1}\|_{\mathcal{Y}} \leq \|e_k\|_{\mathcal{Y}}, \quad for\ all\ k \geq 0, \quad (Monotonicity), \tag{4.15}$$

and

$$\lim_{k \to \infty} e_k = e_\infty = P_{ker[G^*]} e_0 \quad (Error\ Convergence) \tag{4.16}$$

where $P_{ker[G^*]}$ is the orthogonal projection operator onto the kernel of G^*.

Consequently, if e_0 lies in the closure $\overline{\mathcal{R}[G]}$ of $\mathcal{R}[G]$,

$$\lim_{k \to \infty} e_k = 0 \quad (Error\ Convergence\ to\ Zero). \tag{4.17}$$

Furthermore if $ker[G^*] = 0$, i.e., $\overline{\mathcal{R}[G]} = \mathcal{Y}$, then for any initial tracking error e_0 zero tracking error can be achieved.

Remark 4.5 The use of the closure $\overline{\mathcal{R}[G]}$ in the above is necessary only when the signal space is infinite dimensional. In practice, this only occurs in continuous signal tracking problems where $\overline{\mathcal{R}[G]} \neq \mathcal{Y}$ as, for example, the output of a strictly proper continuous system will be continuous whereas the signal space contains discontinuous functions that cannot be tracked without introducing impulses into the control input.

Remark 4.6 The result is also true for iteration-dependent gains $\{\beta_{k+1}\}_{k \geq 1}$ provided that they all lie in some interval $\delta \leq \beta_{k+1} \|G\|^2 \leq 2 - \delta$ for some $\delta > 0$. The proof of this result is not included in the text.

Proof To show monotonic convergence, note that, if $0 < \beta \|G^*\|^2 < 2$,

$$-I < (1 - \beta \|G^*\|^2) I \leq I - \beta GG^* \leq I,$$

from which $\|L\| \leq 1$ and thus the tracking error *norm* converges monotonically as $\|e_{k+1}\|_{\mathcal{Y}} \leq \|L\| \|e_k\|_{\mathcal{Y}}$.

To show convergence, write the initial tracking error e_0 into the following form:

$$e_0 = \tilde{e}_0 + \tilde{e}_0^{\perp}$$

where $\tilde{e}_0 \in ker[G^*]$ and $\tilde{e}_0^{\perp} \in ker[G^*]^{\perp} = \overline{\mathcal{R}[GG^*]} = \overline{\mathcal{R}[G]}$. Note that for $\tilde{e}_0$, we have

$$L\tilde{e}_0 = (I - \beta GG^*)\tilde{e}_0 = \tilde{e}_0 - \beta GG^* \tilde{e}_0 = \tilde{e}_0.$$

Therefore the operator L, when restricted to $ker[G^*]$, is just $L = I$.

For $\tilde{e}_0^{\perp} \in ker[G^*]^{\perp}$, Owens [8] provides a general analysis. For the purposes of the proof in this text, an eigenvalue approach is taken. Suppose that GG^* has a complete set of orthonormal eigenvectors $\{v_j\}$, corresponding to *non-zero* (necessarily positive) eigenvalues $\{\sigma_j^2\}$, that span $ker[G^*]^{\perp}$. The non-unit eigenvalues of L are $\{1 - \beta\sigma_j^2\}_{j \geq 0}$. Also, $0 < \sigma_j^2 \leq \|GG^*\| = \|G\|^2$, so that $|1 - \beta\sigma_j^2| < 1, \ j \geq 1$.

Expanding $\tilde{e}_0^{\perp}$ in terms of these eigenvectors in the form,

$$\tilde{e}_0^{\perp} = \sum_{j \geq 1} a_j v_j \qquad with \quad \sum_{j \geq 1} a_j^2 < \infty,$$

gives, for any integer $q \geq 1$,

$$L^k \tilde{e}_0^{\perp} = \underbrace{\sum_{j=1}^{q} (1 - \beta\sigma_j^2)^k a_j v_j}_{\rightarrow 0 \ as \ k \rightarrow \infty} + \underbrace{\sum_{j > q} (1 - \beta\sigma_j^2)^k a_j v_j}_{norm \ squared \ \leq \sum_{j>q} a_j^2},$$

so that

$$\limsup_{k \rightarrow \infty} \|L^k \tilde{e}_0^{\perp}\|_{\mathcal{Y}} \leq \sum_{j > q} a_j^2 \rightarrow 0 \ \ (q \rightarrow \infty),$$

which proves that $\lim_{k \rightarrow \infty} e_k = \tilde{e}_0$ which is the orthogonal projection of e_0 onto $ker[G^*]$ as required.

Finally, when $e_0 \in \overline{\mathcal{R}[G]}$, we have $e_0 \in ker[G^*]^{\perp}$ as $ker[G^*]^{\perp} = \overline{\mathcal{R}[G]}$ so that $P_{ker[G^*]} e_0 = 0$ and thus $e_\infty = 0$. ■

If the reference can be tracked exactly, there exists at least one input such that $r = y$ which can be written as $e_0 \in \mathcal{R}[G]$. In this case the tracking error will converge to zero. The next result shows that, in this case, the input signal will also converge to a limit with a very nice property.

Theorem 4.2 (Input Convergence Property) Using the notation and assumptions of Theorem 4.1, suppose that $e_0 \in \mathcal{R}[G]$. Then the input sequence $\{u_k\}_{k \geq 0}$ in $\mathcal{U}$ converges to the unique solution of the (minimum norm) optimization problem

$$u_\infty = \arg\min_{u\in\mathcal{U}}\{\ \|u-u_0\|^2_{\mathcal{U}} : r = Gu + d\}. \tag{4.18}$$

Moreover, it converges monotonically,

$$\|u_\infty - u_{k+1}\|^2_{\mathcal{U}} \le \|u_\infty - u_k\|^2_{\mathcal{U}}, \quad k \ge 0. \tag{4.19}$$

Proof The input update equation can be written in the form

$$(\tilde{u}_\infty - u_{k+1}) = (\tilde{u}_\infty - u_k) - \beta G^* G(\tilde{u}_\infty - u_k) = L_0(\tilde{u}_\infty - u_k), \quad 0 < \beta\|G\|^2 < 2,$$

where $\tilde{u}_\infty$ is any input that generates r exactly and $L_0 = I - \beta G^* G : \mathcal{U} \to \mathcal{U}$ is the operator defining the input evolution. This equation has the same structure as the error evolution (with G^*G replacing GG^*) so the previous theorem can be used to prove monotonicity and also that

$$\lim_{k\to\infty}(\tilde{u}_\infty - u_k) = P_{ker[G]}(\tilde{u}_\infty - u_0).$$

This equation also holds if $\tilde{u}_\infty$ is replaced by u_∞. The result now follows from the fact that the unique solution of (4.18) must satisfy the orthogonality condition $\langle\tilde{u}_\infty - u_\infty, u_\infty - u_0\rangle_{\mathcal{U}} = 0$ for any choice of $\tilde{u}_\infty$. With this condition $P_{ker[G]}(u_\infty - u_0) = 0$ and hence u_∞ is the limit of the input sequence $\{u_k\}_{k\ge 0}$. ■

Note 4.1 The above theorem shows that the Algorithm 4.1 (Gradient-Based ILC Algorithm) not only achieves monotonic tracking error norm convergence to zero, but also guarantees input convergence to a point that is closest (in norm) to the initial input u_0. When $u_0 = 0$ then the final converged input is just the (so-called) *minimum control energy solution.*

The performance of the Gradient ILC Algorithm is illustrated by the following example.

Example 4.1 Consider the same setting as in Example 3.2, i.e., the following system:

$$G(s) = \frac{s+1}{s^2+5s+6},$$

sampled at intervals of $T_s = 0.1$ using zero-order hold (resulting in a discrete-time plant model $S(A, B, C)$). The system is required to track a reference signal $r(t) = sin(2\pi t/6)$ defined over the interval $t \in [0, 6]$. For this example, $N = 60$. Choose $Q(t) = I$, $R(t) = I$, $t = 0, 1, \ldots, N$ and thus, in supervector form, $G^* = G^T$. Several learning gain choices $\beta = 0.1/\|G\|^2, 0.3/\|G\|^2, 0.7/\|G\|^2$, respectively, (all choices meeting the convergence requirement of β) are considered. The resultant performance of these Gradient-based ILC designs is shown in Fig. 4.1. As can be seen, the tracking error norm is converging, monotonically, to exact tracking for all three learning gain choices as expected.

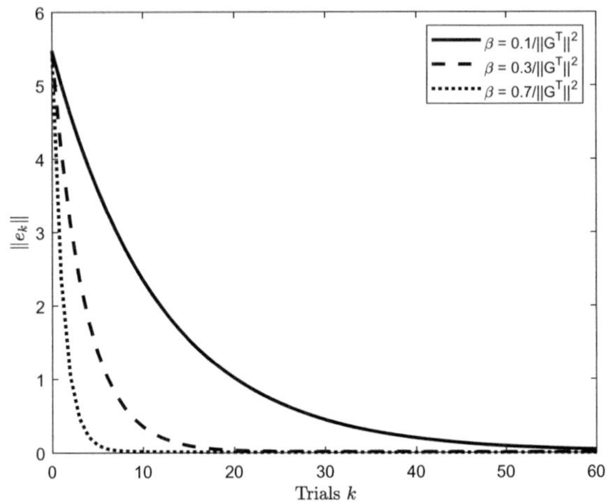

Fig. 4.1 Convergence behaviour of the Gradient-Based ILC Algorithm with different step sizes

4.3 Robustness of the Algorithm

To analyse the robustness of the Gradient-Based ILC Algorithm, suppose G_0 is the nominal model but a *right multiplicative model uncertainty* U is present, i.e., the true plant G is given by

$$G = G_0 U \quad (Right\ Multiplicative\ Uncertainty),$$

where $U : \mathcal{U} \to \mathcal{U}$ is a linear, bounded operator. Note that the case with no model mismatch (modelling error) is recovered by setting $U = I$.

Note 4.2 For discrete-time, time-invariant, ℓ-input, m-output, state-space models the above will be taken to have the frequency domain representation

$$G(z) = G_0(z)U(z), \quad (Transfer\ Function\ Matrix\ Description)$$

where the $\ell \times \ell$ transfer function matrix $U(z)$ is stable.

The error evolution equation is derived from the input update $u_{k+1} = u_k + \beta G_0^* e_k$ and takes the form

$$e_{k+1} = (I - \beta G_0 U G_0^*)e_k, \quad k \geq 0\,,\ (Error\ Evolution). \tag{4.20}$$

The following theorem states a sufficient condition for *Robust Monotone Convergence* of the error, i.e., conditions that ensure that

$$\|e_{k+1}\|_{\mathcal{Y}}^2 < \|e_k\|_{\mathcal{Y}}^2 \quad for\ all\ \ k \geq 0, \tag{4.21}$$

in the presence of the perturbation U.

Theorem 4.3 (Robust Monotone Error Convergence) The Algorithm 4.1 (Gradient-Based ILC Algorithm) is robust monotonically convergent in the tracking error to zero in the presence of multiplicative modelling error U if (a sufficient condition) the induced norm,

$$\|I - \beta G_0 U G_0^*\| < 1. \tag{4.22}$$

In addition, a necessary condition for this relation to hold true is that $ker[G^*] = \{0\}$.

Proof The tracking error evolution with the presence of modelling mismatch U is given by

$$\begin{aligned} e_{k+1} &= r - y_{k+1} = r - Gu_{k+1} - d = r - G_0 U u_{k+1} - d \\ &= r - G_0 U(u_k + \beta G_0^* e_k) - d = r - G_0 U u_k - d - \beta G_0 U G_0^* e_k \\ &= e_k - \beta G_0 U G_0^* e_k = (I - \beta G_0 U G_0^*) e_k. \end{aligned}$$

It follows that a reduction in norm and ultimate convergence to zero error will be achieved if $\|I - \beta G_0 U G_0^*\| < 1$. Finally, note that if $ker[G^*] \neq \{0\}$, there will exist a non-zero initial error e_0 such that $G^* e_0 = 0$ and hence that $e_1 = e_0$, i.e., $\|I - \beta G_0 U G_0^*\| \geq 1$. ■

Remark 4.7 The above robustness monotonic convergence condition is quite general but does not cover all possibilities. For example, consider tracking of a signal, $r(t),\ 0 \leq t \leq T$, for a linear, *strictly proper*, continuous-time, state-space system from zero initial conditions. Very high frequency sinusoidal inputs will produce infinitesimal outputs from $G_0 U G_0^*$, i.e., $\|I - \beta G_0 U G_0^*\| \geq 1$ for all U.

An alternative approach that avoids some of the technical difficulties is to find more detailed conditions that ensure error norm reduction. More precisely, the condition

$$(I - \beta G_0 U G_0^*)^*(I - \beta G_0 U G_0^*) < I,$$

using simple algebraic operations, reduces to

$$G_0 \left(U + U^* - \beta U^* G_0^* G_0 U\right) G_0^* > 0.$$

This gives the following result.

Theorem 4.4 With the notation stated above, robust monotonic reductions in norm will always be achieved on each and every iteration if $ker[G_0^*] = \{0\}$ (i.e., $\overline{\mathcal{R}[G_0]} = \mathcal{Y}$) and

$$U + U^* > \beta U^* G_0^* G_0 U \quad (Operator\ Inequality\ for\ Robustness). \tag{4.23}$$

If U has a bounded inverse $\hat{U}$ on $\mathcal{U}$, this condition can be replaced by

$$\hat{U} + \hat{U}^* > \beta G_0^* G_0 \quad (Inverse\ Operator\ Inequality). \tag{4.24}$$

Note the following:

1. The above again (as for the inverse ILC algorithm) implies that the modelling error needs to be positive real in order to achieve monotonic convergence in the tracking error norm.
2. The nominal model (when $U = I$) satisfies the condition if, and only if, $2I > \beta G_0^* G_0$, an inequality that is always satisfied for the range of gains β permitted.
3. In general, the inequality is more likely to be satisfied if $\beta > 0$ is small enough, a statement that is consistent with the intuition that robustness will require caution in the algorithm. Here, caution implies slower convergence.

For discrete-time, linear state-space systems, the above theorem also has a frequency domain interpretation. Using the constructions used for robustness analysis of the Inverse ILC Algorithm, it can be shown that when $U(z)$ is asymptotically stable and the weighting matrices $Q(t) = Q$, $R(t) = R$, $t = 0, 1, \ldots, N$ are independent of time, the robustness condition can be replaced by

$$\underbrace{RU(z) + U^T(z^{-1})R > \beta U^T(z^{-1})G_0^T(z^{-1})QG_0(z)U(z),}_{\textit{Frequency Domain Condition for Robust Monotonicity}} \quad \forall |z| = 1. \tag{4.25}$$

For single-input–single-output (SISO), discrete systems, the above reduces to

$$Re[U^{-1}(z)] > \frac{\beta}{2}\frac{Q}{R}|G_0(z)|^2, \quad \forall |z| = 1, \tag{4.26}$$

which can be checked graphically if $U(z)$ is known.

The above results are illustrated in the following example.

Example 4.2 Consider the same setting as in the previous example where the nominal underlying continuous model is given by

$$G_0(s) = \frac{s+1}{s^2 + 5s + 6}.$$

Consider two cases. For the first, consider the case where the model mismatch is given by

$$U(s) = \frac{s+1}{s+5},$$

which can be verified to be positive real. The performance of Gradient-based ILC design with learning gain choice $\beta = 0.1/\|G\|^2, 0.3/\|G\|^2, 0.7/\|G\|^2$, respectively, is shown in Fig. 4.2. Note that for these parameter choices, the robustness monotonic convergence condition is satisfied. As can be seen, the tracking error norm converges monotonically to zero tracking for all three learning gain choices even with the presence of model mismatches as expected.

Fig. 4.2 Robust Monotonic Convergence behaviour of the Gradient-Based ILC Algorithm

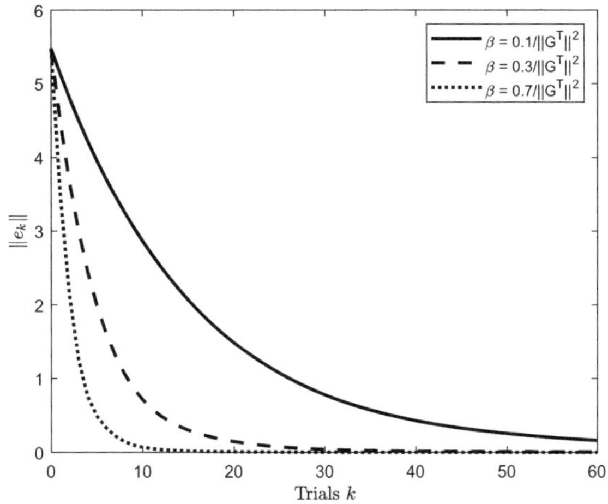

For the second case, the model mismatch is given by

$$U(s) = \frac{1}{s^2 + 5s + 6},$$

which is not positive real. Again, the performance of the Gradient Algorithm with various parameter choices is shown in Fig. 4.3. Note that for these parameter choices, a consideration of the frequency domain form of the robustness criterion indicates that the robustness monotonic convergence condition is not satisfied. As expected, the tracking error norms eventually diverge after an initial reduction.

Fig. 4.3 Divergent behaviour of the Gradient Algorithm

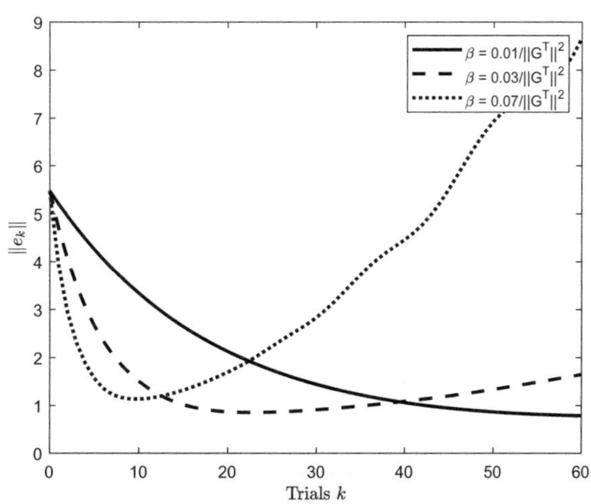

4.4 The Effect of Non-minimum Phase Zeros

Iterative learning control can have performance difficulties in some circumstances. As an example, the inverse-based ILC will have problems as the system inverse will be unstable due to the existence of the non-minimum phase (NMP) zeros. Whilst gradient ILC does not suffer from this particular problem, this section aims to show that the inherent nature of NMP systems can lead to some *slow convergence behaviour*. For simplicity, a linear, discrete-time, time-invariant, SISO system is considered and its behaviour analysed on a discrete interval $0 \le t \le N$. Two simplifications are made, the reader being referred to the references Owens and Chu [9], Owens et al. [12], and Chu and Owens [2] for an analysis of the more general case of multiple NMP zeros (as well as that for linear continuous-time systems):

1. The analysis is simplified by assuming that the system has precisely one NMP zero $|z_1| > 1$.
2. Furthermore, we choose inner products in $\mathcal{Y}$ and $\mathcal{U}$ with scalar weights, $Q(t) = 1$, $R(t) = 1$, $t = 0, 1, \ldots, N$ and thus the system adjoint has supervector matrix $G^* = G^T$.

Note 4.3 Note that, in this case, the squared norm $\|G\|^2$ is equal to the largest eigenvalue of GG^T (largest singular value of G) which has an upper bound $\max_{|z|=1} |G(z)|^2$ (see Sect. 2.14.6) where $G(z)$ denotes the transfer function of the system.

4.4.1 System Description

Suppose the non-minimum phase system $G(z)$ considered has one NMP zero, i.e.,

$$G(z) = \left(\frac{(z - z_1) \prod_{i=1}^{i_0} (z - b_i)}{\prod_{i=1}^{n} (z - a_i)} \right) \tag{4.27}$$

where $|b_j| < 1, 1 \le j \le i_0$, but the remaining zero z_1, $|z_1| > 1$, is a NMP zero lying outside the unit circle.

Remark 4.8 The rare case when a zero lies on the unit circle is technically more complex and is not considered in this text.

$G(z)$ can be decomposed into minimum phase/all-pass decomposition described by the factorization,

$$G(z) = G_m(z) G_a(z), \tag{4.28}$$

achieved by adding a cancelling pole–zero pair at the point $z = z_1^{-1}$. More precisely,

$$G_m(z) = \frac{(1 - z_1 z) \prod_{i=1}^{i_0} (z - b_i)}{\prod_{i=1}^{n} (z - a_i)} \tag{4.29}$$

is minimum phase and

$$G_a(z) = \frac{z - z_1}{1 - z_1 z} \tag{4.30}$$

is asymptotically stable, NMP, and *all-pass* (i.e., $|G_a(z)| \equiv 1, \forall |z| = 1$). In matrix form, this can be written as

$$G = G_m G_a,$$

where G, G_m, and G_a are the supervector matrix representations of $G(z), G_m(z), G_a(z)$, respectively.

Remark 4.9 The matrices G_a and G_m operate on the associated time series vectors on $0 \le t \le N$ in a manner described by the transfer functions $G_a(z)$ and $G_m(z)$. However, the matrices G and G_m will be singular if the system is strictly proper. This technical difficulty in what follows is removed by noting that the only inputs that affect the outputs on this interval are $u(t),\ 0 \le t \le N - k^*$ where the *relative degree* $k^* = n - i_0 - 1$ (excess of poles over zeros). Similarly, the only outputs affected by these inputs are $y(t),\ k^* \le t \le N$. This observation makes it possible to create a description based on $(N + 1 - k^*) \times (N + 1 - k^*)$ matrices rather than the original $(N + 1) \times (N + 1)$. The transfer function mapping the reduced input sequence into the reduced output sequence is obtained by replacing $G_m(z)$ by $z^{k^*} G_m(z)$. This has a supervector description with matrix representation G_m obtained by removing the first k^* rows and last k^* columns from the original supervector description of $G_m(z)$. As $G_a(z)$ has relative degree zero the supervector description on $0 \le t \le N - k^*$ is obtained by removing the last k^* rows and columns. Once this is done, all matrices are then invertible.

4.4.2 Singular Value Distributions of an All-Pass System

To analyse the slow convergence behaviour, we first need to present some results on the singular value distribution of the NMP system G_a. Note that $G_a(z)$ can be written in the form

$$G_a(z) = -z_1^{-1}\left(1 + (z_1^2 - 1)\frac{1}{1 - z z_1}\right) = -z_1^{-1}\left(1 + (1 - z_1^2)\sum_{j \ge 1}(z z_1)^{-j}\right), \tag{4.31}$$

which produces the supervector description, with $a = z_1^{-1} - z_1$,

$$\underbrace{G_a = \left(\frac{-1}{z_1}\right)\begin{bmatrix} 1 & 0 & \cdots & & \cdots & 0 \\ a & 1 & 0 & \cdots & & 0 \\ z_1^{-1}a & a & 1 & \cdots & & 0 \\ z_1^{-2}a & z_1^{-1}a & a & 1 & \cdots & 0 \\ \vdots & & & & & \vdots \\ & & & \cdots & 1 & 0 \\ z_1^{N-k^*-1}a & \cdots & & \cdots & a & 1 \end{bmatrix}}_{Supervector\ Description\ of\ the\ All-Pass\ System}. \tag{4.32}$$

Two statements follow from the above and Sect. 2.14.6,

$$\underbrace{\|G_aG_a^T\| = \|G_a\|^2 = \|G_a^T\|^2 \leq \max_{|z|=1} |G_a(z)|^2 = 1,}_{Norm\ bounds}$$
$$and \quad \underbrace{\det[G_aG_a^T] = \frac{1}{z_1^{2(N-k^*+1)}}}_{Product\ of\ eigenvalues}. \tag{4.33}$$

This shows that all eigenvalues of GG^T are less than, or equal to, one. However, in this case it is possible to calculate the eigenvalues and eigenvectors of $G_aG_a^T$ (i.e., the singular values and singular vectors of G^T) exactly. To do this, note the following.

Lemma 4.1 (Eigenstructure of the All-Pass System) Suppose that $N - k^* \geq 1$, then:

1. $G_aG_a^T$ has exactly one eigenvalue equal to $z_1^{-2(N-k^*+1)}$ with the remaining $N - k^*$ eigenvalues being equal to unity. The eigenspace corresponding to these unit eigenvalues is the span of the $N - k^*$ vectors $\{v_j\}_{2\leq j\leq N-k^*+1}$.
2. The remaining eigenvalue is equal to $z_1^{-2(N-k^*+1)}$ with eigenvector v_1, which is orthogonal to the $\{v_j\}_{2\leq j\leq N-k^*+1}$ in the Euclidean topology, i.e., $v_1^Tv_j = 0,\ 2 \leq j \leq N - k^* + 1$.

In the above, the linearly independent vectors are defined by

$$v_1 = \begin{bmatrix} 1 \\ z_1^{-1} \\ z_1^{-2} \\ \vdots \\ \\ z_1^{-(N-k^*-1)} \\ z_1^{-(N-k^*)} \end{bmatrix},\ v_2 = \begin{bmatrix} 1 \\ -z_1 \\ 0 \\ \vdots \\ 0 \\ 0 \\ 0 \end{bmatrix},\ v_3 = \begin{bmatrix} 0 \\ 1 \\ -z_1 \\ \vdots \\ 0 \\ 0 \\ 0 \end{bmatrix},\ \cdots,\ v_{N-k^*+1} = \begin{bmatrix} 0 \\ 0 \\ 0 \\ \vdots \\ 0 \\ 1 \\ -z_1 \end{bmatrix}. \tag{4.34}$$

Proof The proof is by construction using the definitions and the special structure of $G_aG_a^T$. Firstly note that the set $\{v_j\}_{j\geq 2}$ is linearly independent and a simple calculation yields the orthogonality property of v_1. Next, consider the example computation,

$$G_a G_a^T v_2 = (\frac{-1}{z_1}) G_a \begin{bmatrix} 1-az_1 \\ -z_1 \\ 0 \\ \vdots \\ 0 \\ 0 \\ 0 \end{bmatrix} = \frac{1}{z_1^2} \begin{bmatrix} 1-az_1 \\ a(1-az_1)-z_1^{-1} \\ 0 \\ \vdots \\ 0 \\ 0 \\ 0 \end{bmatrix} = \begin{bmatrix} 1 \\ a-z_1^{-1} \\ 0 \\ \vdots \\ 0 \\ 0 \\ 0 \end{bmatrix} = \begin{bmatrix} 1 \\ -z_1 \\ 0 \\ \vdots \\ 0 \\ 0 \\ 0 \end{bmatrix} = v_2.$$

That is, v_2 is an eigenvector of $G_a G_a^T$ with eigenvalue equal to one. Similar calculations can be undertaken for $G_a G_a^T v_j, \; j > 2$ with the same outcome so that there must be $N - k^*$ unit eigenvalues with eigenspace spanned by the $\{v_j\}_{j \geq 2}$. The product of the eigenvalues of $G_a G_a^T$ is equal to its determinant, which, from the above, must be equal to the final eigenvalue. As $G_a G_a^T$ is symmetric, its eigenvector can be found by computing a vector orthogonal to the $\{v_j\}_{j \geq 2}$. The vector v_1 satisfies this requirement. The proof is complete. ■

An important consequence of this result is that $G_a G_a^T$ has a decomposition

$$G_a G_a^T = V \Lambda V^T = V \underbrace{\begin{bmatrix} z_1^{-2(N-k^*+1)} & 0 \\ 0 & I_{N-k^*} \end{bmatrix}}_{\Lambda} V^T \approxeq V \underbrace{\begin{bmatrix} 0 & 0 \\ 0 & I_{N-k^*} \end{bmatrix}}_{\Lambda_a} V^T = V \Lambda_a V^T, \tag{4.35}$$

where V is an orthogonal matrix $VV^T = V^T V = I$ and I_{N-k^*} is a unit matrix of the indicated dimension. V is computed by setting the second to last columns equal to an orthonormal basis for the span of $\{v_j\}_{j \geq 2}$ and a first column being v_1 normalized to unit Euclidean norm, i.e.,

$$v_1 \mapsto \frac{v_1}{\sqrt{v_1^T v_1}}, \qquad where \quad v_1^T v_1 = \frac{1 - z_1^{-2(N-k^*+1)}}{1 - z_1^{-2}}.$$

Remark 4.10 The approximation of $G_a G_a^T$ is important in the later analysis of NMP Gradient ILC performance and the characterization of slow convergence phenomena. The important observation is that, whereas both G and G_m are nonsingular, the structure of G_a suggests that G is almost singular for reasons associated not only with the, typically, low system gains at high frequency (due to low-pass characteristics, which are largely independent of N) but also due to the additional effect of the NMP-induced eigenvalue $z_1^{-2(N-k^*+1)} \to 0+$ if N is large (the normal situation in practice).

Remark 4.11 The argument put forward in the previous remark is supported by the observation, using norm inequalities and the frequency domain-based lower bound on the minimum eigenvalue of $G_m G_m^T$ (and hence $G_m^T G_m$),

$$min_{\|u\|=1}\left(\frac{u^T G_m^T G_m u}{u^T u}\right) \geq \left(\max_{\|v\|=1}\left(\frac{v^T (G_m G_m^T)^{-1} v}{v^T v}\right)\right)^{-1}$$
$$= \frac{1}{\|(G_m G_m^T)^{-1}\|} = \frac{1}{\max_{|z|=1} |G^{-1}(z)|^2} > 0,$$

which is independent of N. That is, the system G_m has a non-zero, N-independent, lower bound on its eigenvalues.

4.4.3 Slow Convergence Phenomena

Gradient algorithms applied to linear, discrete-time, time-invariant, NMP systems exhibit a very particular slow convergence property that is discussed in this section. It can be illustrated by the following example.

Example 4.3 Consider the discrete-time system with one NMP zero

$$G(z) = \frac{(z-0.1)(z-1.1)}{(z+0.2)(z-0.8)}.$$

The sampling time $T_s = 0.05$, the trial length is $0 \leq t \leq 6$, and $N = 121$. Zero initial conditions are assumed and the reference signal is

$$r(t) = \begin{cases} 5cos(t) & \text{if } 0 \leq t \leq \pi/2, \\ 0 & \text{if } pi/2 < t \leq 6 - \pi/2, \\ 5sin(t + \pi/2 - 6) & \text{if } 6 - \pi/2 < t \leq 6. \end{cases}$$

The initial input is chosen to be $u_0 = 0$ and the learning gain is chosen to be $\beta = 0.3$. The example has all the properties that, by Theorem 4.1, guarantees error convergence to zero.

This example illustrates a special, and problematic, form of such convergence for NMP systems that does not look like convergence to the user!

A simulation of the algorithm over 50 iterations gives the results shown in Fig. 4.4 where error norm is seen to reduce monotonically, but then, apparently stays, to all intents and purposes, constant, *flat-lining* on a "*plateau*" where very little norm reduction is achieved. This phenomenon has already been seen in Parameter Optimal ILC in Chap. 3, but the reasons for the behaviour are very different for reasons explained in the following discussion.

Remark 4.12 Included in the graph is the norm reduction exhibited by an approximate model that is illustrated in this section and verifies the use of the approximation in describing NMP-induced behaviour and also in deriving a theoretical prediction of the value of the error norm on the plateau.

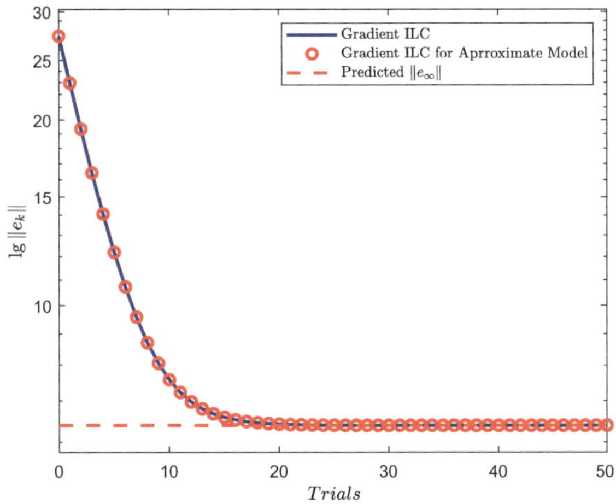

Fig. 4.4 Convergence of a Non-minimum phase system using the Gradient Algorithm with fixed learning gain $\beta = 0.3$ showing the apparently non-zero limit error norm

The purpose of what follows is to throw light on the source of the issues seen in the example. To do this consider now the application of the Algorithm 4.1 (Gradient-Based ILC Algorithm) to a SISO, linear, time-invariant, discrete-time, asymptotically stable system with exactly one NMP zero z_1.

The content leans heavily on the developments of the previous Sect. 4.4.2.

Remark 4.13 Many of the properties described in this section are inherited by other optimization-based Iterative Learning Control algorithms, including the Norm Optimal Iterative Learning Control (NOILC) Algorithm described later in this text.

4.4.3.1 A Qualitative Discussion

The error evolution of the gradient algorithm takes the supervector form

$$e_{k+1} = (I - \beta G G^T) e_k = (I - \beta G_m \underbrace{G_a G_a^T} G_m^T) e_k. \tag{4.36}$$

As G (and hence G^T) is nonsingular, the convergence Theorem 4.1 ensures that the error converges to zero if $k \to \infty$ for $0 < \beta \|G\|^2 < 2$. In this sense, NMP zeros do not affect the ultimate convergence property.

However, the presence of a NMP zero introduces an arbitrarily small eigenvalue of $G_a G_a^T$ as N increases, this eigenvalue going to zero as $N \to \infty$. This small eigenvalue will translate into an arbitrarily small eigenvalue of GG^T, denoted σ_{NMP}^2 for simplicity, an eigenvalue that will go to zero as N becomes large. As a consequence, the component of the initial error supervector e_0 that lies in the subspace corresponding to the associated eigenvector, w_{NMP}, will converge following the law,

$$(1 - \beta \sigma_{NMP}^2)^k w_{NMP} \quad (\textit{Slowly Convergent Component}).$$

For large enough values of N, the situation can arise that all other eigenvector components decay to zero, possibly rapidly, leaving only this component which is converging to zero, but extremely slowly. As ILC iterations, in practice, are finite in number, the error sequence will be monotonically decreasing but appear to be essentially constant (appear to be "flat-lining") as error reductions on each iteration will be infinitesimally small and probably not worth the physical effort, time, and expense of undertaking more plant iterations. Graphically, the error norm $\|e_k\| = \sqrt{e_k^T e_k}$ will reduce monotonically over a finite, relatively small, number of iterations, but then remain essentially constant.

Remark 4.14 The value of ILC iterations will clearly depend on the value of the error norm when the infinitesimally slow convergence "begins". If this value is large, then Gradient ILC, in the form described, has limited benefit but if small, it may have value in applications where tracking improvements are sought but very close tracking is not required.

4.4.3.2 Characterization of the Error Norm Plateau

An estimate of the value of the error norm where the flat-lining/plateauing of the error norm begins can be obtained by making the approximation

$$GG^T = G_m V \Lambda V^T G_m^T \rightarrow G_m V \Lambda_a V^T G_m^T, \tag{4.37}$$

an approximation that will be very accurate over a finite number of iterations.

As Λ_a is symmetric and satisfies $\Lambda_a = \Lambda_a^2 = \Lambda_a \Lambda_a^T$, it follows that the approximation is equivalent to applying the Gradient Algorithm to the process $G_m V \Lambda_a V^T$ which is singular. As a consequence, this approximate gradient algorithm will generate a sequence of errors $\{e_k^a\}_{k \geq 0}$ that converges to a non-zero limit e_∞^a, a limit whose norm will be representative of the error norm on the plateau of the original problem. Using Theorem 4.1, this limit takes the form of the orthogonal projection,

$$e_\infty^a = P_{ker[V \Lambda_a V^T G_m^T]} e_0$$

where $P_{ker[V \Lambda_a V^T G_m^T]}$ is the orthogonal projection operator mapping $\mathcal{U}$ onto $ker[V \Lambda_a V^T G_m^T]$. The next step is to recognize that this kernel is described by $ker[V \Lambda_a V^T G_m^T] = (G_m^T)^{-1} ker[V \Lambda_a V^T]$, which is spanned by the single vector $(G_m^T)^{-1} v_1$. The orthogonal projection operator can be represented by the expression

$$P_{ker[V \Lambda_a V^T G_m^T]} = \left(\frac{1}{v_1^T (G_m^T G_m)^{-1} v_1} \right) (G_m^T)^{-1} v_1 v_1^T G_m^{-1}, \tag{4.38}$$

so that the estimated value of the error and its associated norm on the plateau are

$$\underbrace{e_{\infty}^{a} = \left(\frac{1}{v_1^T (G_m^T G_m)^{-1} v_1} \right) (G_m^T)^{-1} v_1 v_1^T G_m^{-1} e_0}_{\textit{Estimated Error on Plateau}}$$
$$\underbrace{\|e_{\infty}^{a}\|^2 = \frac{(v_1^T G_m^{-1} e_0)^2}{v_1^T (G_m^T G_m)^{-1} v_1}}_{\textit{Estimated Plateau Norm Value}} . \tag{4.39}$$

These are relatively complex representations on the face of it, but, given $e_0(t)$ on $k^* \le t \le N$, they depend only on the single vector $(G_m^T)^{-1} v_1$. This makes a useful approximation possible.

4.4.3.3 Approximation of the Plateau Estimate

The expression (4.39) is not necessarily one that needs to be evaluated in applications. Its value here is that of providing some insight into situations when the plateau is likely to take smaller, and possibly small, values. The key to this is to note the dependence of the plateau on both z_1 and the initial error e_0 and to develop a simplification of the formula, valid when N is large.

The important quantity in (4.39) is $(G_m^T)^{-1} v_1$. It is difficult to make any precise statements about this, but it is possible to gain insight into the situation when N is large. This can be done using the following argument. First note that this signal can be written using the time reversal operator $\mathcal{T}$ (Sect. 2.8), in the form

$$(G_m^T)^{-1} v_1 = \mathcal{T} G_m^{-1} (\mathcal{T} v_1) .$$

This can be evaluated using state-space solution techniques using the observation that the input to G_m^{-1} is just

$$\mathcal{T} v_1 = z_1^{-(N-k^*)} [1, \ z_1, z_1^2, \cdots, \ z_1^{N-k^*}]^T .$$

Note that $(z^{k^*} G_m(z))^{-1}$ has an asymptotically stable realization $S(A_m, B_m, C_m, D_m)$ with $D_m \neq 0$. This gives

$$G_m^{-1} \mathcal{T} v_1 = (z_1^{k^*} G_m(z_1))^{-1} \underbrace{\frac{1}{z_1^{N-k^*}} \begin{bmatrix} 1 \\ z_1 \\ \vdots \\ z_1^{N-k^*} \end{bmatrix}}_{\mathcal{T} v_1} - \underbrace{\frac{1}{z_1^{N-k^*}} \begin{bmatrix} C_m \\ C_m A_m \\ \vdots \\ C A_m^{N-k^*} \end{bmatrix} (z_1 I - A_m)^{-1} B}_{\textit{decaying sequence}}$$
$$= \underbrace{(z_1^{k^*} G_m(z_1))^{-1} \mathcal{T} v_1}_{O(z_1^{-1})} + O(z_1^{N-k^*+1}).$$

As $|z_1| > 1$, it follows that, as N becomes large, this yields the close approximation

$$(G_m^T)^{-1} v_1 \approx \underbrace{(z_1^{k^*} G_m(z_1))^{-1}}_{a\ scalar} v_1 \quad (when\ N\ is\ sufficiently\ large). \tag{4.40}$$

This yields the approximate value for the error and error norm on the plateau (4.39) in the form,

$$Estimated\ Plateau\ Error \quad e_\infty^a = \left(\frac{v_1^T e_0}{v_1^T v_1}\right) v_1 \quad with\ Norm$$

$$\|e_\infty^a\|^2 \approx \frac{(v_1^T e_0)^2}{v_1^T v_1}, \quad i.e., \quad \|e_\infty^a\| \approx \frac{|v_1^T e_0|}{\sqrt{v_1^T v_1}}, \tag{4.41}$$

$$where \quad v_1 = [1\ z_1^{-1}\ z_1^{-2}\ \cdots\ z_1^{-(N-k^*)}]^T.$$

This formula is valid if N is large, is amenable to easy evaluation, and reveals a very simple relationship between the plateau value and the initial error signal e_0, the time series v_1 and the zero z_1. More precisely, as v_1 is the supervector for the time series $\{1,\ z^{-1},\ z^{-1},\ \cdots,\ z^{-(N-k^*)}\ \}$, and $e_0 = \{e_0(k^*),\ e_0(k^*+1),\ \cdots, e_0(N)\}$ in the chosen representation, the important quantities are

$$v_1^T e_0 = e_0(k^*) + e_0(k^*+1) z_1^{-1} + e_0(k^*+2) z_1^{-2} + \cdots + e_0(N) z_1^{-(N-k^*)}$$

$$and \quad v_1^T v_1 = \frac{1 - z_1^{-2(N-k^*+1)}}{1 - z_1^{-2}} \approx \frac{1}{1 - z_1^{-2}} \quad if\ N\ is\ large$$

It follows that the error on the plateau will tend to take the form of a scalar multiple of the decaying time series v_1 whose magnitude depends upon e_0 and z_1. The corresponding plateau error norm will have small values if:

1. Either the NMP zero z_1 is large when $v_1 \to [1\ 0\ 0 \cdots\ 0]^T$ so that $v_1^T v_1 \approx 1$ and $v_1^T e_0 \approx e_0(k^*)$which will normally be small if the reference signal can be tracked.

Note 4.4 Conversely, if z_1 is much closer to unity than the eigenvalues of the plant, the formula (4.41) suggests that the value of $\|e_k\|$ along the plateau will be in the region of the *standard deviation/variance* of the time series e_0.

2. More generally, the plateau will occur at small norm values if the error time series e_0 takes small values in a large enough interval about the point $t = k^*$. This is because the times series v_1 decays along the time interval. If, for example, the initial condition is zero and the initial control iterate is $u_0 = 0$, then $e_0 = r$ and the reduction in the plateau value will be achieved for references that start at $r(0) = 0$ and grow slowly in the vicinity of $t = 0$. This suggests the following:

   a. References, in this situation, that exhibit rapid movements in the initial phase of the motion are likely to lead to larger plateau values.

b. Reference signals describing "slowly" moving reference values will tend to produce lower plateau values.

4.5 Extension to Point-to-Point Tracking Tasks

The Algorithm 4.1 (Gradient-Based ILC Algorithm) can be applied easily to solve the point-to-point tracking problems, in a similar way as that in the previous chapter for the inverse ILC algorithm. To recap the setting, for point-to-point tracking tasks, the system is required to track a reference defined on M, given intermediate times $t_1, t_2, \ldots, t_M$. This applies to both discrete-time and continuous-time linear state-space systems but most of what follows here uses discrete-time systems to describe algorithm properties and behaviour.

In vector form, the reference is given by

$$r^p = \left[r^T(t_1) \;\; r^T(t_2) \;\; \cdots \;\; r^T(t_M)\right]^T,$$

so that the output space $\mathcal{Y}$ is finite, and possibly low, dimensional. The dimension is Mm. As a consequence, the operator GG^* is represented by an $mM \times mM$ matrix, often of much lower dimension than GG^*.

Remark 4.15 This is easily generalized to include "partial" reference signals, i.e., given a set of reference outputs $\{r(t_j)\}_{1 \le j \le M}$, only part of this reference becomes the reference signal for the ILC algorithm, giving

$$r^p = \left[(F_1 r(t_1))^T \;\; (F_2 r(t_2))^T \;\; \cdots \;\; (F_M r(t_M))^T\right]^T.$$

Here, each F_j is $f_j \times m$, representing the idea that the output at t_j is required to satisfy f_j linear equations. The details are left for the reader to consider.

The relationship between the system output, y^p, at these time instants, and the input u can be described as

$$y^p = G^p u + d^p,$$

where the superscript p denotes the signal values at these intermediate points.

Note 4.5 For discrete-time systems in lift/supervector matrix form, y^p, G^p, and d^p just consist of the t_1th, t_2th, ..., t_Mth (block) rows of y, G, d, respectively.

The Algorithm 4.1 (Gradient-Based ILC Algorithm) for point-to-point tracking tasks now becomes

$$u_{k+1} = u_k + \beta G^{p*} e_k^p \quad where \quad e_k^p = r^p - y_k^p$$

is the point-to-point tracking error.

Remark 4.16 The input update can also be written as $u_{k+1} = u_k + \beta_{k+1} G^{p*} e_k^p$ if iteration-dependent gains are used. Such variations can be beneficial in improving algorithm performance but is not considered here. See Owens [8] for more details of possibilities including the idea of spectral annihilation.

The above Gradient-based ILC design for point-to-point tasks has the following properties, inherited from the convergence Theorem 4.1 for the general gradient algorithm:

Theorem 4.5 The Gradient-based Point-to-Point/Intermediate Point ILC algorithm has the following error evolution law:

$$e_{k+1}^p = (1 - \beta G^p G^{p*}) e_k^p.$$

Suppose $0 < \beta \|G^p\|^2 < 2$. The tracking error norm then converges monotonically, i.e.,

$$\|e_{k+1}^p\|_{\mathcal{Y}} \leq \|e_k^p\|_{\mathcal{Y}}, \quad k \geq 0, \tag{4.42}$$

and

$$\lim_{k\to\infty} e_k^p = e_\infty^p = P_{ker[G^{p*}]} e_0^p \tag{4.43}$$

where $P_{ker[G^{p*}]}$ is the orthogonal projection operator onto the kernel of G^{p*}.

Consequently, if $e_0^p \in \mathcal{R}[G^p]$, then

$$\lim_{k\to\infty} e_k^p = 0.$$

Furthermore if $ker[G^{p*}] = 0$, i.e., $\mathcal{R}[G^p] = \mathcal{Y}$, then for any initial tracking error e_0^p zero tracking error can be achieved.

For the above cases, the input sequence $\{u_k\}_{k\geq 0}$ in $\mathcal{U}$ converges to the unique solution of the (minimum norm) optimization problem

$$u_\infty = \arg\min_{u\in\mathcal{U}} \{ \|u - u_0\|_{\mathcal{U}}^2 : r^p = G^p u + d^p \}. \tag{4.44}$$

Proof This proof is identical to that for the full trajectory tracking problem and is omitted here for brevity. ■

To illustrate the performance of the above algorithm consider the following example.

Example 4.4 Consider the same system and the setting as in previous examples. Instead of tracking the reference defined over the whole trial interval, the system is now only required to track the reference at intermediate time instants at $t = 1s, 2s, 3s, 4s, 5s, 6s$. A zero initial input u_0 is chosen. The convergence performance of the algorithm with learning gain choices $\beta = 0.1/\|G^{p^T}\|^2, 0.3/\|G^{p^T}\|^2, 0.7/\|G^{p^T}\|^2$, respectively, is shown in Fig. 4.5.

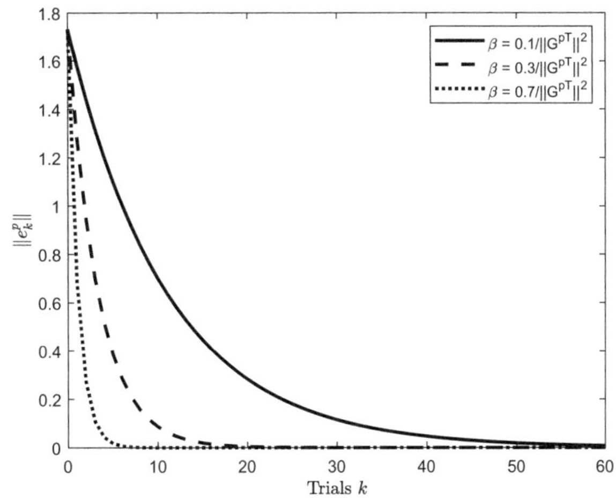

Fig. 4.5 Convergent behaviour of the point-to-point gradient algorithm

As can be seen, the point-to-point tracking error norm converges monotonically to zero as the theory predicts. The system output on the last trial, together with the point-to-point reference, are shown in Fig. 4.6. As can be seen, perfect tracking of the point-to-point reference is achieved. Note that the outputs for different gain choices are on top of each other in the figure and not distinguishable.

Finally, there are two points to note about point-to-point tracking tasks:

1. There are an infinite number of inputs that can produce perfect tracking as there is more than one input that can take one intermediate point to the next exactly. The final converged input for different step sizes, as well as the minimum control energy input u^*, are shown in Fig. 4.7. As can be seen they all converge

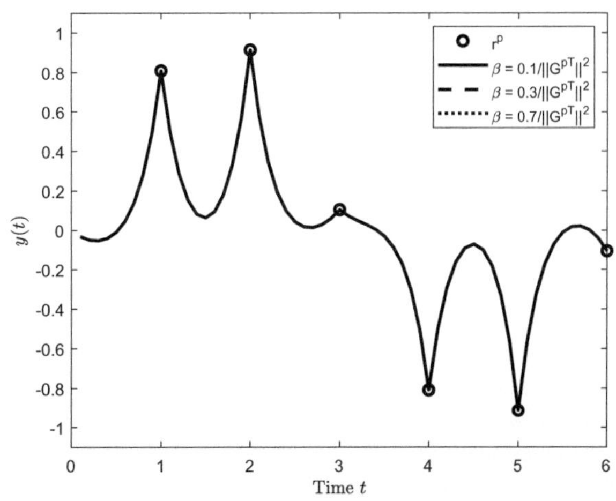

Fig. 4.6 System trajectory for the limit of the point-to-point gradient algorithm

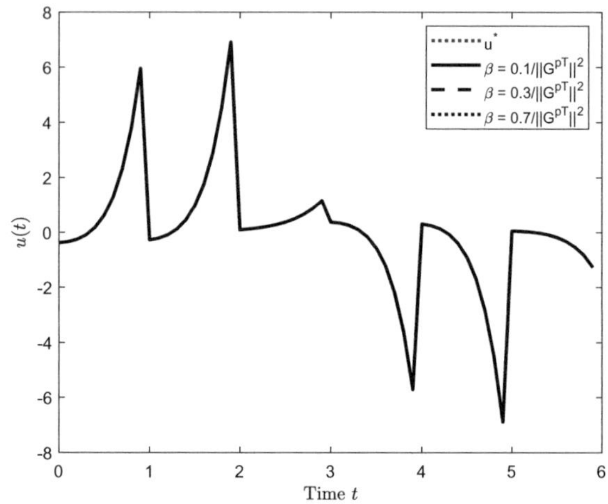

Fig. 4.7 Converged input for the point-to-point gradient algorithm with superimposed minimum energy solution u^*

to the minimum energy solution (lines are on top of each other and thus not distinguishable).

2. The final converged input will depend upon the choice of initial input u_0 but in all cases perfect tracking at the intermediate points is achieved. This suggests that the choice of initial input may have value in achieving a converged input with characteristics that meet the needs of the application.

4.6 Gradient ILC Based on Feedback Design Concepts

In the Algorithm 4.1 (Gradient-Based ILC Algorithm), the convergence behaviour relies heavily on the properties of GG^*. A natural question is to ask if additional modifications to the plant/algorithm can modify the error dynamics by leading to GG^* being replaced, usefully, by a term of the form TT^*. If possible, then the "new" algorithm will inherit the convergence properties of the original algorithm plus any additional properties added by the nature of the operator T.

The approach described in this section uses a methodology based on *output feedback control design techniques*. This will be shown, not only to add useful benefits to algorithm error dynamics, but also to provide a link between the skills needed in ILC and those needed for effective feedback design.

Remark 4.17 Classical closed-loop control systems condition the plant, produce good tracking characteristics, and, subject to stability constraints, achieve improved signal tracking if the closed-loop transfer function becomes close to unity over a wide frequency bandwidth. If a closed-loop controller has already been incorporated into the plant model G, then controller-induced characteristics will not eliminate errors in one iteration but may help the convergence of the Gradient Algorithm by making

GG^* approximate the identity. Even if this has already been implemented, there may still be benefits to using feedback techniques to design the ILC algorithm to add to the benefits of the embedded feedback loops.

Given the above discussion, suppose that a unity, negative, feedback controller design is undertaken, off-line, for the system G (which may already contain embedded control loops) resulting in the choice of forward path compensator K_c (in a unit output feedback loop). This compensator can be used to create a Gradient algorithm with potentially improved convergence properties.

The key to a possible design solution is to consider the closed-loop operator T generated by K_c, and the associated sensitivity S,

$$T = GK_c(I + GK_c)^{-1} = (I + GK_c)^{-1}GK_c = GK_cS = SGK_c,$$
$$where \quad S = I - T = (I + GK_c)^{-1}.$$

Now consider the choice of input updating law,

$$u_{k+1} = u_k + \beta K_c S T^* e_k, \qquad (Modified\ Input\ Update\ Law), \tag{4.45}$$

where G^* has been replaced by the operator $K_c ST^*$. This leads to the error evolution

$$e_{k+1} = (I - \beta TT^*)e_k, \qquad (Error\ Evolution), \tag{4.46}$$

which is identical to the originally developed Gradient Algorithm but with G replaced by T. Theorem 4.1 then yields the following results:

1. if $0 < \beta\|T^*\|^2 < 2$, then the error converges, monotonically, to the projection $P_{ker[T^*]}e_0$ where
2. the kernel $ker[T^*] = ker[(GK_c)^*]$.

If the feedback controller K_c is designed in such a way that the closed-loop system T is asymptotically stable with good tracking characteristics, then, intuitively, $T \approxeq I$. For example, for SISO, tracking problems for discrete systems

$$\|TT^*\| = \|T\|^2 \le \max_{|z|=1} |T(z)|^2 \approx 1.$$

As a consequence, TT^* can be expected to have a very large number of its significant eigenvalues close to 1, the remaining eigenvalues being smaller. The gain range permitted is approximately $0 < \beta\|T\|^2 < 2$ with the choice of $\beta = 1$ eliminating almost all the components of tracking error in just one iteration, properties that were observed for the Inverse-Based ILC Algorithm in the previous chapter.

In reality, designing a feedback law such that $T = I$ is not possible. Balancing gains between different frequency ranges has to be carefully considered, as well as the desired robustness properties against model uncertainties. There are a range of methods in the control design literature on this, e.g., see the text Skogestad and

Postlethwaite [13] and Owens [6, 7]. More detailed discussion can also be found in the companion Optimal Iterative Learning Control text Owens [8].

Remark 4.18 Given e_k and u_k, the computation of the input update requires the calculation of the signal $\Delta_{k+1} = K_c S T^* e_k$. Given models for G and K_c, models for T and $K_c S$ can be created together with a representation of the adjoint T^*. Δ_{k+1} is then found by using simulation software to find $T^* e_k$, followed by the evaluation of $K_c S(T^* e_k)$.

4.7 Discussion and Further Reading

This chapter has introduced the Algorithm 4.1 (Gradient-Based ILC Algorithm) in its general operator form. Despite this generality, the computations needed for the control update are straightforward in principle, requiring:

1. the choice of inner products and norms in associated spaces $\mathcal{U}$ and $\mathcal{Y}$,
2. plus the use of model-based simulations driven by plant data.

The simulations are generated using a representation of the adjoint operator G^*, which, for discrete- and continuous-time state-space systems, can be represented by an adjoint state-space model. The algorithm has well-defined convergence and robustness properties that depend crucially on the gains chosen as well as the structure of the plant dynamics. The gains could be chosen by trial and error but the theory provides a suitable range if an estimate of the operator norm $\|G\|^2$ is computed.

Plant dynamics, for reasons of safe operation, are often designed to incorporate control loops that ensure adequate closed-loop responses to inputs. In this case the input signal to the plant G seen by ILC will be a "reference signal" for the output y (not to be confused with the reference signal for the ILC algorithm). If this was possible, then G already possesses some tracking capability that ILC is expected to improve.

Issues will arise, often, for non-minimum phase (NMP) systems that will affect ILC algorithm performance. A rigorous analysis for the Gradient ILC Algorithm shows that NMP systems may exhibit limitations in what can be achieved in practice as, after an initial number of iterations, convergence will slow to an extremely slow rate that could make further iteration too expensive in time and resources to continue. The value of error norm that can be achieved for NMP systems, therefore, is limited although improvements in performance can be designed into the algorithm if the reference signal and input and the resultant initial error e_0 satisfy slow change requirements at the beginning of the time interval. Gradient ILC is not the only algorithm to suffer from this complication as the Inverse Algorithm and the Norm Optimal Iterative Learning Control (NOILC) Algorithm exhibit very similar characteristics.

The general, operator-based presentation allows the algorithm to be applied in a wide range of circumstances, the main difference being in the form of the underlying model, the operator G, and hence the adjoint operator G^*. The extension to point-to-point (intermediate point) tracking tasks is one such example. Convergence is typically quite rapid for these situations as the reference "signal" is much simpler than that requires in full signal tracking and is finite dimensional. The companion text Owens [8] considers this problem in more detail showing how the use of variable gains $\{\beta_{k+1}\}_{k\geq 0}$ can be used to eliminate the tracking error in a finite number of iterations using spectral annihilation concepts.

Finally, although control loops may be included in the underlying plant (and hence G), feedback ideas can be used to improve the performance of Gradient ILC in a more direct manner, namely by a modification of the input update law based on a feedback design for G. This adds some computational complexity to the algorithm but offers potential benefits to convergence performance, yet to be explored fully in the literature.

Gradient-based ILC has a long history. It was first considered in Togai and Yamano [14] and Furuta and Yamakita [3] and more recently in Kinoshita et al. [5], Hatonen et al. [4], Owerns et al. [11], Owens and Chu [10], and Chu [1]. This chapter is quite general in its underlying potential but has focused primarily on discrete-time state-space systems—more details on gradient ILC design for continuous-time systems can be obtained in the companion book Owens [8], as well as robustness analyses for the case where $m \neq \ell$ and the use of trial/iteration varying gains β_{k+1}.

References

1. Chu B (2016) Iterative learning control for performance optimization. In: 2016 American Control Conference (ACC). IEEE, pp 2129–2134
2. Chu B, Owens DH (2013) Singular value distribution of non-minimum phase systems with application to iterative learning control. In: 52nd IEEE Conference on Decision and Control. IEEE, pp 6700–6705
3. Furuta K, Yamakita M (1987) The design of learning control systems for multivariable systems. In: Proceedings of the IEEE International Symposium on Intelligent Control, pp 371–376
4. Hatonen J, Freeman CT, Owens DH, Lewin PL, Rogers E (2006) A gradient-based repetitive control algorithm combining ILC and pole placement. Eur J Control 12(3):278–292
5. Kinoshita K, Sogo T, Adachi N (2002) Adjoint-type iterative learning control for a single-link flexible arm. IFAC Proc Vol 35(1):223–228
6. Owens DH (1978) Feedback and multivariable systems. Peter Peregrinus, Stevenage
7. Owens DH (1981) Multivariable and optimal systems. Academic Press, London
8. Owens DH (2015) Iterative learning control—an optimization paradigm. Springer
9. Owens DH, Chu B (2010) Modelling of non-minimum phase effects in discrete-time norm optimal iterative learning control. Int J Control 83(10):2012–2027
10. Owens DH, Chu B (2014) Combined inverse and gradient iterative learning control: performance, monotonicity, robustness and non-minimum-phase zeros. Int J Robust Nonlinear Control 24(3):406–431
11. Owens DH, Hatonen J, Daley S (2009) Robust monotone gradient-based discrete-time iterative learning control. Int J Robust Nonlinear Control 19(6):634–661
12. Owens DH, Chu B, Rogers E, Freeman CT, Lewin PL (2014) Influence of nonminimum phase zeros on the performance of optimal continuous-time iterative learning control. IEEE Trans Control Syst Technol 22(3):1151–1158

13. Skogestad S, Postlethwaite I (2005) Multivariable feedback control: analysis and design. John Wiley & Sons
14. Togai M, Yamano O (1985) Analysis and design of an optimal learning control scheme for industrial robots: a discrete system approach. In: Anonymous (ed) Proceedings of the 24th IEEE Conference on Decision Control. IEEE, New York, NY, USA, pp 1399–1404

Chapter 5
Norm Optimal Iterative Learning Control

Gradient methods aim to reduce the magnitude of the tracking error from iteration to iteration, with the ultimate objective of minimizing that tracking error. They are strongly related to the idea of optimization regarded as the minimization of some objective function subject to defined constraints. This chapter describes an approach to Iterative Learning Control (ILC) and monotonicity using optimization problems that ensure monotonic reductions of errors from iteration to iteration.

The chapter is the first chapter to use optimization methodologies in algorithm development for ILC. The chapter is technical in nature to ensure that (a) the problem definition is clear and quite general and (b) to provide techniques that make the analysis of performance characteristics of the proposed algorithm possible. The operator methods used in the derivation and analysis of algorithm properties are the most efficient and effective tool for those purposes but, for state-space systems, *Linear Quadratic Optimal Control theory* plays a central role in the required computational toolbox. For linear, discrete, or continuous, state-space systems, implementable control laws have two possible forms, namely *feedforward* and *state feedback* representations related to the structures seen in the discussion of quadratic optimal control presented in Chap. 2.

The central sets of data in control systems are signals in the form of time series and/or functions describing measured or computed signals including inputs, outputs, states, errors, etc. The chapter uses models and algorithms expressed in operator notation, regarding these signals as points in defined Hilbert spaces. These Hilbert spaces are, typically, spaces of functions or time series and are endowed with *inner products* and associated *norms*. The approach is notationally compact and can be used to provide a clear, geometrical interpretation of algorithm behaviour in a form that is reminiscent of more familiar matrix theory and transfer function notation. It has wide applicability and demonstrates that the algorithms apply to a wide class of model types and tracking objectives including, for example, time/transport delays

B. Chu and D. H. Owens, *Optimal Iterative Learning Control*, Advances in Industrial Control, https://doi.org/10.1007/978-3-031-80236-2_5

or a mixture of differential and algebraic equations. Whilst the mathematical techniques selected are ideal for analysis and formulation of algorithms and apply quite generally, the text focuses on very specific state-space problems for which Optimal Control computational solutions exist. In these applications, the details of the underlying abstract theory play no practical role in developing and applying the *computational* structure.

The power of the approach will be seen in the following sections and chapters and relies heavily on the computational framework provided by the Optimal Control Problem described and solved in the Section on Quadratic Optimal Control earlier in this text (Chap. 2). The problem addressed is designed to achieve a combination of several objectives, namely [16]:

1. *exact tracking* of a specified reference signal at each point on a finite time interval and
2. finding a suitable input signal to ensure that the plant output "passes through" specified points (that is, takes required values) at selected points on the time interval of interest (the, so-called, *Intermediate Point* and *End-Point* problems).

The two ideas briefly described above can be combined to solve *Multi-task Tracking* problems where:

1. together with the need for the outputs to pass through isolated intermediate points,
2. they are required to track defined reference signals on one or more subintervals, and
3. plant motion may be free on other subintervals but control action is still needed to provide the desired improvement in behaviours at intermediate points and tracking on the subintervals.

All of these control objectives fit into the operator framework and can be considered in isolation or in combination.

The reader will see that this chapter describes a benchmark solution to many linear tracking problems but does not necessarily include all design requirements seen in applications. The addition of more complex objectives such as satisfying "hard" or "soft" input/output constraints is left for later chapters (Chap. 6).

5.1 Norm Optimal Problem Formulation

This section introduces the idea of Norm Optimal Iterative Learning Control (NOILC), using an optimization problem to generate ILC algorithms that offer the possibility of using more familiar Optimal Control methods as part of the computational structure and implementations.

The major application of the theory will be to the control of systems with linear state-space models but the algorithms will also apply to more general linear system descriptions. To incorporate this degree of generality into the development, input–output plant dynamics will be described by the relationship

$$y = Gu + d \quad with\ input \ \ u \in \mathcal{U} \ \ and\ output\ y \in \mathcal{Y} \tag{5.1}$$

where $\mathcal{U}$ and $\mathcal{Y}$ are real Hilbert spaces, the bounded linear operator $G : \mathcal{U} \to \mathcal{Y}$, and the term $d \in \mathcal{Y}$ represents initial conditions and similar repeatable contributions to plant dynamics.

Three basic requirements underpin algorithm development, namely:

1. *The control objective* is to track a specified reference signal $r \in \mathcal{Y}$ using an Iterative Control procedure that can be implemented using a combination of off-line calculations and experiments to assess and generate new data to be used in the next iteration.

Remark 5.1 As in matrix theory, r can contain several components including tracking of signals or sequences on subintervals of a time interval and/or desired values for output signals at specified intermediate points. For linear, discrete-time, state-space systems, operating on an interval $0 \le t \le N$, the reader may find it helpful to think of the reference as a time series of values for the output $y(t)$ at a subset of points on $0 \le t \le N$.

2. *The incorporation of data obtained from plant experiments is essential* to make sure that any algorithm is using and responding to the real responses of the plant rather than those predicted by a model.

Remark 5.2 The Inverse ILC Algorithm and the Gradient ILC Algorithm both have the required property as the online measured error data is used directly in the input update law.

3. *The operator G should represent a stable system* as experiments cannot be undertaken in the real world on unstable processes.

Remark 5.3 An unstable system will almost always have stabilizing control loops added to make plant operation safe, etc. The plant model and hence the associated operator G will normally include these control loops, thus ensuring asymptotic stability of the representation.

The tracking error is $e = r - y$ and, as in previous chapters, signals f on iteration k are denoted by f_k. Iterations are initiated by an input signal u_0 generating an initial error $e_0 = r - y_0$ with output $y_0 = Gu_0 + d$ being obtained either from simulation or from experiments on the physical plant.

5.1.1 The Objective Function and Monotonic Errors

The approach to iterative algorithm development is to consider the first iteration. More precisely, the algorithm should use a mechanism that ensures that the magnitude of a non-zero, initial error e_0 is guaranteed to be reduced by the change $u_1 - u_0$ in

u_0. Gradient methodologies produce error reduction (expressed in terms of the norm in $\mathcal{Y}$) by choosing a suitable descent direction and ensuring that the magnitude of the change in input signal is not too large.

A systematic choice of the size of input change can be achieved by replacing the gradient paradigm by an optimization problem and minimizing an objective function of the form,

$$\underbrace{J(u, u_0) = \|e\|_{\mathcal{Y}}^2 + \epsilon^2 \|u - u_0\|_{\mathcal{U}}^2}_{(\ Objective \quad Function\ )}, \quad with \quad e = r - y \quad \& \quad \epsilon^2 > 0\ , \tag{5.2}$$

subject to the dynamical system constraint (5.1).

The objective function contains error magnitudes, $\|e\|_{\mathcal{Y}}$, and the magnitude $\|u - u_0\|_{\mathcal{U}}$ of the change in input signal. It is expressed in a quadratic form because this will lead to solutions that depend linearly on data (u_0, e_0). The parameter ϵ^2 is added to provide a simple mechanism, familiar from Optimal Control Theory, for varying the relative contributions of $\|e\|_{\mathcal{Y}}^2$ and $\|u - u_0\|_{\mathcal{U}}^2$ to the objective function. Intuitively, if it is large, then the optimization will place greater emphasis on making u very similar to u_0. If it is small, greater variation in u is permitted, allowing the minimizing input to concentrate on achieving a greater reduction in the error norm $\|e\|_{\mathcal{Y}}$.

The formal way of representing this problem is to write the minimizing input u_1 as

$$u_1 = \arg\min_{u \in \mathcal{U}} \{\ J(u, u_0)\ :\ e = r - y, \quad y = Gu + d\} \quad . \tag{5.3}$$
$$(Underlying\ Optimization\ Problem\ in\ NOILC)$$

An important property of u_1 is that it *guarantees monotonic reductions in the norm of the error signal* as measured by the norm in the space $\mathcal{Y}$. This is proved by noting that, from optimality and the positivity of the norm $\|u_1 - u_0\|_{\mathcal{U}}^2$,

$$\|e_1\|_{\mathcal{Y}}^2 + \epsilon^2 \|u_1 - u_0\|_{\mathcal{U}}^2 = J(u_1, u_0) \le J(u_0, u_0) = \|e_0\|_{\mathcal{Y}}^2. \tag{5.4}$$

That is,

$$\|e_1\|_{\mathcal{Y}}^2 \le \|e_0\|_{\mathcal{Y}}^2, \quad (Monotonic\ Error\ Reduction\ Property), \tag{5.5}$$

equality holding if, and only if, $\|u_1 - u_0\|_{\mathcal{U}} = 0$, i.e., $u_1 = u_0$.

5.1.2 Minimizing Input: A Derivation

The purpose of this section is to show that the formal representation of the minimizing input for (5.3) is

$$\underbrace{u_1 = u_0 + \epsilon^{-2} G^* e_1} \qquad (NOILC\ Optimal\ Input) \tag{5.6}$$

where G^* is the adjoint operator of G. In particular, $u_1 = u_0$ if, and only if, $G^* e_1 = 0$, i.e., $e_1 \in ker[G^*]$.

To prove (5.6), compute the value of the objective function (performance index) when the input is replaced by $\tilde{u} = u_1 + \Delta$ where $\Delta \in \mathcal{U}$ is arbitrarily chosen. If $\tilde{e}$ is the the error generated by $\tilde{u}$, then $\tilde{e} = e_1 - G\Delta$ and $\tilde{u} - u_0 = u_1 - u_0 + \Delta = \epsilon^{-2} G^* e_1 + \Delta$. Using the definition of inner products and norms gives

$$\begin{aligned} J(\tilde{u}, u_0) &= \|e_1 - G\Delta\|_{\mathcal{Y}}^2 + \epsilon^2 \|u_1 - u_0 + \Delta\|_{\mathcal{U}}^2 \\ &= J(u_1, u_0) - 2\langle e_1, G\Delta\rangle_{\mathcal{Y}} + \|G\Delta\|_{\mathcal{Y}}^2 \\ &\quad +2\langle G^* e_1, \Delta\rangle_{\mathcal{U}} + \epsilon^2 \|\Delta\|_{\mathcal{U}}^2 \end{aligned} \tag{5.7}$$

so that, using the property $\langle e_1, G\Delta\rangle_{\mathcal{Y}} = \langle G^* e_1, \Delta\rangle_{\mathcal{U}}$ of the adjoint G^*,

$$J(\tilde{u}, u_0) = J(u_1, u_0) + \underbrace{\|G\Delta\|_{\mathcal{Y}}^2 + \epsilon^2 \|\Delta\|_{\mathcal{U}}^2}_{\geq 0}. \tag{5.8}$$

The positivity of the norms then implies that $J(\tilde{u}, u_0) \geq J(u_1, u_0)$ for all Δ, i.e., u_1 is, indeed, the solution of the NOILC problem. A by-product of the analysis is that $J(\tilde{u}, u_0) = J(u_1, u_0)$ if, and only if, $\Delta = 0$, reaffirming previously noted properties.

5.2 The NOILC Algorithm: General Form

Replacing u_0 and e_0 by the new signals u_1 and e_1, application of the process described above to minimize $J(u, u_1)$ will lead to a new input u_2 and a further reduction in error norm, i.e., $\|e_2\|_{\mathcal{Y}}^2 \leq \|e_1\|_{\mathcal{Y}}^2$. This process can be continued indefinitely and leads to the following general algorithm and associated properties.

Algorithm 5.1 (Norm Optimal Iterative Learning Control) Using the notation and terminology of the preceding discussion, the Norm Optimal Iterative Learning Control algorithm initiated by the input u_0 generates a sequence of inputs $\{u_k\}_{k\geq 0} = \{u_0, u_1, u_2, \cdots\}$ (and associated errors $\{e_k\}_{k\geq 0} = \{e_0, e_1, e_2, \cdots\}$) by computing

$$u_{k+1} = \arg\min_{u \in \mathcal{U}} \{\ J(u, u_k)\ :\ e = r - y,\ \ y = Gu + d\} \quad for \quad k \geq 0. \tag{5.9}$$

That is, the solution,

$$u_{k+1} = u_k + \epsilon^{-2} G^* e_{k+1}, \quad (Minimizing\ Input\ on\ Iteration\ k+1) \tag{5.10}$$

is the input that satisfies the model equations and also minimizes

$$J(u, u_k) = \|e\|_{\mathcal{Y}}^2 + \epsilon^2 \|u - u_k\|_{\mathcal{U}}^2, \quad with \quad e = r - y. \tag{5.11}$$

5.2.1 Simple Illustrative Examples

Despite its abstract form, the practical meaning of the algorithm is that an Optimal Control problem must be solved on each iteration by whatever means are appropriate to the situation met by the engineer. Any solution procedure is equivalent to calculation of the solution u_{k+1} of the two equations,

$$y_{k+1} = Gu_{k+1} + d \quad \& \quad u_{k+1} = u_k + \epsilon^{-2} G^* e_{k+1} \quad with \quad e_{k+1} = r - y_{k+1}, \tag{5.12}$$

which are strongly related to the ideas of plant and costate equations (respectively) in classical Optimal Control methods for state-space models.

Remark 5.4 In practice, the formal calculation of G^* and definitions of $\mathcal{U}$ and $\mathcal{Y}$ are often easily avoided by focussing on the minimization of the objective function and using the input update equations associated with the chosen computational procedure.

To illustrate possible forms of the problem, consider the system, G, described by the simple first-order system

$$\tfrac{dy(t)}{dt} = -y(t) + u(t), \quad y(0) = 0, \quad 0 \le t \le 5. \tag{5.13}$$

Illustrative NOILC problems are now stated in the following two examples.

Example 5.1 Suppose that it is required to find, iteratively, the input that ensures that the output exactly matches the trajectory $r(t) = t$ on the subinterval $0 < a \le t \le b < 5$ and satisfying the terminal/end condition $r^e(5) = y(5) = 1$. Note that the specified values of $r(t)$ outside the subinterval $[a, b]$ are irrelevant to the problem and its solution and that the reference signal has two components, namely the specified values $r(t) = t,\ a \le b$ on the subinterval *plus* the specified end value $r^e(5) = y(5) = 1$.

For completeness, the Hilbert spaces for the outputs and inputs can be taken to be

$$\mathcal{Y} = L_2[a, b] \times \mathcal{R} \quad \& \quad \mathcal{U} = L_2[0, 5], \tag{5.14}$$

with norms of $(w_1, w_2) \in \mathcal{Y}$ and $u \in \mathcal{U}$ defined by

$$\|(w_1, w_2)\|^2_{\mathcal{Y}} = Fw_2^2 + \int_a^b w_1^2(t)dt \quad \& \quad \|u\|^2_{\mathcal{U}} = \int_0^5 u^2(t)dt \tag{5.15}$$

but it is easier to move immediately to the optimization problem. More precisely, a possible objective function for the NOILC problem has the quadratic form

$$J(u, u_k) = \underbrace{\underbrace{F(1 - y(5))^2}_{terminal\ error} + \underbrace{\int_0^5 Q(t)(r(t) - y(t))^2 dt}_{tracking\ error}}_{\|e\|^2_{\mathcal{Y}}} + \epsilon^2 \underbrace{\int_0^5 R(u(t) - u_k(t))^2 dt}_{\|u-u_k\|^2_{\mathcal{U}}} \tag{5.16}$$

with $R = 1$ on $0 \le t \le 5$, whilst $Q(t) = 1$ on $a \le t \le b$ and is zero elsewhere. The weights $F > 0$ and $\epsilon^2 > 0$ are parameters available to the design.

Note that $J \ge 0$ and equal to zero if, and only if, the system output generated by u_k tracks the two components of the reference signal exactly.

Example 5.2 Setting $a = 0$ and $b = 5$ makes the stated end condition unattainable and it is removed from the problem. The ILC problem in this case could be the tracking of the signal $r(t) = t$ on $[0, 5]$ by iteratively minimizing

$$J(u, u_k) = \underbrace{\int_0^5 (r(t) - y(t))^2 dt}_{\|e\|^2_{\mathcal{Y}}} + \epsilon^2 \underbrace{\int_0^5 (u(t) - u_k(t))^2 dt}_{\|u-u_k\|^2_{\mathcal{U}}} . \tag{5.17}$$

In both examples the minimization of the objective functions subject to the plant dynamics are classical Optimal Control problems and can be approached using computational tools from that subject area. This is made clearer in the more general state-space example discussed in the next section.

To complete this section, a computational study of the second example above is used to verify the monotonic error norm property of NOILC and indicates convergence of the algorithm. Choosing the initial input $u_0(t) \equiv 0$, then $y_0(t) \equiv 0$ and $e_0 = r$. Using Optimal Control methodologies in the minimization of J for the $(k+1)^{th}$ iteration provides the three equations that define the input u_{k+1},

$$\begin{aligned} \dot{y}_{k+1}(t) &= -y_{k+1}(t) + u_{k+1}(t), \quad y_{k+1}(0) = 0 \\ \dot{p}(t) &= p(t) - (r(t) - y_{k+1}(t)), \quad p(5) = 0 . \\ u_{k+1}(t) &= u_k(t) - \epsilon^{-2} p(t) \end{aligned} \tag{5.18}$$

Using Riccati methodologies to solve the equations produces the plot of $\|e_k\|$ against k shown in Fig. 5.1, verifying monotonicity, indicating convergence, and also verifying the intuition that reducing values of ϵ^2 will tend to increase convergence rates. That is, as ϵ^2 gets smaller, the error norm is reduced to a value below any chosen fraction of $\|e_0\|_{\mathcal{Y}}$ in fewer iterations.

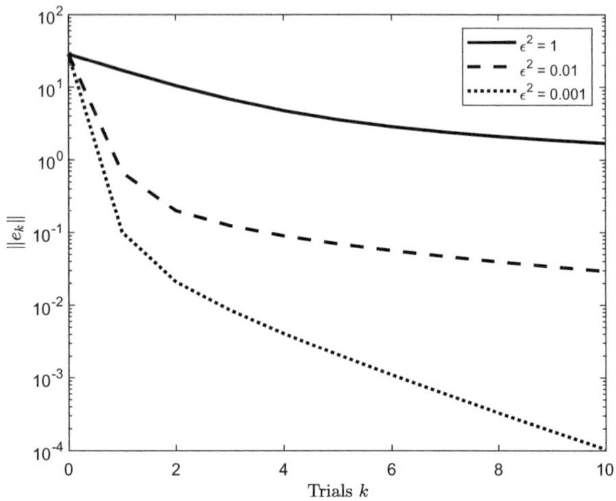

Fig. 5.1 An Example of the Convergence Behaviour of NOILC with different ϵ^2 values

5.2.2 NOILC for Linear State-Space Systems

State-space models are commonly used in control studies. This section reformulates the linear quadratic optimal control problem described in previous chapters from the NOILC perspective. The presentation focuses on the case of continuous systems but the ideas carry over to the discrete-time case in a straightforward manner. The details of the discrete-time case are left for the reader to consider.

Consider an ℓ-input, m-output, linear, time-varying or time-invariant, strictly proper, state-space system of the form

$$\begin{aligned} \dot{x}(t) &= A(t)x(t) + B(t)u(t), \quad x(0) = x_0 \\ y(t) &= C(t)x(t). \end{aligned} \tag{5.19}$$

It is possible to combine the different aspects of NOILC into one objective function. These aspects include the following:

1. trajectory tracking on a finite interval (the *tracking problem*) or
2. control to ensure that outputs pass through a finite number of points/subspaces at specified times (*intermediate and/or end-point problems*) or
3. tracking on several subintervals with no constraint in the gaps between intervals (*tracking on subintervals*) or
4. a combination of these objectives (*multi-task problems*).

Let $t_0 < t_1 < t_2 < \cdots < t_M = T$ be intermediate and/or end points on the interval of control. In NOILC jargon the error has two components, namely,

$$e(t) = r(t) - y(t), \quad e^e(t_j) = r^e(t_j) - F_j y(t_j), \tag{5.20}$$

where $r(t)$ is an $m \times 1$ continuous reference signal, the F_j are $f_j \times m$ matrices, and the $f_j \times 1$ vectors $r^e(t_j)$, $1 \le j \le M$, are intermediate reference values of interest.

The performance index/objective function for the problem has three components (representing the two components in the NOILC objective function)

$$J(u, u_k) = \underbrace{J_f(u) + J_x(u)}_{\|e\|^2_{\mathcal{Y}}} + \underbrace{J_u(u)}_{\|u-u_k\|^2_{\mathcal{U}}} . \tag{5.21}$$

More precisely, the objective function is defined by

$$\begin{aligned} J_f(u) &= \sum_{j=1}^{M} (e^e(t_j))^T Q_j e^e(t_j), \\ J_x(u) &= \int_0^T e^T(t) Q(t) e(t) dt, \quad and \\ J_u(u) &= \int_0^T (u(t) - u_k(t))^T R(t)(u(t) - u_k(t)) dt \end{aligned} \tag{5.22}$$

where the Q_j, $1 \le j \le m$ are symmetric, positive definite $f_j \times f_j$ matrices; $Q(t)$ is $m \times m$, symmetric, and positive semi-definite; and $R(t)$ is $\ell \times \ell$, symmetric, and positive definite.

In order to represent the objective function in a NOILC problem, the reference signals, intermediate times, and weight $Q(t)$ must satisfy consistency conditions to guarantee that there is an underlying Hilbert space structure and to ensure that the problem has a solution in an acceptable sense. These can be illustrated by the following example, namely:

1. The reference signal $r(t)$ may not need to be defined on the whole of the interval. If trajectory tracking is only required on defined, non-intersecting, closed subintervals, $\mathcal{I}_j$, $1 \le j \le N_{\mathcal{I}}$, then values of $r(t)$ are not defined elsewhere.
2. In particular, the intermediate times $\{t_j\}$ must either be outside the subintervals or, if within a subinterval, must satisfy the continuity condition $r^e(t_j) = F_j r(t_j)$.
3. The weight matrix $Q(t)$ should satisfy the requirement that it is positive definite on each subinterval $\mathcal{I}_j$, and zero elsewhere.

5.3 Properties of NOILC

The NOILC Algorithm has a number of important and interesting properties described in the following sections.

5.3.1 Input Formula and Error Evolution

The input on iteration $k + 1$ is given by the formula

$$u_{k+1} = u_k + \epsilon^{-2} G^* e_{k+1} \quad (Input\ Update\ Formula). \tag{5.23}$$

As $G(u_{k+1} - u_k) = -e_{k+1} + e_k$, it follows that the errors on each iteration are related by

$$e_{k+1} = Le_k \ , \ k \geq 0 \quad (Error\ Evolution). \tag{5.24}$$

Expressions of this form govern the behaviour of many ILC algorithms and the operator $L : \mathcal{Y} \to \mathcal{Y}$ is one of the most important objects in the study of ILC. For NOILC, it is defined by the expression

$$L = (I + \epsilon^{-2} G G^*)^{-1} \quad (the\ Learning\ Operator). \tag{5.25}$$

Remark 5.5 This relation has the same structure as the error evolution for the Gradient Algorithm, where $L = I - \beta G G^*$. The different form of L has an effect on the operator's properties.

In NOILC, L has the properties of being

1. self-adjoint and positive definite (as GG^* is self-adjoint and positive semi-definite) and
2. if I denotes the identity mapping in $\mathcal{Y}$,

$$0 < \frac{1}{1 + \epsilon^{-2}\|G\|^2} I \leq L \leq I \quad (Upper\ and\ Lower\ Bounds). \tag{5.26}$$

Proof The statement $L \leq I$ follows from the monotonicity of the error norm. The proof is completed by writing $H = \epsilon^{-2} G G^*$, $\|H\| = \epsilon^{-2}\|G\|^2$ and noting that

$$(I + H)^{-1} - (1 + \|H\|)^{-1} I = L(\|H\| I - H)(1 + \|H\|)^{-1} \geq 0 \tag{5.27}$$

as $\|H\| I - H \geq 0$ and L and H commute. ■

3. In particular $ker[L] = \{0\}$ in all cases and, in addition, $Le = e$ (no change in error) if, and only if, $G^* e = 0$, i.e., $e \in ker[G^*]$.

5.3.2 General Properties

In addition to the above:

1. The error sequence $\{e_k\}_{k \geq 0}$ has norms that satisfy the error monotonicity conditions,

$$\|e_{k+1}\|^2_{\mathcal{Y}} + \epsilon^2 \|u_{k+1} - u_k\|^2_{\mathcal{U}} \leq \|e_k\|^2_{\mathcal{Y}}, \tag{5.28}$$

and hence,

$$\|e_{k+1}\|_{\mathcal{Y}}^2 \leq \|e_k\|_{\mathcal{Y}}^2 \tag{5.29}$$

with equality holding if, and only if, $u_{k+1} = u_k$ and hence $e_{k+1} = e_k$. In this situation $Le_k = e_k$ and hence $e_k \in ker[G^*]$.

2. If equality holds for some value of $k = \tilde{k}$, then $u_k = u_{\tilde{k}}$ for all $k \geq \tilde{k}$ and the algorithm converges in a finite number of iterations with $\lim_{k\to\infty} e_k = e_\infty = e_{\tilde{k}}$.
3. As the sequence $\{\|e_k\|_{\mathcal{Y}}\}_{k\geq 0}$ is positive and reducing, the limit

$$\lim_{k\to\infty} \|e_k\|_{\mathcal{Y}} = E_\infty \geq 0 \tag{5.30}$$

exists. The error converges to zero if, and only if, $E_\infty = 0$.

4. If $ker[G^*] = \{0\}$ and $e_0 \neq 0$, then $0 < \|e_{k+1}\|_{\mathcal{Y}}^2 < \|e_k\|_{\mathcal{Y}}^2$ for all $k \geq 0$. That is, reductions in error norm are guaranteed on each and every iteration.

Proof If $\|e_{k+1}\|_{\mathcal{Y}}^2 = \|e_k\|_{\mathcal{Y}}^2$ holds for some index k, then $u_{k+1} = u_k$ and hence $G^*e_{k+1} = 0$. It follows that $e_{k+1} = e_k = 0$ and hence $u_k = u_{k-1}$ which leads to the conclusion that $e_{k-1} = 0$. An inductive argument then leads to the conclusion that $e_j = 0, \ 0 \leq j \leq k+1$ which contradicts the assumption that $e_0 \neq 0$. ■

5. Writing $\|e_k\|_{\mathcal{Y}}^2 - \|e_{k+1}\|_{\mathcal{Y}}^2 \geq \epsilon^2 \|u_{k+1} - u_k\|_{\mathcal{U}}^2$ for all $k \geq 0$ and adding give

$$\|e_0\|_{\mathcal{Y}}^2 - E_\infty^2 \geq \epsilon^2 \sum_{k=0}^{\infty} \|u_{k+1} - u_k\|_{\mathcal{U}}^2, \tag{5.31}$$

so that $\lim_{k\to\infty} \|u_{k+1} - u_k\|_{\mathcal{U}} = 0$. That is, ultimately, the changes in input signal become infinitesimally small.

6. Finally, the minimum value of $J(u, u_k)$ is $J(u_{k+1}, u_k)$ which takes the form

$$\begin{aligned} J(u_{k+1}, u_k) &= \langle e_k, (I + \epsilon^{-2}GG^*)^{-1}e_k\rangle_{\mathcal{Y}} \\ &= \langle e_k, Le_k\rangle_{\mathcal{Y}} = \langle e_0, L^{2k+1}e_0\rangle_{\mathcal{Y}}. \end{aligned} \tag{5.32}$$

Proof The derivation uses the defining relationships, the algebra

$$J(u_{k+1}, u_k) = \|e_{k+1}\|_{\mathcal{Y}}^2 + \epsilon^2\|\epsilon^{-2}G^*e_{k+1}\|_{\mathcal{U}}^2 = \langle e_{k+1}, (I + \epsilon^{-2}GG^*)e_{k+1}\rangle_{\mathcal{Y}}$$

and the substitution $e_{k+1} = Le_k$ and $e_k = L^k e_0$. ■

5.3.3 Convergence of the Error Sequence

The primary objective of any ILC algorithm is convergence of the error sequence $\{e_k\}_{k\geq 0}$. Although practical application of ILC is necessarily only for a finite number of iterations, this convergence is essential to ensure that behaviour over a finite number of iterations will, increasingly, produce more accurate tracking.

5.3.3.1 Convergence Properties

The error evolution $e_{k+1} = Le_k,\ k \geq 0$ is the important relationship in proving error convergence. In NOILC, the important properties of L are the facts that it is self-adjoint and satisfies the bounds (5.26). This section will demonstrate the proof of the following statement, namely that:

1. For all $e_0 \in \mathcal{Y}$ the NOILC Algorithm error sequence has the convergence property in $\mathcal{Y}$,
$$\lim_{k\to\infty} e_k = e_\infty, \quad (NOILC\ Error\ Convergence) \tag{5.33}$$
where e_∞ is the unique vector (in the kernel $ker[I - L]$) in the representation $e_0 = e_\infty + \tilde{e}_0,\ \tilde{e}_0 \in \overline{\mathcal{R}[I - L]}$. By construction, it is orthogonal to the range $\mathcal{R}[I - L]$.

Remark 5.6 This relation can be represented by

$$e_\infty = P_{ker[I-L]}e_0, \tag{5.34}$$

where $P_{ker[I-L]}$ is the self-adjoint, orthogonal projection operator mapping $\mathcal{Y}$ onto $ker[I - L]$.

Simple manipulations give

$$I - L = \epsilon^{-2}LGG^* = \epsilon^{-2}GG^*L \ \Rightarrow\ ker[I - L] = ker[G^*]. \tag{5.35}$$

2. Convergence to zero for all e_0 is guaranteed if $ker[G^*] = \{0\}$, i.e.,
$$ker[G^*] = \{0\} \ \Rightarrow\ \lim_{k\to\infty} \|e_k\|_{\mathcal{Y}} = e_\infty = 0. \tag{5.36}$$
Note the following:
   a. If $\mathcal{Y}$ is infinite dimensional, the condition $ker[G^*] = \{0\}$ is equivalent to the statement that *almost every* reference r can be generated by a suitable choice of input $u \in \mathcal{U}$.
   b. In particular, if $\mathcal{Y}$ is finite dimensional then the condition $ker[G^*] = \{0\}$ is equivalent to the statement that *every* reference r can be generated by a suitable choice of input $u \in \mathcal{U}$.
   c. If the bound (5.26) on L is strengthened by the statement $\|L\| < 1$, then the proof follows easily from the bound $\|e_k\| \leq \|L\|^k \|e_0\|$ and is left as an exercise for the reader.

3. Finally, if the initial error e_0 lies in the range $\mathcal{R}[I - L] = \mathcal{R}[GG^*]$ of $I - L$, then the series
$$\sum_{k=0}^{\infty} \|e_k\|^2_{\mathcal{Y}} < \infty \quad (Error\ Norm\ Summability) \tag{5.37}$$
converges. In this case the error e_k clearly converges to zero.

In effect the NOILC algorithm will converge whatever the choice of reference $r \in \mathcal{Y}$. This is surprising as clarified in the following comments:

1. For continuous state-space systems, the state vector and hence the output signal will be continuous. The convergence statements above imply that tracking of all square integrable references can be achieved! If $r(t)$ has points of discontinuity, then the input needed to track the signal must contain impulses (Delta function), i.e., error convergence does not necessarily imply convergence of the input to a square integrable signal. The issue of input convergence will be discussed in a later section.
2. For discrete state-space systems and any NOILC problem where the output space is finite dimensional, the problems that could be seen in the continuous state-space systems case will not arise as the notion of discontinuity has no practical value in these cases.

5.3.3.2 Error Convergence Proof

Turning now to the proof of the convergence statements.

Proof The first step is to note that $\mathcal{Y}$ can be decomposed into the direct sum of two disjoint, closed, orthogonal subspaces,
$$\mathcal{Y} = ker[I - L] \bigoplus \overline{\mathcal{R}[I - L]} = ker[G^*] \bigoplus \overline{\mathcal{R}[GG^*]}, \tag{5.38}$$
and write, uniquely, $e_0 = e_\infty + \tilde{e}_0$ with $\tilde{e}_0 \in \overline{\mathcal{R}[I - L]}$. Note that $Le_\infty = e_\infty$ and $L\tilde{e}_0 \in \overline{\mathcal{R}[GG^*]}$ so that $L^k e_0 = e_\infty + L^k \tilde{e}_0,\ k \geq 0$. The general convergence result (5.33) will then follow if $L^k \tilde{e}_0 \to 0,\ k \to \infty$.

Turning to a proof of the summability condition. If $e_0 \in \mathcal{R}[I - L]$, then using the positivity of L it is possible to write $e_0 = (I - L)w_0$ and $w_0 = (I + L)w_1$ (and hence $e_0 = (I - L^2)w_1$). Using $e_k = L^k e_0$, $\|L\| \leq 1$, the Cauchy–Schwarz inequality, and the fact that L is positive and self-adjoint then gives, for all $M \geq 0$,
$$\begin{aligned} 0 \leq \textstyle\sum_{k=0}^{M} \|e_k\|^2_{\mathcal{Y}} = \sum_{k=0}^{M} \langle L^k e_0, L^k e_0 \rangle_{\mathcal{Y}} = \sum_{k=0}^{M} \langle L^{2k} e_0, e_0 \rangle_{\mathcal{Y}} \\ \textstyle\sum_{k=0}^{M} \langle L^{2k}(I - L^2)w_1, e_0 \rangle_{\mathcal{Y}} = \langle (I - L^{2(M+1)})w_1, e_0 \rangle_{\mathcal{Y}} \leq 2\|w_1\|_{\mathcal{Y}} \|e_0\|_{\mathcal{Y}} < \infty \end{aligned} \tag{5.39}$$
for all M, which proves the summability condition.

Next, suppose that e_0 is arbitrary and note that $\mathcal{R}[I-L]$ is dense in $\overline{\mathcal{R}[I-L]}$. The signal $\tilde{e}_0$ can be written as a vector $\hat{e}_0 \in \mathcal{R}[I-L]$ plus a small correction of arbitrarily small magnitude δ. As $\|L\| \le 1$, it follows that $\|L^k e_0 - e_\infty\| \le \|L^k \hat{e}_0\| + \delta$ so that $\limsup_{k\to\infty} \|L^k e_0 - e_\infty\| \le \delta$. Convergence to e_∞ must therefore be true as δ is arbitrary. Finally, if $ker[I-L] = \{0\}$, then $e_\infty = 0$ and (5.36) follows. ■

5.3.4 *Convergence of the Input Sequence*

The example of continuous state-space systems attempting to track a discontinuous reference signal indicates that the tracking error e_k still converges to zero in $\mathcal{Y}$. This can only happen if the input sequence $\{u_k\}_{k\ge 0}$ develops impulsive components, i.e., the limiting form of the input does not lie in the input space $\mathcal{U}$. In one sense, this is a technical observation, but it is also a practical observation as the inputs u_k in the iterative sequence will necessarily tend to contain increasingly large magnitude "spikes" at points of discontinuity of r. Such spikes are unlikely to be desirable in applications.

A natural consequence of the discussion is that an important part of NOILC design is to choose a reference r that can be achieved by an acceptable input $u_\infty \in \mathcal{U}$. The following sections provide a formal statement and proof of sufficient conditions for input convergence plus a bonus, namely that the converged solution has a "minimum energy" property identical to that noted for the Gradient ILC Algorithm of Chap. 4.

5.3.4.1 Input Convergence and "Minimum Energy" Properties: Overview

Suppose that there exists at least one input that generates the desired reference r.

Remark 5.7 An alternative statement is that $r - d \in \mathcal{R}[G]$ or that $e_0 \in \mathcal{R}[G]$, a natural assumption if the aim is to converge to zero tracking error!

Using the same approach as the error convergence analysis in Sect. 5.3.3 then indicates the following:

1. The error sequence converges to zero as desired.
2. In addition, the following discussion will demonstrate that the NOILC input sequence $\{u_k\}_{k\ge 0}$ then converges to the unique input u_∞ that simultaneously tracks the desired reference r whilst minimizing the error norm (squared) $\|u - u_0\|^2_{\mathcal{U}}$. That is,

$$u_\infty = argmin\{\ \|u - u_0\|^2_{\mathcal{U}}\ ;\ r = Gu + d\ \}, \quad (Minimum\ Energy\ Limit). \tag{5.40}$$

3. u_∞ exists and is unique and satisfies

$$u_\infty - u_0 \in \overline{\mathcal{R}[G^*]} = (ker[G])^\perp . \tag{5.41}$$

As a consequence, the limit u_∞ has the property

$$P_{ker[G]}(u_\infty - u_0) = 0, \tag{5.42}$$

where $P_{ker[G]}$ is the self-adjoint, positive semi-definite, orthogonal projection operator mapping $\mathcal{U}$ onto $ker[G]$.

The proof of these assertions is provided in the next section.

In practice NOILC will only be applied for a finite number of iterations so input convergence is not as important as error convergence. It does, however, add confidence in the paradigm if input convergence is, indeed, possible using NOILC methodologies. The following comments do, however, suggest that:

1. If $\mathcal{R}[G]$ is finite dimensional, every vector $r - d$ can be tracked.
2. In infinite dimensional (function) spaces, $\mathcal{R}[G]$ is only dense in its closure so, although almost all signals $r - d \in \overline{\mathcal{R}[G]}$ are included in this discussion, some are not.
3. The assumption must, therefore, limit the details of the choice of r. For strictly proper, continuous, state-space systems, the output will be continuous and possess several degrees of differentiability. For example, for a system with transfer function $G(s) = s^{-2}$, all responses must be continuous and, at least, twice differentiable.
4. The minimum energy terminology for the limit input is borrowed from Optimal Control Theory which aims to find the optimal input that transfers, for a state-space system, an initial state x_0 to a specified final state x_f on a specified finite time interval $[0, t_f]$ whilst minimizing a performance index $J(u) = \|u\|^2_{\mathcal{U}}$ in the form

$$J(u) = \int_0^{t_f} u^T(t)R(t)u(t)dt, \tag{5.43}$$

where $R(t)$ is symmetric and positive definite.

5.3.4.2 An Input Evolution Equation and a Proof of the Input Convergence and Minimum Energy Properties

With the notation and assumptions of the previous section, write the input update formula (5.23) in the form

$$\begin{aligned} u_\infty - u_{k+1} &= u_\infty - u_k - \epsilon^{-2}G^* e_{k+1} \\ &= u_\infty - u_k - \epsilon^{-2}G^*(r - y_{k+1}) \\ &= u_\infty - u_k - \epsilon^{-2}G^* G(u_\infty - u_{k+1}), \end{aligned} \tag{5.44}$$

so that

$$u_\infty - u_{k+1} = L_0(u_\infty - u_k), \;\; k \geq 0 \;\;\; (Input\ Evolution) \tag{5.45}$$

where $L_0 : \mathcal{U} \to \mathcal{U}$ has the form

$$L_0 = (I + \epsilon^{-2} G^* G)^{-1}, \quad (Input\ Evolution\ Operator). \tag{5.46}$$

A number of comments of interest are summarized as follows:

1. As L_0 has a similar structure to L, it follows that it is self-adjoint and positive as an operator from $\mathcal{U}$ onto itself. In addition

$$I - L_0 = (I + \epsilon^{-2} G^* G)^{-1} \epsilon^{-2} G^* G$$

and hence L_0 satisfies the same norm bounds (5.26) as L.

2. The same techniques used for error evolution and convergence indicate the parallel result that

$$\lim_{k \to \infty} (u_\infty - u_k) = P_{ker[I - L_0]}(u_\infty - u_0) = P_{ker[G]}(u_\infty - u_0) = 0, \tag{5.47}$$

that is,

$$\lim_{k \to \infty} u_k = u_\infty = argmin\{\ \|u - u_o\|^2_{\mathcal{U}}\ ;\ r = Gu + d\ \} \tag{5.48}$$

as required.

3. In particular, as $\|L_0\| \leq 1$, the following monotonicity property must be true:

$$\|u_\infty - u_{k+1}\|_{\mathcal{U}} \leq \|u_\infty - u_k\|_{\mathcal{U}}, \quad (Input\ Monotonicity), \tag{5.49}$$

i.e., the input iterates get increasingly closer to the limit as iterations progress.

5.3.5 *Characterization of Convergence Rates Using Spectral Analysis of L*

The convergence of the error is important but the rate of convergence is equally important in practical applications as each iteration is expected to be a possibly time consuming, mix of off-line computation and online experimentation to generate new plant data. Intuition suggests that convergence rates will increase as the input weighting ϵ^2 is reduced. Convergence rates will also depend on the details of the reference signal, the choice of norms is $\mathcal{Y}$ and $\mathcal{U}$, and the plant dynamics, as described by G. An exact characterization of convergence is probably not possible but insight can be obtained by *spectral analysis* of the Learning Operator L.

5.3.5.1 An Eigenstructure for GG^*

The easiest way to obtain a useful insight is to use matrix-like methodologies. More precisely:

1. Assume that GG^* has a spectrum that consists only of eigenvalues of $\sigma_0^2 = 0$ (if $ker[G^*] \neq \{0\}$) and (a finite or infinite number of) positive values $\sigma_j^2 > 0$, ordered in the natural ascending way,
$$0 < \cdots \leq \sigma_{j+1}^2 \leq \sigma_j^2 \leq \cdots \leq \sigma_2^2 \leq \sigma_1^2. \tag{5.50}$$
2. The operator norm has an exact relationship to the largest eigenvalue,
$$\sigma_1^2 = \|GG^*\| = \|G\|^2 = \|G^*\|^2, \quad (Operator\ Norm\ and\ Eigenvalues). \tag{5.51}$$
3. The span of the eigenvectors of GG^* corresponding to zero eigenvalues is precisely $ker[GG^*] = ker[G^*]$.
4. Suppose also that GG^* has a complete (finite or infinite) set of eigenvectors $\{w_j\}_{j\geq 1}$ spanning $\overline{\mathcal{R}[GG^*]}$. As GG^* is self-adjoint, they can be assumed to be orthonormal, namely, they satisfy
$$\begin{array}{c} GG^*w_j = \sigma_j^2 w_j, \quad j \geq 1, \qquad \& \qquad \langle w_j, w_p\rangle_{\mathcal{Y}} = \delta_{j,p} \\ (Eigenvector\ and\ Orthonormality\ Equations\ for\ GG^*) \end{array} \tag{5.52}$$
where $\delta_{j,p}$ is the Kronecker delta. In addition, $GG^*w_0 = 0$ for all $w_0 \in ker[GG^*]$. All problems with finite dimensional signal space $\mathcal{Y}$ satisfy these requirements automatically, notably those associated with discrete-time state-space models.
5. The set of eigenvectors form a basis for $\overline{\mathcal{R}[GG^*]}$. Again, all problems with finite dimensional signal space $\mathcal{Y}$ satisfy this requirement. In this circumstance, any error, for example, can be expanded in terms of the eigenvectors
$$\begin{array}{c} e_k = a_{0,k}w_0 + \sum_{j\geq 1} a_{j,k}w_j, \quad (Eigenvector\ Expansion) \\ w_0 \in ker[G^*], \quad \|w_0\|_{\mathcal{Y}} = 1, \end{array} \tag{5.53}$$
by suitable choice of the coefficients $a_{j,k}$. The norm of e_k can then be expressed in the form
$$\|e_k\|_{\mathcal{Y}}^2 = \sum_{j\geq 0} a_{j,k}^2, \tag{5.54}$$
so the coefficients are square summable.

Finally, the number of eigenvalues in an ILC problem can be small and finite, very large (but finite), or infinite. In all cases the eigenvalues lie in the interval $0 \leq \sigma^2 \leq \|G\|^2$. Little can be said, in general, about the values or distribution within that interval other than the intuitive observation that strictly proper, time-invariant, state-space systems (described by a convolution operator G) have an infinite number of eigenvalues, including an infinite number of small eigenvalues associated with:

1. high frequency behaviour when output responses to high frequency inputs become extremely small,
2. small outputs for oscillatory inputs of frequency close to system zeros that are very close to the imaginary axis of the complex plane, and
3. non-minimum phase TFM characteristics.

5.3.5.2 An Eigenstructure for L and Convergence Rates

The operator L has the same eigenvectors as GG^* and satisfies the equation

$$Lw_j = \ell(\sigma_j^2)w_j, \quad j \geq 0, \quad (Eigenvector\ Equation\ for\ L) \tag{5.55}$$

with eigenvalues defined by the function

$$\ell(\sigma^2) = \frac{1}{1+\epsilon^{-2}\sigma^2} \quad with \quad \sigma^2 = \sigma_j^2, \quad j \geq 0. \tag{5.56}$$

This function provides the link between eigenstructure, the reference signal, and convergence. It is represented schematically in Fig. 5.2 for several values of ϵ^2.

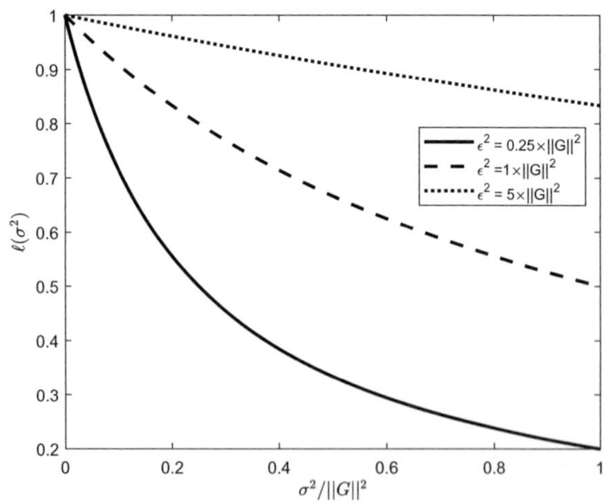

Fig. 5.2 Magnitude of Eigenvalues of L with different ϵ^2 values

Note the following:

1. $\ell(\sigma^2)$ has the value of unity if $\sigma^2 = 0$ and decreases as σ^2 increases, satisfying

$$\frac{1}{1+\epsilon^{-2}\|G\|^2} \le \ell(\sigma^2) \begin{cases} \le 1, & if \;\; ker[G^*] = \{0\} \\ < 1, & otherwise \end{cases} \tag{5.57}$$

   in the range of the eigenvalues of GG^*.
2. It takes its lowest value when $\sigma^2 = \|G\|^2 = \sigma_1^2$.
3. The effect of using low values of ϵ^2 is revealed by

$$\lim_{\epsilon^2 \to 0+} \ell(\sigma^2) = 0, \quad 0 < \sigma^2 \le \|G\|^2. \tag{5.58}$$

The effect of σ_j^2 on NOILC performance can be seen by supposing that the initial error $e_0 = w_j$, when

$$e_k = L^k e_0 = L^k w_j = \ell^k(\sigma_j^2) w_j \tag{5.59}$$

indicating the following:

1. e_k then converges slowly to zero if σ_j^2 is small and non-zero, but producing no change in error if $\sigma_j^2 = 0$. The fastest convergence is seen for the case of $j = 1$.
2. For all $\sigma_j^2 > 0$, the first iterate e_1 becomes arbitrarily small if ϵ^2 becomes very small.

The analysis extends to the more general case when e_0 can be expressed as an expansion $e_0 = \sum_{j \ge 0} a_j w_j$ when

$$e_k = L^k e_0 = \sum_{j \ge 0} a_j \ell^k(\sigma_j^2) w_j, \quad (Error\ Eigenvector\ Expansion). \tag{5.60}$$

Clearly, the observations made above apply to each term of the expansion and hence, in particular:

1. The contribution of each eigenvector $w_j,\ j \ge 1$ to the error reduces monotonically as iterations progress.
2. The contribution of eigenvectors corresponding to larger values of σ_j^2 is reduced more rapidly than others, implying that, after a large number of iterations, the error will be dominated by eigenvectors corresponding to small eigenvalues σ_j^2 and, after an infinite number of iterations, only the contribution of the eigenvector w_0 corresponding to the zero eigenvalue of GG^* will remain, i.e., $e_k \to a_0 w_0 = e_\infty$ as deduced in (5.33).
3. Reducing the value of ϵ^2 increases the reduction of eigenvector contributions to the error e_1, i.e., as $\epsilon^2 \to 0+$, the first error e_1 again approaches $a_0 w_0 = e_\infty$.

In summary, the eigenvector analysis of error evolution in NOILC provides insight into the effect of ϵ^2 and suggests that, although substantial reductions in the contribution of eigenvectors corresponding to larger eigenvalues σ_j^2 can be achieved quickly, those corresponding to small eigenvalues will be more problematic. In frequency domain terms (when they apply to the model and performance objectives) low frequency behaviour of the reference can be quickly tracked but the tracking of higher frequency content of the reference will be much slower.

5.3.5.3 Choosing a Value of ϵ^2

The control weighting factor ϵ^2 must be chosen by the user. The trends of the effect of the choice is clear from the previous section. This opens up the options of choosing by trial and error or by a more systematic approach. The eigenstructure of L analysed in the previous section provides a ballpark value using the function $\ell(\sigma_j^2)$. More precisely, the use of

$$\epsilon^2 = \sigma_1^2 = \|G\|^2 \qquad (A\ Nominal\ Value\ for\ \epsilon^2) \tag{5.61}$$

will reduce the contribution to the error of eigenvectors corresponding to eigenvalues σ_j^2 close to $\|G\|^2$ by approximately one half on every iteration. The reduction of the contribution of other eigenvectors will be less substantial and possibly much slower. Calculation or estimation of the value of the norm $\|G\|$ hence could be a starting point for a subsequent trial and error variation, possibly based on simple facts such as the following.

Note 5.1 (Guidance Note) There is a useful link between convergence and the eigenvalues that could guide the procedure, namely that the contribution of eigenvectors with eigenvalues in the range $\lambda\|G\|^2 \le \sigma^2 \le \|G\|^2$ will more than halve on each iteration if $\epsilon^2 \le \lambda\|G\|^2$.

The computation of the norm $\|G\|$ is not essential and, in many cases, it can be difficult. The easiest case is that of $J_f = 0$, constant positive definite matrices $Q(t) = Q$ and $R(t) = R$, and linear, time-invariant, discrete-time, state-space systems with transfer function matrix $G(z)$. In this case the norm can be bounded using the frequency response using

$$\|G\|^2 \le \sup_{|z|=1} r\left[G(z)R^{-1}G^T(z^{-1})Q\right]. \tag{5.62}$$

This was derived in earlier chapters, the right-hand side being a good approximation to $\|G\|^2$ on long (enough) time intervals.

5.3.5.4 Why Use End-Weighting in J?

In purely intermediate and end-point ILC problems (when $J_x = 0$ in (5.22)), specifying a value of the end point is essential, and part of the definition of the reference r. In cases of full trajectory tracking for state-space systems using $J_f = 0$ in (5.22), the situation is more complex, as follows:

1. In the discrete-time case, placing emphasis on reducing the tracking error at any isolated point t can be achieved by increasing the value of the weight matrix $Q(t)$ at that point. This should increase error reduction at that time relative to others.
2. Using $J_f = 0$ in continuous time, however, does introduce an issue as the Optimal Control formulation gives a change in input linearly dependent on the costate vector $p(t)$. In the absence of final time weighting, the value of the costate at the final time is zero, i.e., the value of $u_k(t)$ (correct or not) at the final time remains unchanged as iterations progress! This does not affect convergence in norm but will result in a rapid (and unnecessary) change in inputs, over an increasingly short time interval, at the final time.

A simple mechanism to introduce changes in input values at the final time is to retain a degree of terminal weighting by setting $M = 1$, t_M equal to the final time, $f_M = m$, $F_M = I$, Q_M symmetric and positive definite, and $r^e(t_M) = r(t_M)$, i.e.,

$$\begin{aligned} J_f(u) &= (r(t_M) - y(t_M))^T Q_M (r(t_M) - r(t_M)) \\ &= e^T(t_M) Q_M e(t_M). \end{aligned} \tag{5.63}$$

In this case, the costate vector at the final time will, in general, be non-zero. Iteration-dependent changes in input at the final time are the inevitable (and possibly beneficial) consequence.

5.3.6 A Numerical Example

In this section, we use a numerical example to illustrate the key properties of NOILC algorithm. The tracking problem has been illustrated in Sect. 5.2.1. In this section we consider the point-to-point tracking problem.

Example 5.3 Consider the same system and the setting as in the previous chapter, i.e.,

$$G(s) = \frac{s+1}{s^2 + 5s + 6}$$

sampled at intervals of $T_s = 0.1$ using zero-order hold (resulting in a discrete-time plant model $S(A, B, C)$). The system is required to track a reference at intermediate time instants at $t = 1s, 2s, 3s, 4s, 5s, 6s$. Zero initial input is chosen. The convergence performance of the NOILC algorithm with $\epsilon^2 = 1, 0.01, 0.001$, respectively, is shown in Fig. 5.3. As can be seen, the point-to-point tracking error norm converges

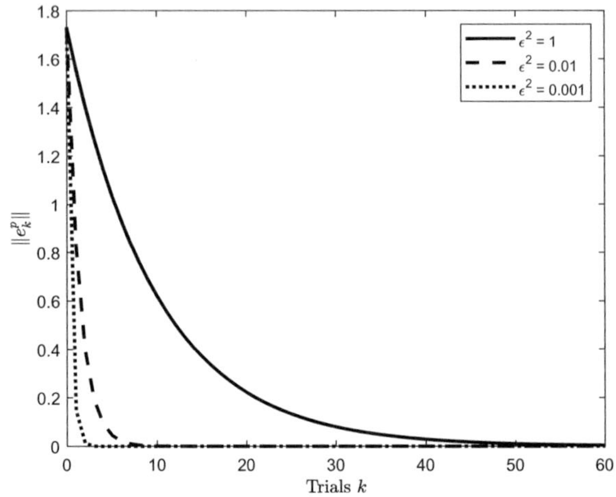

Fig. 5.3 Error Convergence of NOILC for Point-to-Point tracking with different ϵ^2 values

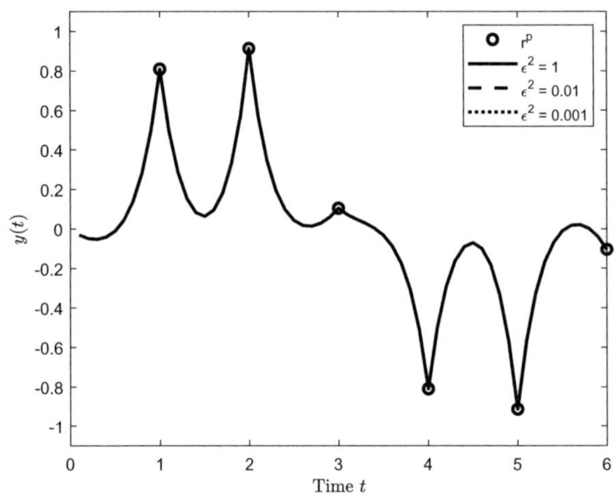

Fig. 5.4 Converged trajectory of NOILC for point-to-point tracking with different ϵ^2 values

monotonically to zero as the theory predicts. Furthermore, as expected a smaller ϵ^2 will lead to faster convergence speed and high accuracy being achieved in all three cases although the number of iterations needed to achieve a given accuracy is much reduced as ϵ^2 becomes smaller. The system output on the last trial together with the point-to-point reference are shown in Fig. 5.4. As can be seen, perfect tracking of the point-to-point reference is achieved.

For point-to-point tracking tasks, there are generally an infinite number of inputs that can produce perfect tracking. The final converged input for different ϵ^2 values, as well as the minimum control energy input u^*, are shown in Fig. 5.5. As can be seen they all converge to the minimum energy solution. Note that in Figs. 5.4 and 5.5, lines for system outputs and inputs for different ϵ^2 values are on top of each other and thus not distinguishable.

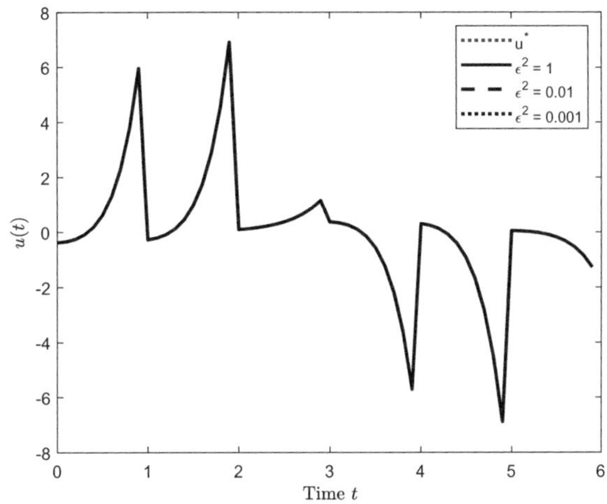

Fig. 5.5 Converged Input for NOILC for Point-to-Point tracking with different ϵ^2 values

5.4 Feedback and Feedforward Implementations

The abstract input characterization (5.23) on each NOILC iteration expresses u_{k+1} as being dependent on e_{k+1}. For continuous or discrete, linear state-space systems G, the adjoint G^* is not causal so that the value of the input at any specific time t will depend on values of the (as yet unknown) error e_{k+1} at times $t' > t$. This lack of causality can be removed in two ways, namely, by using a *feedback* implementation or by using a *feedforward* implementation. In both cases the error e_k on the previous iteration is the measured plant signal that drives the tracking on iteration $k + 1$.

To illustrate the idea, consider the linear, continuous-time, time-varying, state-space NOILC problem described in Sect. 5.2.2 with tracking of a reference signal $r(t)$ (possibly on subintervals) whilst passing through defined intermediate and end points at times $\{t_j\}$.

5.4.1 Online Feedback Implementations

The form of the feedback implementation is described in previous chapters in terms of a state feedback control law. The relevant details can be found in the section on the solution of the TPBVP for linear quadratic optimal control problems (Sect. 2.12). For the NOILC problem considered, the control update law is suitable for online implementation in the form,

$$u_{k+1}(t) = u_k(t) - \underbrace{\epsilon^{-2}R^{-1}(t)B^T(t)K(t)x_{k+1}(t)}_{online\ state\ feedback} + \underbrace{\epsilon^{-2}R^{-1}(t)B^T(t)\xi_{k+1}(t)}_{Computed\ Off-Line} \quad (5.64)$$

with measured state $x_{k+1}(t)$ and an additional signal $\xi_{k+1}(t)$ computed off-line. The defining equations are

$$\begin{aligned} \dot{K}(t) + K(t)A(t) + A^T(t)K(t) - K(t)B(t)\epsilon^{-2}R^{-1}(t)B^T(t)K(t) \\ +C^T(t)Q(t)C(t) = 0 \\ \& \\ \dot{\xi}_{k+1}(t) = -\left(A^T(t) - K(t)B(t)\epsilon^{-2}R^{-1}(t)B^T(t)K(t)\right)\xi_{k+1}(t) \\ -C^T(t)Q(t)r(t) + K(t)B(t)u_k(t). \end{aligned} \tag{5.65}$$

These equations must also satisfy the costate end and jump conditions,

$$\begin{aligned} K(t_j+) - K(t_j-) = -C^T(t_j)F_f^T Q_j F_j C(t_j), \ 1 \le j \le M, \\ with \quad K(t_M+) = 0, \\ \& \\ \xi_{k+1}(t_j+) - \xi_{k+1}(t_j-) = C^T(t_j)F_f^T Q_j r^e(t_j), \ 1 \le j \le M, \\ with \quad \xi(t_M+) = 0. \end{aligned} \tag{5.66}$$

Both $K(t)$ and $\xi_{k+1}(t)$ can be computed using simulation in reverse time from the final time back to $t = 0$.

Remark 5.8 Implementation will require the addition of a state estimator/observer to the control scheme to compute $x_{k+1}(t)$ from measured output data $y_{k+1}(t'),\ 0 \le t' \le t$. This additional complexity (and uncertainty as the incorporation of the observer is not fully understood at this time) is compounded by the absence of e_k from the formulation. The calculations used in the feedforward formulation will alleviate this problem as it uses e_k directly. It does not suffer from these issues and is consistent with the implementation procedures adopted in many application areas.

5.4.2 *Off-Line Computations and a Feedforward Implementation*

The error evolution equation $e_{k+1} = Le_k$ describes the expected evolution of the error whatever the chosen implementation. Substituting into the input update Eq. (5.23) leads to the *Feedforward* version of the update formula, namely,

$$\begin{aligned} u_{k+1} = u_k + \underbrace{\epsilon^{-2}G^* e_{k+1}, \quad e_{k+1} = Le_k,}_{two\ off-line\ calculations} \\ (Feedforward\ Formula\ For\ Input\ Updates) \end{aligned} \tag{5.67}$$

regarded as a computational approach to finding u_{k+1} directly from the error data e_k.

For practical purposes, this formula reduces to the off-line solution of a two step modified optimal control problem based on the measured error e_k obtained from plant experiments on iteration k. More precisely:

1. Using the formula $e_{k+1} = Le_k$ the first step is to compute a predicted error e_{k+1} off-line by writing $e_{k+1} = e_k - z_{k+1}$, $z_k = 0$ and regarding e_k as the reference signal for a modified optimization problem

$$\begin{aligned} z = Gv, \quad J_z(v) &= \|e_k - z\|^2_{\mathcal{Y}} + \epsilon^2 \|v\|^2_{\mathcal{U}} \\ v_{k+1} = argmin\{\, J_z(v)\ ;\ z &= Gv \,\}, \quad z_{k+1} = Gv_{k+1}. \end{aligned} \tag{5.68}$$

   Next set the *computed/predicted* error $\hat{e}_{k+1} = e_k - z_{k+1}$. Note that $z = Gv$ is just the plant model equations with the zero initial condition ($d = 0$).
2. Finally, use a representation of the adjoint system to calculate the input update $u_{k+1} = u_k + \epsilon^{-2} G^* \hat{e}_{k+1}$ and use this input as an open-loop control for the plant.

The procedure is seen to be *feedforward* in the sense that only the data e_k, u_k is *fed forward* (or passed) from iteration k to be used for the input update.

Computational approaches to the two steps are summarized as follows:

1. For state-space systems, the objective function $J_z(v)$ is precisely $J(v, 0)$ with reference defined by $r(t) \mapsto e_k(t)$ and $r^e(t_j) \mapsto e^e_k(t_j)$ and input u_k replaced by zero.
2. The off-line computation of v_{k+1} (and hence z_{k+1}) is then obtained using the representation

$$v_{k+1}(t) = \underbrace{-\epsilon^{-2} R^{-1}(t) B^T(t) K(t) x_{k+1}(t)}_{model\ state\ feedback} + \underbrace{\epsilon^{-2} R^{-1}(t) B^T(t) \xi_{k+1}(t)}_{generated\ by\ e_k} \tag{5.69}$$

   with computed state $x_{k+1}(t)$ and additional signal $\xi_{k+1}(t)$. The defining equations and jump conditions for $K(t)$ are unchanged, but the equations for ξ_{k+1} become

$$\begin{aligned} \dot{\xi}_{k+1}(t) = &- \left(A^T(t) - K(t)B(t)\epsilon^{-2} R^{-1}(t) B^T(t) K(t)\right) \xi_{k+1}(t) \\ &- C^T(t) Q(t) \underbrace{e_k(t)}, \end{aligned} \tag{5.70}$$

   with jump and end conditions,

$$\begin{aligned} \xi_{k+1}(t_j+) - \xi_{k+1}(t_j-) &= C^T(t_j) F_f^T Q_j e^e_k(t_j), \ 1 \le j \le M, \\ with \quad &\xi(t_M+) = 0. \end{aligned} \tag{5.71}$$

3. Given $\hat{e}_{k+1} = e_k - z_{k+1}$, the input update $u_{k+1} = u_k + \epsilon^{-2} G^* \hat{e}_{k+1}$ is computed from a computation of $G^* \hat{e}_{k+1}$. This can be approached using a state-space representation of the adjoint operator G^*, leading to the reverse time simulation,

$$\begin{aligned} (G^* \hat{e}_{k+1})(t) = &\ - R^{-1}(t) B^T(t) p(t) \\ \dot{p}(t) = &\ - A^T(t) p(t) + Q(t) \hat{e}_{k+1}(t), \\ &\ for\ \ t \ne t_j, \ 1 \le j \le M \\ p(t_j+) - p(t_j-) = &\ C^T(t_j) F_f^T Q_j \hat{e}^e_{k+1}(t_j), \ 1 \le j \le M, \\ &\ \&\ \ p(t_M+) = 0. \end{aligned} \tag{5.72}$$

5.5 Robustness Conditions

Robustness of a control strategy is normally taken to mean that a property of a controlled system is retained in the presence of a degree of modelling error. The properties derived and summarized in this chapter show that NOILC has many excellent characteristics (Sect. 5.3) with some insight available to support parameter choice. The most necessary property is that of convergence of the error which is well-defined. All this will work well if the model is an exact representation of the plant, a situation that will be met very rarely. In reality, the plant dynamics will differ from those of the model, possibly by a substantial and unknown amount. It is therefore natural to ask whether or not the NOILC algorithm with off-line calculations that use data obtained from plant experiments will retain good convergence properties if the model dynamics differ from those of the plant. That is, convergence of the error is *robust* to the presence of plant uncertainty.

The answer to this question is very complex and depends on the nature of the difference between the plant and its model. This can take many forms but progress can be made by considering a special form of uncertainty. Suppose that the plant model is described by the operator G but that the plant is linear but described by the operator GU which includes a *right multiplicative perturbation* $U : \mathcal{U} \to \mathcal{U}$. Plant dynamics will be described by a relation of the form

$$y = GUu + d', \quad (Plant\ Dynamics). \tag{5.73}$$

Remark 5.9 The nominal value of U is the identity I when the plant GU and model G are identical. More generally, $U \neq I$ will affect convergence characteristics and could create a situation of divergence.

In what follows, conditions are derived that characterize perturbations that retain a form of monotonic error convergence, i.e.,

$$\|e_{k+1}\|_0 < \|e_k\|_0, \quad k \geq 0, \tag{5.74}$$

where $\| \cdot \|_0$ is a modified, additional norm added to $\mathcal{Y}$ and as defined below.

Two assumptions underpin what follows:

1. Using state-space concepts, the plant and model will be assumed to be asymptotically stable to make the collection of data from the plant viable. It is natural therefore to also assume that U is asymptotically stable.
2. It is assumed that a feedforward implementation of NOILC is to be used (see Sect. 5.4).

Starting from the input update Eq. (5.67), and using (5.73), gives the error evolution

$$e_{k+1} = (I - \epsilon^{-2} GUG^* L)e_k. \quad (Error\ Evolution\ with\ Uncertainty) \tag{5.75}$$

Next, $\mathcal{Y}$ is given an additional inner product

$$\langle w, z\rangle_0 = \langle w, Lz\rangle_{\mathcal{Y}} \text{ and associated norm } \|e\|_0^2 = \langle e, Le\rangle_{\mathcal{Y}}. \tag{5.76}$$

Remark 5.10 As L is positive definite and self-adjoint, the definitions satisfy the properties required of an inner product and norm.

Write, using the properties of inner products and adjoint operators,

$$\begin{aligned} \|e_k\|_0^2 - \|e_{k+1}\|_0^2 &= \|e_k\|_0^2 - \|(I - \epsilon^{-2}GUG^*L)e_k\|_0^2 \\ &= 2\epsilon^{-2}\langle\psi, U\psi\rangle_{\mathcal{U}} - \epsilon^{-4}\langle\psi, U^*G^*LGU\psi\rangle_{\mathcal{U}}, \\ \psi = G^*Le_k &= \epsilon^2(u_{k+1} - u_k) \in \mathcal{R}[G^*] \subset \mathcal{U}\,. \end{aligned} \tag{5.77}$$

This yields the following condition for robust monotonicity:

$$\begin{aligned} U + U^* > \epsilon^{-2}U^*G^*LGU \;\; on \;\; \mathcal{R}[G^*]\,. \\ (Inequality\ for\ Robust\ Monotonicity\ (a)) \end{aligned} \tag{5.78}$$

If the range of G^* presents computational problems, this can be replaced by the sufficient condition,

$$\begin{aligned} U + U^* > \epsilon^{-2}U^*G^*LGU \;\; on \;\; \mathcal{U}. \\ (Operator\ Inequality\ for\ Robust\ Monotonicity\ (b)) \end{aligned} \tag{5.79}$$

If, in addition, U has a bounded inverse operator $\hat{U}$, (5.79) becomes

$$\begin{aligned} \hat{U} + \hat{U}^* > \epsilon^{-2}G^*LG \;\; on \;\; \mathcal{U}. \\ (Inverse\ Operator\ Inequality\ for\ Robust\ Monotonicity) \end{aligned} \tag{5.80}$$

Note 5.2 (Important Note) The inequalities (5.79) and (5.80) are inequalities defined with respect to the inner product in $\mathcal{U}$.

5.5.1 General Properties

Condition (5.79) leads to a number of simple operator conditions for robust monotonicity:

1. Condition (5.79) reveals the necessary positivity property for U,

$$U + U^* > 0\,. \tag{5.81}$$

Note 5.3 This is the main constraint on the form of U tolerated in NOILC. Although the theory presented gives only sufficient conditions, it is intuitively obvious that some form of positivity condition will be needed in any, more detailed, analysis. This is believed because eigenvalues of L close to unity will be affected and, to ensure monotonicity, can only move "into" (a directional constraint) the unit circle in the presence of U. If an eigenvalue is perturbed sufficiently to move "out of" the unit circle, the norm will increase and convergence will be lost.

2. Robustness is retained if (5.79) is strengthened to

$$\begin{aligned} U + U^* \geq \epsilon^{-2} U^* G^* L G U + \alpha I, \; for \; some \;\; \alpha > 0. \\ (Strong \; Operator \; Inequality \; for \; Robust \; Monotonicity). \end{aligned} \tag{5.82}$$

In particular, this suggests a stronger, positivity condition on U,

$$U + U^* \geq \alpha' I, \quad \alpha' > \alpha\,, \tag{5.83}$$

when large enough values of ϵ^2 will retain robust monotonicity.

3. Alternatively, if U has a bounded inverse denoted $\hat{U}$, a stronger condition than (5.80) takes the form,

$$\hat{U} + \hat{U}^* \geq \epsilon^{-2} G^* L G + \alpha I, \; for \; some \;\; \alpha > 0. \tag{5.84}$$

4. More generally, writing $\epsilon^{-2} G^* L G = \epsilon^{-2} G^* G L_0 = I - L_0$ where the input evolution operator $L_0 = (I + \epsilon^{-2} G^* G)^{-1}$, then L_0 satisfies the bounds (5.26) so that

$$I - L_0 \leq \frac{\epsilon^{-2} \|G\|^2}{I + \epsilon^{-2} \|G\|^2} I. \tag{5.85}$$

It follows that (5.79) and/or (5.80) are satisfied if

$$\begin{aligned} (a) \;\; U + U^* > \;\; & \beta U^* U, \;\; \beta = \tfrac{\epsilon^{-2}\|G\|^2}{I + \epsilon^{-2}\|G\|^2}, \\ or, \; (b) \;\; \hat{U} + \hat{U}^* > \;\; & \beta I, \end{aligned} \tag{5.86}$$

or, in strengthened form, for some scalar $\alpha > 0$, if

$$\begin{aligned} (a) \;\; U + U^* \geq \;\; & \beta U^* U + \alpha I, \;\; \beta = \tfrac{\epsilon^{-2}\|G\|^2}{I + \epsilon^{-2}\|G\|^2}, \\ or, \; (b) \;\; \hat{U} + \hat{U}^* \geq \;\; & \beta I + \alpha I. \end{aligned} \tag{5.87}$$

Note that these conditions are dependent on ϵ^2 but independent of the model G other than the presence of its norm $\|G\|^2$ in β. The parameter β plays an important role in robustness. It takes values,

$$0 < \beta < 1, \; with \;\; \lim_{\epsilon \to 0+} \beta = 1, \;\; and \;\; \lim_{\epsilon \to \infty} \beta = 0. \tag{5.88}$$

Remark 5.11 This suggests that the robustness of the NOILC algorithm will improve as ϵ^2 increases.

5. Condition (5.86)(a) can be written in the form

$$(\beta^{-1}I - U^*)(\beta^{-1}I - U) < \beta^{-2}I. \tag{5.89}$$

A sufficient condition for this to hold is the norm condition,

$$\|\tfrac{1}{\beta}I - U\| < \tfrac{1}{\beta}. \quad \text{(A Norm Condition for Robustness)} \tag{5.90}$$

6. Using (5.88) gives the operator norm inequality,

$$\|\frac{1}{\beta}I - U\| \le \|I - U\| + \|\frac{1}{\beta}I - I\| = \|I - U\| - 1 + \frac{1}{\beta} \tag{5.91}$$

which reveals that (5.90) is true for all β if

$$\|I - U\| < 1 \quad (Robustness\ independent\ of\ \beta). \tag{5.92}$$

That is, NOILC has then the defined property of robust monotonicity for all β satisfying (5.88). In particular, robustness is present for all choices of $\epsilon^2 > 0$ and, suprisingly, also for all dynamics G satisfying the conditions used in the analysis.

The conditions are the simplest computational approaches to checking for robustness if the operator norm can be computed or usefully bounded. This is the case in the next section.

5.5.2 A Note on Boundedness and Convergence

The monotonicity condition (5.74) would guarantee the boundedness of the error norm sequence $\{\|e_k\|_0\}_{k\ge 0}$ and, using (5.26), the inequalities

$$\frac{1}{1+\epsilon^{-2}\|G\|^2}\|e\|_{\mathcal{Y}} \le \|e\|_0 \le \|e\|_{\mathcal{Y}} \tag{5.93}$$

then prove the boundedness (but not necessarily the monotonicity) of $\{\|e_k\|_{\mathcal{Y}}\}_{k\ge 0}$. In addition, the sequence $\|e_k\|_0 \to 0$ if, and only if, $\|e_k\|_{\mathcal{Y}} \to 0$, i.e., convergence with respect to one norm implies convergence with respect to the other.

To prove the possibility of convergence of the error to zero, the strengthened condition (5.87)(a) can be used in (5.77) to give

$$\|e_k\|_0^2 - \|e_{k+1}\|_0^2 \ge \epsilon^2 \alpha \|u_{k+1} - u_k\|_{\mathcal{U}}^2, \quad k \ge 0 \tag{5.94}$$

from which

$$\begin{array}{c} \|e_0\|_0^2 \ge \epsilon^2 \alpha \sum_{k \ge 0} \|u_{k+1} - u_k\|_{\mathcal{U}}^2, \ \ i.e., \ \ \lim_{k\to\infty} \|u_{k+1} - u_k\|_{\mathcal{U}}^2 = 0, \\ (Input\ Square\ Summability - Property), \end{array} \tag{5.95}$$

a property that has already been seen, (5.31), for the NOILC algorithm.

The consequence of this is that, if G has range dense in $\mathcal{Y}$, the error sequence $e_k \to 0$ in the *weak sense* that

$$\begin{array}{c} \lim_{k\to\infty} \langle \gamma, e_k \rangle_{\mathcal{Y}} = 0, \quad for\ all\ \gamma \in \mathcal{Y}, \\ (Weak\ Convergence\ of\ the\ Error\ to\ Zero) \end{array}. \tag{5.96}$$

If $\mathcal{Y}$ is finite dimensional, the convergence is therefore true in the sense of the norm $\lim_{k\to\infty} \|e_k\|_{\mathcal{Y}} = 0$.

Proof Equation (5.95) implies that $\lim_{k\to\infty}\langle \gamma', u_{k+1} - u_k\rangle_{\mathcal{U}} = 0$ for all $\gamma' \in \mathcal{U}$. As a consequence

$$0 = \lim_{k\to\infty} \langle \gamma', G^* L e_k \rangle_{\mathcal{U}} = \lim_{k\to\infty} \langle LG\gamma', e_k \rangle_{\mathcal{Y}}. \tag{5.97}$$

As $\mathcal{R}[LG] = \mathcal{R}[GL_0]$ is dense in $\mathcal{Y}$, the weak convergence condition is true for a dense choice of γ. That this choice can be extended to all $\gamma \in \mathcal{Y}$ follows from the fact that the sequence $\|e_k\|_{\mathcal{Y}}$ is bounded. ■

Remark 5.12 This result provides some reassurance that convergence will occur in practice. Unfortunately, in infinite dimensions, weak convergence does not imply convergence in norm. All discrete-time state-space tracking problems will converge in norm as they are finite dimensional as are intermediate point/point-to-point problems for continuous-time state-space problems.

5.5.3 Discrete-Time Frequency Domain Conditions

Suppose that both plant, model, and uncertainty can be described by multi-input–multi-output (MIMO), linear, proper, time-invariant, discrete-time, state-space models with transfer function matrices $G(z)U(z)$, $G(z)$, and $U(z)$ of dimension $m \times \ell$, $m \times \ell$, $\ell \times \ell$, respectively. $U(z)$ is assumed to have an inverse $\hat{U}(z)$ at all but a finite number of points z.

Suppose that the inner product in $\mathcal{U}$ is given by

$$\langle u, v \rangle_{\mathcal{U}} = \sum_{t=0}^{N} u^T(t) R v(t), \quad with\ \ell \times \ell \ \ R = R^T > 0. \tag{5.98}$$

Note 5.4 R is the weight matrix used in the NOILC computations and $R^{-1}U^T(z^{-1})R$ is the TFM associated with U^*.

With these definitions, condition (a) of (5.86) and the analysis of previous chapters yield the condition

$$RU(z) + U^T(z^{-1})R > \frac{1}{\beta}U^T(z^{-1})RU(z), \quad \forall |z| = 1, \tag{5.99}$$

in the sense of positivity of $\ell \times \ell$ complex matrices. In contrast, condition (b) of (5.86) yields the condition

$$R\hat{U}(z) + \hat{U}^T(z^{-1})R > \beta R, \quad \forall |z| = 1. \tag{5.100}$$

This is not necessarily easy to check computationally. A better approach comes from using the norm evaluation results of previous chapters which imply that (5.90) is satisfied if

$$sup_{|z|=1} r\left[\left(\frac{1}{\beta}I - R^{-1}U^T(z^{-1})R\right)\left(\frac{1}{\beta}I - U(z)\right)\right] < \left(\frac{1}{\beta}\right)^2, \tag{5.101}$$

where $r[\cdot]$ is the spectral radius of its argument.

The single-input–single-output (SISO) case ($m = \ell = 1$) simplifies this condition, reducing it to checking that

$$\begin{aligned} &either \ (a) \quad \left|\tfrac{1}{\beta} - U(z)\right| < \tfrac{1}{\beta}, \quad \forall \ |z| = 1, \\ &or \ (b) \quad 2Re\left[\hat{U}(z)\right] > \beta, \quad \forall \ |z| = 1. \end{aligned} \tag{5.102}$$

These conditions have a graphical interpretation, namely:

1. For (a), that the plot, in the complex plane of the frequency response $U(z)$, $|z| = 1$, stays within the circle of radius $\frac{1}{\beta}$ centred at the point $(\frac{1}{\beta}, 0)$. This circle lies within the right-half complex plane and contains the point $z = 1$.
2. For (b), the requirement is that the plot of the inverse transfer function $\hat{U}(z)$ for $|z| = 1$ lies to the right of the vertical line in the right-half complex plane passing through the point $(\frac{1}{2}\beta, 0)$.

5.5.3.1 Deliberate Model Approximation

The robustness analysis described above indicates that there is, indeed, a degree of robustness implicit in NOILC, the main constraint being the need to satisfy a positivity condition, (5.81), on U. If U is unknown, then the conditions cannot be checked. In situations when a detailed model and plant are in good agreement, it may be possible to replace a known, detailed model GU by a simplified/reduced order model G for the purposes of a feedforward NOILC algorithm. In this situation U is known. The biggest constraint on the application of the robustness theory is the need for U to satisfy the positivity condition (5.81). Once chosen, the robustness criteria can be applied to the use of the approximate model G to assess the range of ϵ^2 that might retain monotonicity.

To illustrate the ideas, consider a SISO, discrete system operating on the interval $0 \leq t \leq 20$ with transfer function,

$$G(z)U(z) = \frac{0.1(z-a)}{(z-0.9)(z-0.8)} \tag{5.103}$$

with model G and approximation error U,

$$G(z) = \frac{0.1}{(z-0.9)} \quad \& \quad U(z) = \frac{(z-a)}{(z-0.8)}. \tag{5.104}$$

The problem considered is the tracking of a time series $r(t),\ 0 \leq t \leq 20$, using a NOILC algorithm in feedforward form based on the objective function J, with unity weights $Q(t) = R(t) = 1$,

$$J(u, u_k) = \sum_{t=0}^{20} \left((r(t) - y(t))^2 + \epsilon^2 (u - u_k)^2 \right). \tag{5.105}$$

Using the relation $\|G\| \leq \sup_{|z|=1} |G(z)| = 1$ yields the robustness condition (5.102)(a) in the form

$$\left| \frac{1}{\beta} - \frac{(z-a)}{(z-0.8)} \right| < \frac{1}{\beta},\ \forall\ |z| = 1\ ,\ \beta = \frac{\epsilon^{-2}}{1+\epsilon^{-2}}, \tag{5.106}$$

whilst condition (5.102)(b) takes the form

$$2Re\left[\frac{(z-0.8)}{(z-a)} \right] > \beta\ \ \forall\ |z| = 1. \tag{5.107}$$

These expressions provide a range of values of the parameter for which robust monotonic convergence is present. The reader is invited to explore this range either using numerical, algebraic, or graphical methods. For example:

1. If $a = 0$, then the condition reduces to $\beta < 0.4$ so that $\epsilon^2 > 1.5$ is required.
2. Further insight can be gained from a simple observation, namely that, as $\beta < 1$ the above expression with $a = 0.8$ reduces to $2 > \beta$ which is trivially satisfied. It therefore follows that, in an open interval about the point $a = 0.8$, monotonicity will be observed for all values of $\epsilon^2 > 0$! In this situation, the robust monotonicity property is weight independent although the norm $\|e_k\|_0$ is ϵ^2 dependent so the

nature of the monotonicity and convergence may vary. A more detailed insight can be obtained from the discussion leading to (5.92).

5.5.4 Improving Robustness Using Exponentially Weighted Signals

Using the state-space notation and assumptions of the previous section, *a simple way of increasing the robustness of a NOILC algorithm* is to transform all signals into *exponentially weighted* versions. This section provides a theoretical justification of this statement for linear, discrete-time systems.

If $f = \{ f(0), f(1), \cdots, f(N) \}$ is a time series, then the associated exponentially weighted time series f^λ is denoted by

$$f^\lambda = \{ f(0), \lambda f(1), \lambda^2 f(2), \cdots, \lambda^N f(N) \}, \qquad \lambda > 0. \tag{5.108}$$

In what follows, $\lambda < 1$ will, implicitly, be assumed.

Suppose that a discrete, ℓ-input, m-output, time-invariant, linear, state-space model with time series x, u, y, and the tentative NOILC objective function, with a reference times series r,

$$J(u, u_k) = \sum_{t=0}^{N} \left((r(t) - y(t))^T Q(r(t) - y(t)) + \epsilon^2 (u - u_k)^T R(u(t) - u_k(t)) \right). \tag{5.109}$$

If this is replaced by an exponentially weighted version,

$$J_\lambda(u, u_k) = \sum_{t=0}^{N} \lambda^{2t} \left((r(t) - y(t)^T Q(r(t) - y(t)) + \epsilon^2 (u - u_k)^T R(u(t) - u_k(t))) \right), \tag{5.110}$$

then the state-space model can be written in terms of the time series $r^\lambda, x^\lambda, u^\lambda, y^\lambda$. To obtain a state-space model relating these time series, the discrete model matrix data (A, B, C, D) is replaced by $(\lambda A, \lambda B, C, D)$. The proof of this is left as an exercise for the reader. The TFM of this model is $G_\lambda(z) = G(z\lambda^{-1})$ with modelling error TFM $U_\lambda(z) = U(z\lambda^{-1})$.

Remark 5.13 For continuous, state-space systems the weighting for r, x, u, y is described by the mapping $f(t) \mapsto f^\lambda(t) \exp(-\lambda t)$, $\lambda > 0$. The model data (A, B, C, D) then becomes $(A - \lambda I, B, C, D)$. An increase in robustness can also be expected using this technique.

The exponentially weighted NOILC problem can then be regarded as the choice of u^λ to track r^λ by sequentially minimizing $J_\lambda(u, u_k)$ written as $J(u^\lambda, u_k^\lambda)$. The robustness theory now applies to the exponentially weighted NOILC problem using the operator norm $\|G_\lambda\|$ given by

$$\begin{aligned} \|G_\lambda\|^2 &= \sup_{|z|=1} r\left[G_\lambda(z)R^{-1}G_\lambda^T(z^{-1})Q\right] \\ &= \sup_{|z|=1} r\left[G(z\lambda^{-1})R^{-1}G_\lambda^T(z^{-1}\lambda^{-1})Q\right] \\ &= \sup_{|z|=\lambda^{-1}} r\left[G(z)R^{-1}G^T(\overline{z})Q\right] \end{aligned} \quad (5.111)$$

with β replaced by β_λ, i.e.,

$$\beta \rightarrow \beta_\lambda = \frac{\epsilon^{-2}\|G_\lambda\|}{1+\epsilon^{-2}\|G_\lambda\|}. \quad (5.112)$$

Note that, for a strictly proper system ($D = 0$),

$$\lim_{\lambda\rightarrow 0+} \|G_\lambda\| = 0 \;\; \& \;\; \lim_{\lambda\rightarrow 0+} \beta_\lambda = 0. \quad (5.113)$$

Using these facts in the robustness criterion

$$\begin{aligned} & R\hat{U}_\lambda(z) + \hat{U}_\lambda^T(z^{-1})R > \beta_\lambda R, \quad |z| = 1, \\ \textit{written as} \quad & R\hat{U}(z\lambda^{-1}) + \hat{U}^T(z^{-1}\lambda^{-1})R > \beta_\lambda R, \quad |z| = 1, \\ \textit{which becomes} \quad & R\hat{U}(z) + \hat{U}^T(\overline{z})R > \beta_\lambda R, \quad |z| = \lambda^{-1}, \end{aligned} \quad (5.114)$$

indicates that robust monotonic convergence will be achieved for all small enough values of λ if

$$R\hat{U}_\infty + \hat{U}_\infty^T R > 0, \quad \textit{where} \;\; U_\infty = \lim_{|z|\rightarrow\infty} U(z), \quad (5.115)$$

if the limit exists and is nonsingular.

Remark 5.14 This suggests that robustness of the exponentially weighted NOILC algorithm, if λ is small enough, is dominated by the high frequency behaviour of the modelling error $U(z)$.

Remark 5.15 An intuitive explanation of the beneficial effects of exponential weighting is suggested by noting that the minimization of the weighted NOILC performance index will, initially, place greater emphasis on eliminating errors at the beginning of a time interval. Once significantly reduced, later times will become more significant and further iterations will put greater emphasis on reducing errors there until they are reduced when .... and so on until the end of the interval.

5.6 The Effect of Non-minimum Phase Zeros on NOILC

The effect of the choice of ϵ^2 has already been considered in previous sections but the nature of the plant dynamics is significant. Chapter 4 has already considered the effects of non-minimum phase (NMP) zeros on the convergence performance of linear, SISO, discrete-time, time-invariant, state-space systems. The presentation in Sect. 4.4 there defines the NMP property as follows:

1. for continuous-time systems, by the presence of, at least one, TFM zero with positive real part, and
2. for discrete-time systems, by the presence of, at least one, TFM zero with modulus greater than or equal to one.

The presentation in Chap. 4 concentrates on the Gradient Algorithm for discrete systems with only one NMP zero tracking a reference time series $\{r(t)\}_{0 \le t \le N}$ on an interval $0 \le t \le N$ of $N+1$ sample points. A more detailed analysis of the more general case is more technically involved and was not included.

In this section, a parallel presentation will show that, under the same circumstances, the NOILC algorithm exhibits the same properties and has the same general characterization of behaviour as that seen in the Gradient Algorithm. The presentation relies heavily on the ideas and notation of Sect. 4.4 and the reader may find it useful to peruse that section before preceding.

5.6.1 Summary of Non-minimum Phase Effects on NOILC

Two simplifications are made, the reader being referred to the references for an analysis of the more general case of multiple NMP zeros as well as for linear continuous-time systems Owens and Chu [17], Owens et al. [21], Chu and Owens [8]:

1. The system is assumed to have precisely one NMP zero $|z_1| > 1$.
2. The inner products in $\mathcal{Y}$ and $\mathcal{U}$ have scalar weights, $Q(t) = 1$, $R(t) = 1$, $t = 0, 1, \ldots, N$ and thus the system adjoint has supervector matrix $G^* = G^T$.

Note 5.5 Note that, in this case, the squared norm $\|G\|^2$ is equal to the largest eigenvalue of GG^T (largest singular value of G) which has an upper bound $\max_{|z|=1} |G(z)|^2$ (see Sect. 2.14.6) where $G(z)$ denotes the transfer function of the system.

Given the above, the system $G(z)$ is assumed to take the form,

$$G(z) = \left(\frac{(z - z_1) \prod_{i=1}^{i_0} (z - b_i)}{\prod_{i=1}^{n} (z - a_i)} \right) \tag{5.116}$$

where the poles $|a_j| < 1, \quad 1 \le j \le n$ and the zeros $|b_j| < 1, 1 \le j \le i_0$. The remaining zero $z_1, \ |z_1| > 1$, is a NMP zero lying outside the unit circle.

$G(z)$ can be decomposed into minimum phase/all-pass decomposition described by the factorization,

$$G(z) = G_m(z) G_a(z), \tag{5.117}$$

achieved by adding a cancelling pole–zero pair at the point $z = z_1^{-1}$. More precisely,

$$G_m(z) = \frac{(1 - z_1 z) \prod_{i=1}^{i_0} (z - b_i)}{\prod_{i=1}^{n} (z - a_i)} \tag{5.118}$$

is asymptotically stable and minimum phase and

$$G_a(z) = \frac{z - z_1}{1 - z_1 z} \tag{5.119}$$

is asymptotically stable, NMP, and *all-pass* (i.e., $|G_a(z)| \equiv 1, \forall |z| = 1$). In matrix form, this can be written as

$$G = G_m G_a,$$

where G, G_m, and G_a are the supervector matrix representations of $G(z)$, $G_m(z)$, $G_a(z)$, respectively.

The following remark is an important adjustment to the presentation that ensures that G and G_m are invertible, a necessary property for the results to be stated easily.

Remark 5.16 (Ensuring Invertibility) The matrices G_a and G_m operate on the associated time series vectors on $0 \le t \le N$ in a manner described by the transfer functions $G_a(z)$ and $G_m(z)$. However, the matrices G and G_m will be singular if the system is strictly proper. This technical difficulty in what follows is removed by noting that the only inputs that affect the outputs on this interval are $u(t),\ 0 \le t \le N - k^*$ where the *relative degree* $k^* = n - i_0 - 1$ (excess of poles over zeros). Similarly, the only outputs affected by these inputs are $y(t),\ k^* \le t \le N$. This observation makes it possible to create a description based on $(N + 1 - k^*) \times (N + 1 - k^*)$ matrices rather than the original $(N + 1) \times (N + 1)$. The transfer function mapping the reduced input sequence into the reduced output sequence is obtained by replacing $G_m(z)$ by $z^{k^*} G_m(z)$. This has a supervector description with matrix representation G_m obtained by removing the first k^* rows and last k^* columns from the original supervector description of $G_m(z)$. As $G_a(z)$ has relative degree zero the supervector description on $0 \le t \le N - k^*$ is obtained by removing the last k^* rows and columns. Once this is done, all matrices are then invertible.

Remark 5.17 The approach used in the preceding remark cannot be applied, in general, for multi-input–multi-output systems as there does not exist a single "relative degree" for many such systems.

The keys to the NOILC NMP analysis are as follows (see Sect. 4.4 and the convergence statements for the Gradient and NOILC Algorithms):

1. the observation that the limit errors of both the Gradient and NOILC algorithms are identical, namely equal to the orthogonal projection of the initial error onto $ker[G^*]$ and
2. if N is large, the NOILC algorithm is almost indistinguishable, on a large, finite, number of iterations, from the same NOILC algorithm applied to an approximate model whose adjoint has kernel spanned by the vector $(G_m^T)^{-1}v_1$ where

$$v_1 = [1 \ z_1^{-1} \ \cdots \ z_1^{N-k^*}]^T.$$

The inevitable conclusion is then identical to that obtained for the Gradient Algorithm, namely that the convergence has two phases and at least one undesirable property, namely:

1. an initial phase when the error norm $\|e_k\|_{\mathcal{Y}}$ decreases, possibly rapidly, as the contribution of eigenvectors corresponding to larger eigenvalues is reduced, followed by
2. an infinite number of iterations where the error norm decreases extremely slowly as the algorithm slowly eliminates the high frequency content (usually small) and finally the eigenvectors associated with the system zeros. In practice, when iterations will be limited in number, this phenomenon will appear, on a graph of $\|e_k\|_{\mathcal{Y}}$ against iteration index k, to be initial monotonic reductions followed by flat-lining (or plateauing) at a non-zero value.
   From the above, it follows that the plateau is characterized, if N is large, by the estimated error and error norm given in (4.41),

$$\begin{aligned} &Estimated\ Plateau\ Error \quad e_\infty^a = \left(\frac{v_1^T e_0}{v_1^T v_1}\right) v_1 \quad with\ Norm \\ &\|e_\infty^a\|^2 \approx \frac{(v_1^T e_0)^2}{v_1^T v_1}, \quad i.e., \quad \|e_\infty^a\| \approx \frac{|v_1^T e_0|}{\sqrt{v_1^T v_1}}. \end{aligned} \tag{5.120}$$

These observations have been substantiated in many computational examples. In this section, a discrete-time, linear time-invariant SISO system is used to demonstrate the above observations in the context of a tracking problem with the choice of constant weighting matrices $Q(t) = R(t) = 1, \ t \geq 0$.

5.6.2 A Numerical Study

In this section, numerical examples are provided to verify the results described above. The simulations compare the prediction with real simulations over different time intervals, as well as different initial input choices.

Consider the following non-minimum phase system:

$$G(s) = \frac{5(s-1)}{s^2+2s+1}, \tag{5.121}$$

which is sampled using zero-order hold and a sampling time of 0.1s. The discrete-time model is

$$G(z) = \frac{0.429\,(z-1.1055)}{(z-0.9048)^2}$$

and the NMP discrete zero is $z_1 = 1.1055$ and $k^* = 1$.

Example 5.4 The trial length is taken to be 3s and the reference is the sine-wave shown in Fig. 5.6. The initial input is chosen to be $u_0(t) = 0,\ t = 0, \cdots, N-1$ and the simulation runs over 20 NOILC iterations.

In this case, $N = T/T_s = 30$ and the small eigenvalue induced by the NMP zero is $z_1^{-2N} = 0.0024$. The results are shown in Figs. 5.7 and 5.8, where it can be seen that, although the system and approximate model produce almost identical results in the first few iterations, they diverge from each other at a noticeable rate in later iterations. That is, there is a substantial difference between the approximate model prediction and the simulation results for the real model, explained by the observation that the trial length T is not large compared with the "time constant" associated with the NMP zero. The next example increases the interval length T to verify this by showing closer agreement.

Example 5.5 The trial length is now set at 8s and the reference signal is the sine-wave input shown in Fig. 5.9. The initial input is again chosen to be $u_0(t) = 0,\ t = 0, \cdots, N-1$. The simulation was run over 40 iterations and the results are shown in Figs. 5.10 and 5.11.

Fig. 5.6 The Reference Signal $r(t)$

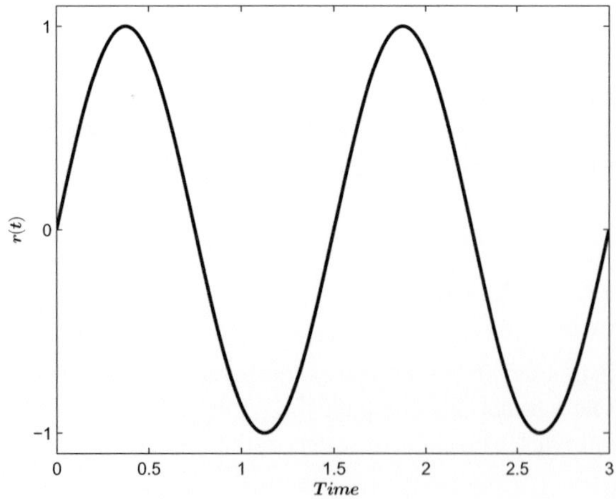

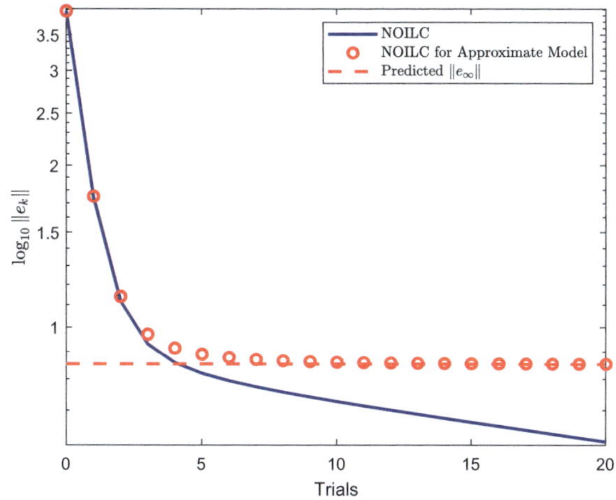

Fig. 5.7 Comparison of Convergence for $T = 3s$

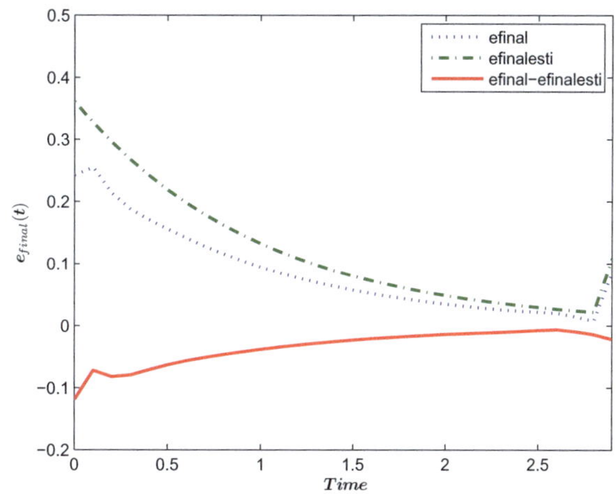

Fig. 5.8 Comparison of the Final Tracking Error $e_{20}(t)$ with its estimate

From the figure, it can be seen that the approximate model provides a good description of the slow convergence phenomenon for many iterations and a good estimate of the value of the error norm on the plateau. The convergence performance is well predicted and the "final" tracking error on the plateau is estimated to a high accuracy.

Finally, different initial inputs are used to verify the analytical predictions. The result is shown in Fig. 5.12, confirming the expected results.

5.7 Discussion and Further Reading

This chapter introduces the key algorithm of the book, namely, the Norm Optimal Iterative Learning Control Algorithm. It was first proposed in Amann et al. [1, 2]

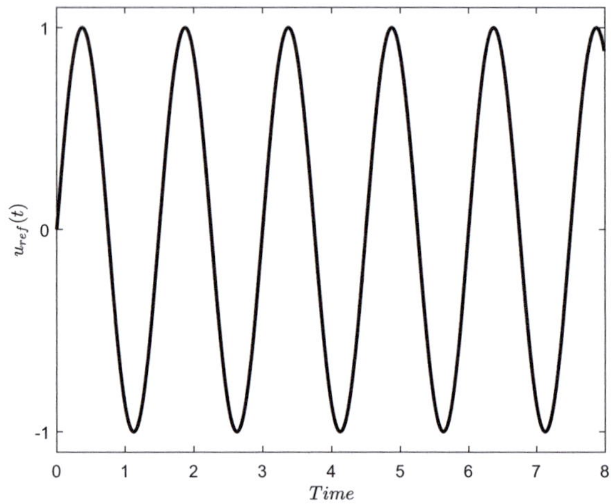

Fig. 5.9 The reference signal $r(t)$

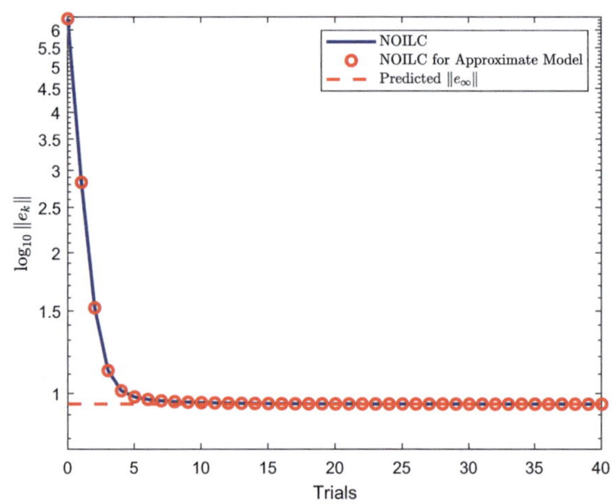

Fig. 5.10 Comparison of convergence showing good agreement between the model and approximate model predictions

and then further developed in Amann et al. [3], Owens et al. [19, 20, 22], Owens and Chu [18], Chu and Owens [5–7], and Chu et al. [10, 11]. An extended version of the presentation can be found in the companion text Owens [16].

NOILC has a range of important properties that make it an attractive option for applications, namely, by producing monotonic reductions in tracking error norm every iteration plus convergence, in norm, of the error to a well-defined value that is always equal to zero if the kernel of the adjoint operator $ker[G^*] = \{0\}$. In addition, the input sequence, if it can converge to a signal that produces zero tracking error, will also have the property that it solves a Minimum Norm Optimal Control Problem (a variant of the so-called Minimum Energy Control Problem).

The nice thing about NOILC is that it has close links with well-established linear optimal control. Because of this, both the theory and the computations can draw

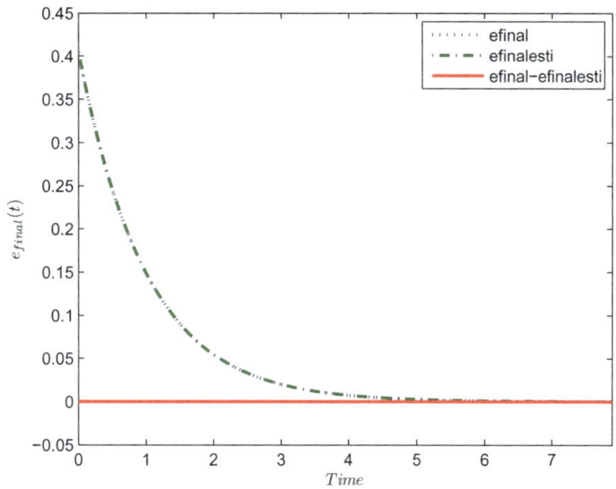

Fig. 5.11 Comparison of the final tracking error $e_{40}(t)$ with its estimate

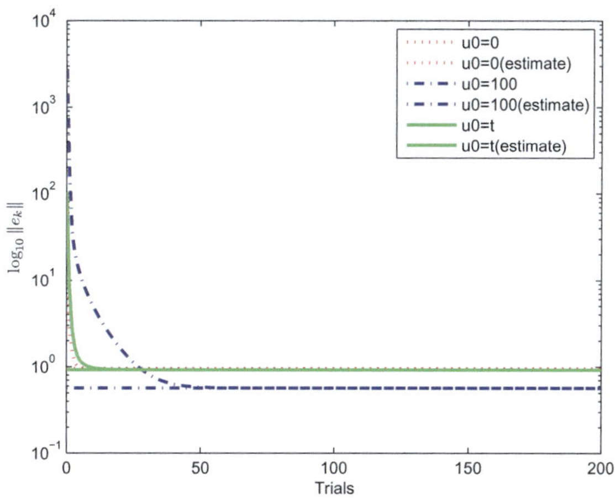

Fig. 5.12 Comparison of convergence for $T = 8s$

heavily on the substantial volume of knowledge that exists on theoretical and computational procedures. For discrete and continuous state-space models, many iterative control tasks can be formulated as a NOILC problem with a quadratic objective function/performance index. Conversely, many linear quadratic tracking problems in Optimal Control can be interpreted as an iterative control problem for some tasks that can be deduced from the form of the weights $Q(t)$, $R(t)$, etc.

NOILC can be implemented either as a feedforward algorithm or in a feedback structure. Most is known about the feedforward approach as the feedback approach presents both theoretical and implementation issues that could challenge researchers and practitioners alike. For feedforward implementations, a form of monotonic convergence can be ensured if the modelling error takes a specified form and satisfies derived robustness conditions. An interesting and relatively unexplored aspect of this robustness theory is the increase in robustness that is predicted if the objective function used in the algorithm contains exponential weighting factors.

In addition to the weighting matrices $Q(t)$ and $R(t)$, which play a similar role as that in the standard linear quadratic control, the NOILC algorithm in the form presented in this chapter also has a scalar weighting parameter ϵ^2. Its impact on the spectrum of the learning operator (and thus convergence rate of the resulting algorithm), as well as consequences on the robustness properties, suggest that the use of larger values of ϵ^2 will lead to lower convergence rates but increased robustness. A ballpark starting value for the choice of ϵ^2 seems to be around $\|G\|^2$.

Insight into the nature of NOILC convergence can be obtained from a spectral (eigenvalue/eigenvector) analysis of the operator GG^* that appears in the operator L describing the error evolution. A simple function and associated graph describe the relationship between the magnitude of the eigenvalue and the rate of convergence of that eigenvector component of the error with the general conclusion that eigencomponents corresponding to small eigenvalues will attenuate less rapidly than those of larger eigenvalues.

Normally, small eigenvalues are associated with, for example, high frequency characteristics in the plant dynamics. Such characteristics can be summarized here as the unwillingness of the system to response to high frequency inputs. Typically, the high frequency content of a reference signal will be small so this is not a great problem in practice. However, for signal tracking for non-minimum phase (NMP) systems, additional small eigenvalues associated with the NMP zeros are introduced.

As mentioned in Chap. 4, non-minimum phase zeros present difficulties for the Gradient Algorithm in the form of infinitesimally slow convergence after an initial phase of more rapid convergence. Insight into this phenomenon is obtained using the analysis introduced in Sect. 4.4 based on a model approximation, valid if the time interval of control is sufficiently large. This analysis is a useful guide to the effects of and relationship between the NMP zero value, the initial tracking error e_0, and hence the reference signal chosen by the user.

NOILC exhibits a very similar slow convergence behaviour to that observed for gradient ILC algorithm. Using the SISO discrete-time system, this slow convergence phenomenon can be modelled and analysed in exactly the same way as the Gradient Algorithm with identical results for the predicted error and error norm during the slow convergence (plateau) phase. Readers are referred to Grenander and Szego [12], Owens and Chu [17], Owens et al. [21], and Chu and Owens [8] for a detailed proof of the general form of the results as well as a treatment of continuous-time state-space systems.

For the rest of the book, we will build on this chapter to develop algorithms to handle situations of practical interest. This includes hard and soft constraint handling, algorithms with faster convergence as well as the use of prediction in NOILC design. In the last three chapters, we will introduce a few case studies to illustrate the application of the NOILC framework.

Remark 5.18 NOILC now has a number of different forms and interpretations, examples of which include introducing a penalty on the norm of the input (in addition or in place of penalizing the input change) [13], the use of basis functions [4], and the use of data-driven method [14], [15], and [9]—the readers are referred to the literature for more results.

References

1. Amann N, Owens DH, Rogers E (1996) Iterative learning control for discrete-time systems with exponential rate of convergence. IEE Proc Control Theory Appl 143(2):217–224
2. Amann N, Owens DH, Rogers E (1996) Iterative learning control using optimal feedback and feedforward actions. Int J Control 65(2):277–293
3. Amann N, Owens DH, Rogers E (1998) Predictive optimal iterative learning control. Int J Control 69(2):203–226
4. Bolder J, Oomen T (2015) Rational basis functions in iterative learning control with experimental verification on a motion system. IEEE Trans Control Syst Technol 23(2):722–729
5. Chu B, Owens DH (2009) Accelerated norm-optimal iterative learning control algorithms using successive projection. Int J Control 82(8):1469–1484
6. Chu B, Owens DH (2010) Iterative learning control for constrained linear systems. Int J Control 83(7):1397–1413
7. Chu B, Owens DH (2011) Accelerated predictive norm-optimal iterative learning control. Proc Inst Mech Eng Part I J Syst Control Eng 225(6):744–759
8. Chu B, Owens DH (2013) Singular value distribution of non-minimum phase systems with application to iterative learning control. In: 52nd IEEE Conference on Decision and Control. IEEE, pp 6700–6705
9. Chu B, Rapisarda P (2023) Data-driven iterative learning control for continuous-time systems. In: 2023 62nd IEEE Conference on Decision and Control (CDC). IEEE, pp 4626–4631
10. Chu B, Freeman CT, Owens DH (2015) A novel design framework for point-to-point ILC using successive projection. IEEE Trans Control Syst Technol 23(3):1156–1163
11. Chu B, Rauh A, Aschemann H, Rogers E, Owens DH (2021) Constrained iterative learning control for linear time-varying systems with experimental validation on a high-speed rack feeder. IEEE Trans Control Syst Technol 30(5):1834–1846
12. Grenander U, Szego G (1984) Toeplitz forms and their applications, 2nd edn. Chelsea Publishing Company, New York
13. Gunnarsson S, Norrlof M (2001) On the design of ILC algorithms using optimization. Automatica 37(12):2011–2016
14. Janssens P, Pipeleers G, Swevers J (2012) A data-driven constrained norm-optimal iterative learning control framework for LTI systems. IEEE Trans Control Syst Technol 21(2):546–551
15. Jiang Z, Chu B (2022) Norm optimal iterative learning control: a data—driven approach. IFAC-PapersOnLine 55(12):482–487
16. Owens DH (2015) Iterative learning control—an optimization paradigm. Springer
17. Owens DH, Chu B (2010) Modelling of non-minimum phase effects in discrete-time norm optimal iterative learning control. Int J Control 83(10):2012–2027
18. Owens DH, Chu B (2024) Error corrected references for accelerated convergence of low gain norm optimal iterative learning control. IEEE Trans Autom Control 69(9):5836–5851
19. Owens DH, Freeman CT, Van Dinh T (2012) Norm-optimal iterative learning control with intermediate point weighting: theory, algorithms, and experimental evaluation. IEEE Trans Control Syst Technol 21(3):999–1007
20. Owens DH, Freeman CT, Chu B (2013) Multivariable norm optimal iterative learning control with auxiliary optimisation. Int J Control 86(6):1026–1045
21. Owens DH, Chu B, Rogers E, Freeman CT, Lewin PL (2014) Influence of nonminimum phase zeros on the performance of optimal continuous-time iterative learning control. IEEE Trans Control Syst Technol 22(3):1151–1158
22. Owens DH, Freeman CT, Chu B (2015) Generalized norm optimal iterative learning control with intermediate point and sub-interval tracking. Int J Autom Comput 12:243–253

Chapter 6
Norm Optimal Iterative Learning Control with Constraints

In many practical applications, the systems are under some constraints due to physical limitations or performance requirements. Hence, the Iterative Learning Control (ILC) design must take these constraints into account. This chapter considers the norm optimal ILC (NOILC) design problem with constraints. Two types of constraints are considered: hard constraints and soft constraints. For the former, the constraints must be satisfied by the control design as otherwise this may cause damages to the system. The design problem is formulated into a successive projection framework where two constrained ILC algorithms are derived. For the latter, some additional design objectives are desired and the resulting design problem is formulated as an auxiliary optimization problem for which two switching algorithms are developed. The convergence properties of the design and implementation procedures are described in detail and illustrated via numerical examples.

6.1 Soft and Hard Constraints

Consider the following familiar operator representation of system dynamics:

$$y = Gu + d \tag{6.1}$$

where G is a bounded linear operator mapping from some input space $\mathcal{U}$ to the output space $\mathcal{Y}$ and d represents the effect of initial condition. The control design objective is to find an input u in $\mathcal{U}$ such that the output y follows a reference trajectory r (*the tracking requirement*).

In practical applications in addition to meeting the above tracking requirement, system constraints that are widely encountered will also need to be satisfied. There are different types of constraints, e.g., input constraints, input rate constraints, and state

B. Chu and D. H. Owens, *Optimal Iterative Learning Control*, Advances in Industrial Control, https://doi.org/10.1007/978-3-031-80236-2_6

or output constraints. Constraints can be divided into two classes: hard constraints and soft constraints. Hard constraints are constraints that must be satisfied, e.g., the input or output limits on actuators. Soft constraints are constraints that should be met as much as possible, e.g., constraints on total energy usage or other performance-related measures. The design problems for both cases are described below.

6.1.1 *Iterative Learning Control with Hard Constraints*

Constraints on the input u are often of the hard category. For simplicity, this chapter only considers the input constraints but output hard constraints can be considered in a similar way. Suppose that the input is constrained to be in a set Ω, taken to be a closed convex set in the input space $\mathcal{U}$. In practice, the set Ω often has a simple form. For example, the following illustrates what might be met in practice:

- $\Omega = \{u \in \mathcal{U} : |u(t)| \leq M(t)\}$;
- $\Omega = \{u \in \mathcal{U} : \lambda(t) \leq |u(t)| \leq \mu(t)\}$;
- $\Omega = \{u \in \mathcal{U} : 0 \leq u(t)\}$,

where $M(t)$, $\lambda(t)$, $\mu(t)$ are known bounds on the input at time t.

The ILC design problem with hard constraints can be stated as follows:

ILC Design Problem for Hard Constraints: *Given a desired reference $r \in \mathcal{Y}$ and an initial input signal $u_0 \in \Omega \in \mathcal{U}$, construct an Iterative Learning Control Algorithm for the plant with dynamics $y = Gu + d$ to find a system input signal in $\mathcal{U}$ such that the output y follows the reference trajectory r as close as possible and that the hard constraints are satisfied on each iteration, i.e.,*

$$u_k \in \Omega, \qquad \forall k \geq 0 .$$

If there are no constraints, the ILC design problem can be solved using one of the many design methods available, for example, the Gradient ILC Algorithm (Chap. 4) or the NOILC Algorithm (Chap. 5). However when constraints are present, the problem becomes more complicated, as one needs to decide how the constraints can be incorporated into the design process whilst retaining known performance properties. In the following sections, the successive projection method proposed by Owens and Jones [22] is used to interpret NOILC algorithm, and a systematic approach for constraints handling in NOILC is then proposed in the form of two new algorithms, together with well-defined convergence properties.

6.1.2 Iterative Learning Control with Soft Constraints

Soft constraints are considered when there are design freedoms, e.g., in point-to-point tracking where the control input generating perfect tracking is non-unique. In this case, the extra design freedom can be exploited to optimize additional performance objectives. To formulate the problem, introduce an auxiliary signal as follows.

An Auxiliary Signal z is selected as a suitable measure of desired control performance and is specified by a linear relation

$$z = G_1 u + d_1 \qquad \text{(Auxiliary Variable Dynamics)} \tag{6.2}$$

where G_1 is a bounded linear operator mapping $\mathcal{U}$ into a real Hilbert space $\mathcal{Z}$ and $d_1 \in \mathcal{Z}$ is an iteration-independent term again describing the effect of initial conditions and exactly repeated disturbances or bias.

The precise ILC tracking problem is defined as follows:

ILC Design Problem with Soft Constraints: *Given a desired reference $r \in \mathcal{Y}$ and an initial input signal $u_0 \in \mathcal{U}$, construct an Iterative Learning Control Algorithm for the plant with dynamics $y = Gu + d$ to find a system input signal in $\mathcal{U}$ that satisfies the tracking requirement $y = r$ where $y = Gu + d$ whilst simultaneously minimizing the objective function*

$$J(z, u) = \|z - z_0\|^2_{\mathcal{Z}} + \|u - u_0\|^2_{\mathcal{U}} \tag{6.3}$$

where (z_0, u_0) is the solution pair of the auxiliary equation for the given input u_0. That is, the ILC algorithm converges to a solution of the constrained optimization problem

$$u_\infty = argmin\{J(z, u) : u \in \mathcal{U}, \; r = Gu + d, \; z = G_1 u + d_1\}, \tag{6.4}$$

i.e., the objective function provides a soft constraint on the auxiliary variables by providing a mechanism for minimizing a combination of (norm) magnitudes of z and u.

Remark 6.1 As a special case, if $G_1 = 0$ and $d_1 = 0$, then $(z_0, u_0) = (0, u_0)$ is a solution of the auxiliary system (for all choices of u_0) and, in this case, the auxiliary optimization problem is the minimization of the input energy measure $\|u - u_0\|^2_{\mathcal{U}}$.

In later sections, two switching algorithms will be developed to solve this problem.

6.2 Hard Constraints: Design Using Successive Projection

In this section, we consider (for simplicity) ℓ-input m-output discrete-time, linear time-invariant systems to illustrate the design method.

Remark 6.2 Continuous- time systems can be treated in a similar way and the details are not included here.

In matrix form the system dynamics can be described as

$$y = Gu + d, \tag{6.5}$$

where G is the operator mapping from some input space $\mathcal{U} = \mathbb{R}^{N\ell}$ to output space $\mathcal{Y} = \mathbb{R}^{Mm}$. Note this describes many situations of practical interests, including but not limited to signal tracking, intermediate point, and end-point tracking. The tracking error on the kth iteration is $e_k = r - y_k$, where u_k, y_k, e_k, r are $N\ell \times 1$ and $Mm \times 1$ supervectors (respectively) and G is the $Mm \times N\ell$ system supervector matrix (see Chap. 2 for more details of supervector methodologies).

6.2.1 *ILC Design Using Successive Projection*

6.2.1.1 ILC Design as Successive Projection

Consider the ILC design problem initially without constraints. The task of the ILC control law is to iteratively find a series of inputs such that $u_k \to u^*$ as k tends to infinity where u^* is an input producing perfect tracking of the reference r, i.e., $r = Gu^* + d$. That is equivalent to iteratively finding a point $(0, u^*) \in H = \mathcal{Y} \times \mathcal{U}$ in the intersection of the following two sets in H:

- $K_1 = \{(e, u) \in H : e = r - Gu - d\}$;
- $K_2 = \{(e, u) \in H : e = 0\}$,

where the inner product in H is taken to be (induced by the inner products in $\mathcal{Y}$ and $\mathcal{U}$)

$$\langle (e, u), (z, v) \rangle = e^T Qz + \epsilon^2 u^T Rv = \langle e, z \rangle_Q + \epsilon^2 \langle u, v \rangle_R \tag{6.6}$$

with $Q > 0$, $R > 0$ symmetric positive definite, $\epsilon^2 > 0$ and the associated induced norm given by

$$\|(e, u)\| = \sqrt{\langle (e, u), (e, u) \rangle} = \sqrt{\|e\|_Q^2 + \epsilon^2 \|u\|_R^2}. \tag{6.7}$$

It is worth noting that both sets are closed and convex in H.

The problem of finding a point in the intersection of two closed, convex sets is a classic one and has a long history [4, 14, 16, 20, 22]. It can be solved using a

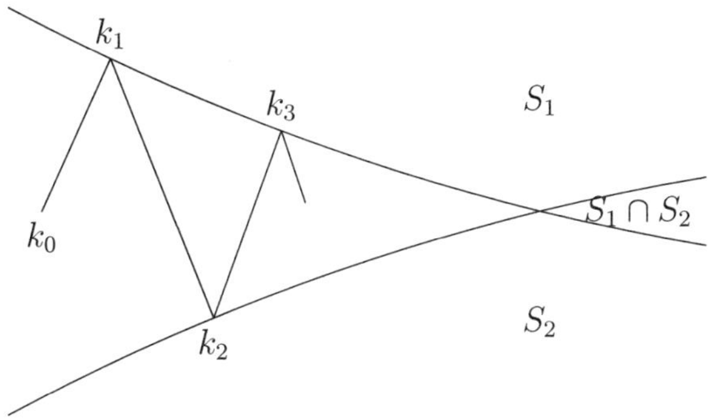

Fig. 6.1 Illustration of the successive projection algorithm

successive projection method as described in the following theorem (and illustrated schematically in Fig. 6.1):

Theorem 6.1 (Owens and Jones [22]) Let $S_1 \subset H$, $S_2 \subset H$ be two closed convex sets in a real Hilbert space H with $S_1 \cap S_2$ non-empty. Define

$$S_j = \begin{bmatrix} S_1, \ j \ odd \\ S_2, \ j \ even \end{bmatrix}.$$

Then, given the initial guess $k_0 \in H$, the sequence $\{k_j\}_{j \geq 0}$ satisfying

$$\|k_j - k_{j-1}\| = \min_{k \in S_j} \|k - k_{j-1}\|, \quad j \geq 1 \tag{6.8}$$

with $k_j \in S_j$, $j \geq 1$, is uniquely defined for each $k_0 \in H$ and satisfies

$$\|k_{j+1} - k_j\| \leq \|k_j - k_{j-1}\|, \quad j \geq 2. \tag{6.9}$$

Furthermore, for any $x \in S_1 \cap S_2$,

$$\|x - k_j\|^2 \geq \|x - k_{j+1}\|^2 + \|k_{j+1} - k_j\|^2 \tag{6.10}$$

so that the sequence continuously gets closer to every point in $S_1 \cap S_2$. In addition

$$\sum_{j=1}^{\infty} \|k_{j+1} - k_j\|^2 \leq \|x - k_1\|^2 \tag{6.11}$$

so that, for each $\epsilon > 0$, there exists an integer N such that for $j \geq N$

$$\inf_{k \in S_{j+1}} \|k - k_j\| < \epsilon, \tag{6.12}$$

that is, the iterates $k_j \in S_j$ become arbitrarily close to S_{j+1}.

Moreover, when $S_1 \cap S_2$ is empty, the algorithm converges in the sense that $\|k_{j+1} - k_j\| \to d(S_1, S_2)$ defining the minimum distance $d(S_1, S_2)$ between the two sets S_1 and S_2.

Using the successive projection method to find the intersection of K_1 and K_2, the well-known NOILC algorithm can be derived. Its convergence properties can also be easily derived. For more details, please refer to Chu and Owens [8, 9] and Chu et al. [10, 11].

6.2.1.2 ILC Design with Hard Constraints

Now consider the constrained ILC problem discussed in Sect. 6.1. When hard constraints are present, they must also be satisfied by the design. The problem can be formulated as finding the intersection of the following closed, convex sets in $H = \mathcal{Y} \times \mathcal{U}$:

- $K_1 = \{(e, u) \in H : e = r - Gu - d\}$;
- $K_2 = \{(e, u) \in H : e = 0\}$,

under the constraint $K_3 = \{(e, u) \in H : u \in \Omega\}$. Note that, it is normal that $K_1 \cap K_2 \cap K_3$ is either a singleton pair $(e, u) = (0, u^*)$ or a linear variety solving the ILC problem or it is the empty set $\emptyset$. In the last case, perfect tracking is not achievable due to the introduction of the input constraint Ω.

There are three sets in the constrained problem. However, set K_3 can be associated with either K_1 (yielding two sets $K_1 \cap K_3$ and K_2) or K_2 (yielding two sets $K_2 \cap K_3$ and K_1) and also notice that the intersection of two closed convex sets is still a closed convex set. Then, the original three-set problem becomes a two-set problem, which is to find the intersection of $S_1 = K_1 \cap K_3$ (resp. K_1) and $S_2 = K_2$ (resp. $K_2 \cap K_3$). This is in a form that can be solved using the original successive projection method described in Theorem 6.1.

The successive projection method hence generates two iterative algorithms for the constrained ILC problem, which are demonstrated in the following two sections. In what follows, we do not specify the exact form of the constraints other than that they are closed and convex.

6.2.2 *Constrained ILC Algorithm 1 (Algorithm 6.1)*

This first algorithm identifies that input constraints with the dynamics, a situation that explains why the algorithm is more computationally intensive than its alternative form (introduced later). The formal construction sets $S_1 = K_1 \cap K_3$ and $S_2 = K_2$ to be the closed convex sets in Theorem 6.1, which can be described as follows:

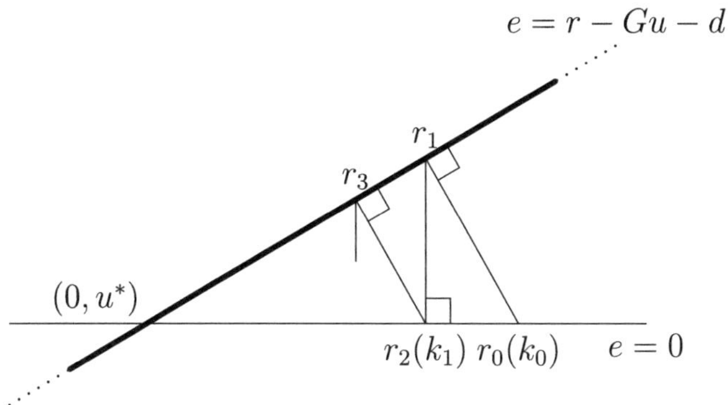

(a) $(K_1 \cap K_3) \cap K_2 \neq \emptyset$ and perfect tracking is possible

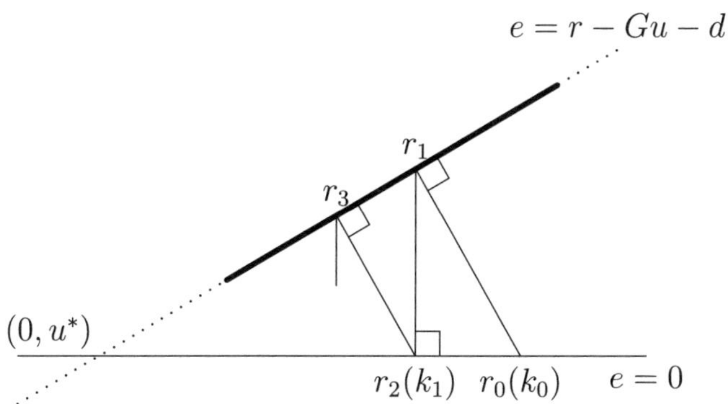

(b) $(K_1 \cap K_3) \cap K_2 = \emptyset$ and perfect tracking is not possible

Fig. 6.2 Illustration of Algorithm 6.1 (Constrained ILC Algorithm 1) using successive projection

- $S_1 = \{(e, u) \in H : e = r - Gu - d, u \in \Omega\}$;
- $S_2 = \{(e, u) \in H : e = 0\}$.

Using Theorem 6.1, the following algorithm can now be constructed and is illustrated schematically in Fig. 6.2a and b, in which the cases when perfect tracking is possible (intersection occurs) or not (the intersection is empty) are shown, respectively.

Algorithm 6.1 (Constrained ILC Algorithm 1) Given any initial input u_0 satisfying the constraint with associated tracking error e_0, the input sequence $u_{k+1}, k = 0, 1, 2, \cdots$, defined by

$$u_{k+1} = \arg\min_{u \in \Omega} \left\{ \|r - Gu - d\|_Q^2 + \epsilon^2 \|u - u_k\|_R^2 \right\} \tag{6.13}$$

also satisfies the constraint and iteratively solves the constrained ILC problem.

To show the algorithm construction process, according to Theorem 6.1, given $r_0 = (0, u_0) \in S_2$, the sequence $\{r_1, r_2, \cdots\}$ given by

$$\|r_i - r_{i-1}\| = \inf_{y \in S_j} \|y - r_{i-1}\|, \tag{6.14}$$

where S_j is defined as

$$S_j = \begin{bmatrix} S_1, \ j \ odd \\ S_2, \ j \ even \end{bmatrix},$$

iteratively finds the intersection of S_1 and S_2. The sub-sequence $\{k_1, k_2, \cdots\} \subset S_2$ (as shown in Fig. 6.2) defined by

$$k_k = r_{2k} \tag{6.15}$$

also iteratively finds the intersection of S_1 and S_2. That is, it solves the ILC problem.

Note that $k_{k+1} = r_{2(k+1)}$ is solved by

$$\|r_{2k+1} - k_k\| = \inf_{y \in S_1} \|y - k_k\| \tag{6.16}$$

and

$$\|k_{k+1} - r_{2k+1}\| = \inf_{y \in S_2} \|y - r_{2k+1}\|. \tag{6.17}$$

Express (6.16) explicitly using the norm definition as the following problem:

$$r_{2k+1} : (e_{k+1}, u_{k+1}) = \arg \min_{(e,u) \in S_1} \left\{ \|e - 0\|_Q^2 + \epsilon^2 \|u - u_k\|_R^2 \right\}, \tag{6.18}$$

which is equivalent to solving

$$u_{k+1} = \arg \min_{u \in \Omega} \left\{ \|r - Gu - d\|_Q^2 + \epsilon^2 \|u - u_k\|_R^2 \right\} \tag{6.19}$$

and (6.17) simply gives $k_{k+1} : (0, u_{k+1})$ which is just the input update in Algorithm 6.1 (Constrained ILC Algorithm 1).

6.2.2.1 Convergence Properties

This section studies the convergence properties of Algorithm 6.1 (Constrained ILC Algorithm 1). As mentioned in Sect. 6.2.1, due to the introduction of the input constraints Ω, there may be no intersection of K_1, K_2, and K_3, which means perfect tracking of the reference signal may be not possible. In this case, the convergence properties may have some differences (from the case where perfect tracking is possible). Hence the convergence results here are presented in two parts: $(K_1 \cap K_3) \cap K_2 \neq \emptyset$ and $(K_1 \cap K_3) \cap K_2 = \emptyset$.

When $(K_1 \cap K_3) \cap K_2 \neq \emptyset$, perfect tracking of the reference signal is possible. Applying Theorem 6.1 to analyse the convergence of Algorithm 6.1 (Constrained ILC Algorithm 1) (by explicitly writing the expression of the sequence generated as shown earlier in the derivation of the algorithm) shows that the algorithm has the highly desirable property as given in the following theorem.

Theorem 6.2 When perfect tracking is possible, Algorithm 6.1 (Constrained ILC Algorithm 1) can achieve monotonic convergence to zero tracking error, that is

$$\|e_{k+1}\|_Q \leq \|e_k\|_Q, \, k = 0, 1, \cdots \tag{6.20}$$

and

$$\lim_{k\to\infty} e_k = 0, \ \lim_{k\to\infty} u_k = u^*. \tag{6.21}$$

Furthermore,

$$\|u_{k+1} - u^*\|_R \leq \|u_k - u^*\|_R, \tag{6.22}$$

i.e., the input iterates approach the converged solution monotonically in norm.

Proof Monotonic convergence can be easily obtained from the successive projection method, as according to Theorem 6.1, the sequence generated satisfies

$$\|r_{2(k+1)} - r_{2k+1}\| \leq \|r_{2k+1} - r_{2k}\| \leq \|r_{2k} - r_{2k-1}\|$$

or equivalently

$$\|k_{k+1} - r_{2k+1}\| \leq \|r_{2k+1} - k_k\| \leq \|k_k - r_{2k-1}\|.$$

Note that k_{k+1} is $(0, u_{k+1})$, r_{2k+1} is (e_{k+1}, u_{k+1}), we have

$$\|k_{k+1} - r_{2k+1}\| = \|e_{k+1}\|_Q.$$

Similarly

$$\|k_k - r_{2k-1}\| = \|e_k\|_Q.$$

Hence we have

$$\|e_{k+1}\|_Q \leq \|e_k\|_Q \tag{6.23}$$

as required.

Zero tracking error is obtained by noting that Algorithm 6.1 (Constrained ILC Algorithm 1) iteratively finds the intersection of $S_1 = K_1 \cap K_3$ and $S_2 = K_2$, which is $(0, u^*)$ when $(K_1 \cap K_3) \cap K_2 \neq \emptyset$, and hence, achieves perfect tracking.

Given that $x = (0, u^*) \in S_1 \cap S_2$, then according to Theorem 6.1

$$\|k_k - x\|^2 \geq \|r_{2k+1} - x\|^2 \geq \|k_{k+1} - x\|^2. \tag{6.24}$$

As x is $(0, u^*)$, k_k is $(0, u_k)$, and k_{k+1} is $(0, u_{k+1})$, it immediately follows that

$$\|u_{k+1} - u^*\|_R \leq \|u_k - u^*\|_R \tag{6.25}$$

as required. That completes the proof. ■

When $(K_1 \cap K_3) \cap K_2 = \emptyset$, perfect tracking is not possible. The algorithm does however compute a "best approximation" of the unconstrained input u^*, as follows:

Theorem 6.3 When perfect tracking is not possible, Algorithm 6.1 (Constrained ILC Algorithm 1) converges to a point u_s^* defined by the following optimization problem:

$$u_s^* \in \arg\min_{u \in \Omega} \left\{ \|e\|_Q^2 : e = r - Gu - d \right\}. \tag{6.26}$$

Moreover, this convergence is monotonic in the tracking error, that is,

$$\|e_{k+1}\|_Q \leq \|e_k\|_Q, k = 0, 1, \cdots. \tag{6.27}$$

Proof According to Theorem 6.1, when $(K_1 \cap K_3) \cap K_2 = \emptyset$, that is, perfect tracking cannot be achieved, Algorithm 6.1 (Constrained ILC Algorithm 1) will converge to point u_s^*, where $r_1 = (e, u) \in S_1$, $r_2 = (0, u_s^*) \in S_2$ defining the minimum distance of the two sets, which is the solution of the following optimization problem:

$$(r_1, r_2) \in \arg \min_{r_1 \in S_1, r_2 \in S_2} \|r_1 - r_2\|^2. \tag{6.28}$$

Recalling the definitions of S_1 and S_2, (6.28) is equivalent to solving

$$\left(u, u_s^*\right) \in \arg \min_{u \in \Omega, u_0} \left\{ \|r - Gu - d\|_Q^2 + \epsilon^2 \|u - u_0\|_R^2 \right\}. \tag{6.29}$$

Hence, Algorithm 6.1 (Constrained ILC Algorithm 1) converges to point u_s^*, which is defined by

$$\begin{aligned} u_s^* &\in \arg \min_{u \in \Omega, u_0} \left\{ \|r - Gu - d\|_Q^2 + \epsilon^2 \|u_0 - u\|_R^2 \right\} \\ &= \arg\min_{u \in \Omega} \left\{ \min_{u_0} \|r - Gu - d\|_Q^2 + \epsilon^2 \|u_0 - u\|_R^2 \right\}. \end{aligned} \tag{6.30}$$

Notice that the inner minimization has the solution

$$u_0 = u. \tag{6.31}$$

Hence, substitute (6.31) into (6.30) and the optimization problem can be transformed into

$$u_s^* \in \arg\min_{u \in \Omega} \|r - Gu - d\|_Q^2. \tag{6.32}$$

Note that the performance index to be minimized is convex. Also notice that the constraint is convex, so that this quadratic programming problem has a solution.

The proof of monotonic convergence is similar to that of $(K_1 \cap K_3) \cap K_2 \neq \emptyset$ and is omitted here. That completes the proof. ■

Algorithm 6.1 (Constrained ILC Algorithm 1) has the appealing property of monotonic convergence of the tracking error norm to the minimum possible value. It requires the solution of the constrained quadratic programming (QP) problem (6.13) which in general does not admit a closed form solution. There are many efficient numerical solvers available. However, when the dimensions of the time series u_k and plant operator G are large, solving the QP problem can be computationally expensive or even becomes unmanageable. This case is discussed in the next section and two methods are given as possible solutions.

6.2.2.2 Efficient Implementation of the Algorithm

Solving a large-scale QP problem is the main potential obstacle in applying Algorithm 6.1 (Constrained ILC Algorithm 1). In this section, two methods are given to address this problem, that is, an iterative solution and a receding horizon method.

There are a number of iterative algorithms in the literature to solve large-scale QP problems; see Goldstein [13], Bertsekas [5], Bierlaire et al. [6], He [17], Bartlett et al. [3], and Gill et al. [12]. Here, the Goldstein–Levitin–Polyak (GLP) method is introduced [5, 13].

The GLP method minimizes a continuously differentiable function $f : H \to R$ over a closed convex set $\Omega \subset H$ using the iterative algorithm

$$x_{k+1} = P_\Omega \left[x_k - a_k \nabla f(x_k)\right], k = 0, 1, \cdots \tag{6.33}$$

where $P_\Omega(z)$ denotes the projection of $z \in H$ onto Ω, $\nabla f(x_k)$ denotes the gradient of f at x_k, and $a_k \geq 0$ is the step size and should satisfy

$$0 < \epsilon \leq a_k \leq \frac{2(1-\epsilon)}{L}, \forall k \geq 0 \tag{6.34}$$

where L is a Lipschitz constant satisfying

$$|\nabla f(x) - \nabla f(y)| \leq L|x - y|, \forall x, y \in \Omega. \tag{6.35}$$

The convergence properties of this algorithm are discussed in Goldstein [13] and omitted here. One appealing property is that $f(x_k)$ will decrease monotonically, which is very useful in an ILC setting.

Algorithm 6.1 (Constrained ILC Algorithm 1) requires the solution of a QP problem, for which the gradient and the Lipschitz constant can be easily obtained. Also note that the constraint is usually simple and the projection can be undertaken conveniently. Hence, the GLP method can be used to solve the QP problem arising in

Algorithm 6.1 (Constrained ILC Algorithm 1). Note that in practice due to the computational/time cost, it is often not possible to wait for too many iterations to get the exact solution. A prescribed maximum iteration or an expected accuracy can be given as criteria to terminate the algorithm in this case.

Another approach to solving the large-scale QP problem is to use the receding horizon method [9, 18]. The main idea of applying the receding horizon method is introduced here, but for more details please refer to textbooks such as Rossiter [26].

On iteration $k+1$, the following problem must be solved:

$$u_{k+1} = \arg\min_{u\in\Omega} \left\{ \|r - Gu - d\|_Q^2 + \epsilon^2 \|u - u_k\|_R^2 \right\}. \tag{6.36}$$

The difficulty with this problem lies in the possible large dimension of u and G. The receding horizon method, however, solves this problem approximately by solving a series of smaller scale QP problems:

Step 1 At time t and for the current state x_t, solve an optimal control problem over a fixed future interval, say $[t, t+N_u-1]$, taking into account the constraints.
Step 2 Apply only the first step in the resulting optimal control sequence.
Step 3 Measure the state reached at time $t+1$.
Step 4 Repeat the fixed horizon optimization at time $t+1$ over the future interval $[t+1, t+N_u]$, starting from the (now) current state x_{t+1}.

Then, the original problem is reduced to a series of QP problems in Step 1:

$$u_{k+1,t}^{opt} = \arg\min_{u_{k+1,t}\in\Omega} \left\{ \sum_{i=t}^{t+N_u-1} \|r(i) - y_{k+1}(i)\|^2 + \epsilon^2 \left\| u_{k+1,t}(i) - u_k(i) \right\|^2 \right\} \tag{6.37}$$

where $u_{k+1,t} = \left[u_{k+1}^T(t)\ u_{k+1}^T(t+1)\ \cdots\ u_{k+1}^T(t+N_u-1) \right]^T$. Note that this problem is of smaller size and can be solved more easily than the full-scale optimization.

The choosing of N_u is very important. If $N_u = N$, (6.37) becomes the original problem (6.36). Large value of N_u will give a more accurate solution at the cost of a large computational load, whilst too small value may result in poor performance. There are a number of results in the literature on how to choose the horizon, e.g., Rossiter [26].

Remark 6.3 In this section, two methods are given to solve the large size constrained QP problem. There are also many other algorithms that can be used. It should be kept in mind that due to the practical restrictions, only an approximate solution of the QP problem can be obtained. Using this solution, the appealing convergence properties of Algorithm 6.1 (Constrained ILC Algorithm 1) may not be maintained, depending on the properties of the methods used.

6.2.2.3 A Note on the Convergence Rate

In this section, the effect of the input weighting ϵ^2 on the convergence rate is discussed. According to (6.13), the input weighting ϵ^2 provides a balance between penalizing the tracking error and the change of input. Intuitively, a smaller ϵ^2 implies a larger acceptable change of input. This, in turn, results in a smaller tracking error and hence leads to a faster convergence rate. Also note that when perfect tracking is not possible, the algorithm will converge to the solution of (6.26) that is independent of the weighting ϵ^2. This implies that the weighting ϵ^2 has no effect on the asymptotic accuracy but only affects the convergence rate. This is illustrated by the following example.

Example 6.1 Consider the following simple second-order system:

$$G(s) = \frac{s-3}{s^2+5s+6}, \tag{6.38}$$

which is sampled using a zero-order hold and a sampling time of 0.1s. The trial length is 20s, zero initial conditions are assumed , and the reference signal is generated by the sine-wave input u^* shown in Fig. 6.3. The constraint is taken to be $|u(t)| \leq 0.8, t = 0, 1, \cdots$, which is violated by u^* so that perfect tracking is not possible. The initial input is chosen to be $u_0 = 0$. The simulation is designed to evaluate the effect of weighting matrices on the performance of Algorithm 6.1 (Constrained ILC Algorithm 1) over 100 iterations. Simulations are run in six cases with weighting chosen to be $Q = R = 1$ and $\epsilon^2 = 3, 1, 0.1, 0.05, 0.01, 0.001$, respectively. Note that this is equivalent to using $Q = 1$ and $R = 3, 1, 0.1, 0.05, 0.01, 0.001 \times I$, and $\epsilon^2 = 1$. The norms of the tracking error for each test are plotted and shown in Fig. 6.4.

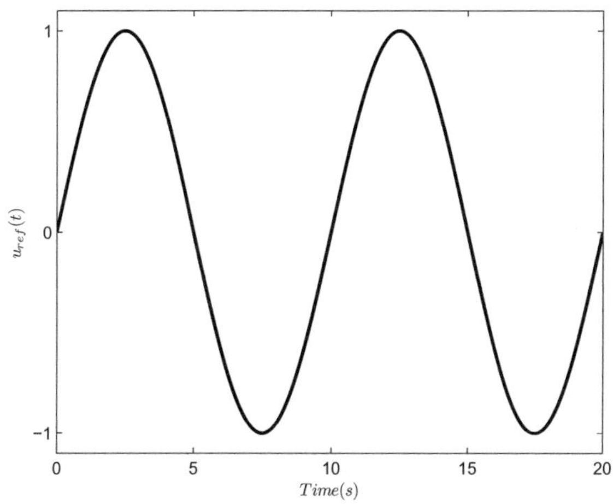

Fig. 6.3 The input signal $u_{ref}(t)$

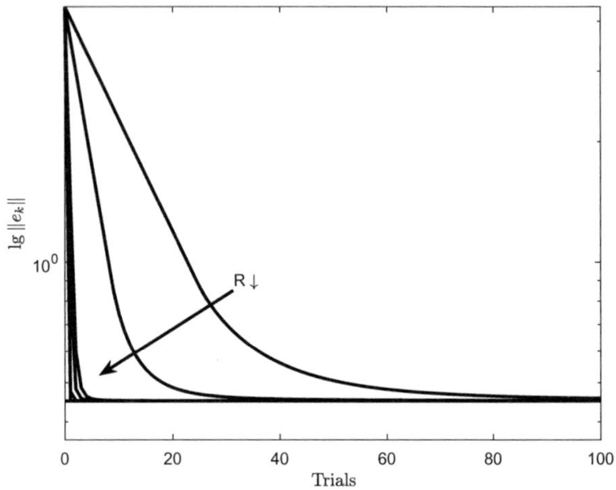

Fig. 6.4 Effect of different weighting ϵ^2 (or R) on the convergence performance

From the figure, it can be seen that the input weighting ϵ^2 (or R) has no effect on the asymptotic accuracy. However, smaller values of ϵ^2 (or R) result in faster convergence, which verifies our expectations.

6.2.3 Constrained ILC Algorithm 2 (Algorithm 6.2)

In this section, an alternative algorithm is given by taking $S_1 = K_1$ and $S_2 = K_2 \cap K_3$ to be the closed, convex sets in Theorem 6.1, which can be expressed as follows:

- $S_1 = \{(e, u) \in H : e = r - Gu - d\}$;
- $S_2 = \{(e, u) \in H : e = 0, u \in \Omega\}$.

Applying Theorem 6.1 in a similar way as in the previous section, the following alternative algorithm to Algorithm 6.1 (Constrained ILC Algorithm 1) can be constructed and is illustrated schematically in Fig. 6.5a and b.

Algorithm 6.2 (Constrained ILC Algorithm 2) Given any initial input u_0 satisfying the constraint with associated tracking error e_0, the input sequence $u_{k+1}, k = 0, 1, 2, \cdots$, defined by the solution of the input unconstrained NOILC optimization problem

$$\tilde{u}_k = \arg\min_u \left\{ \|r - Gu - d\|_Q^2 + \epsilon^2 \|u - u_k\|_R^2 \right\} \tag{6.39}$$

followed by the simple input projection

$$u_{k+1} = \arg\min_{u \in \Omega} \|u - \tilde{u}_k\| \in \Omega \tag{6.40}$$

also satisfies the constraint and iteratively solves the constrained ILC problem.

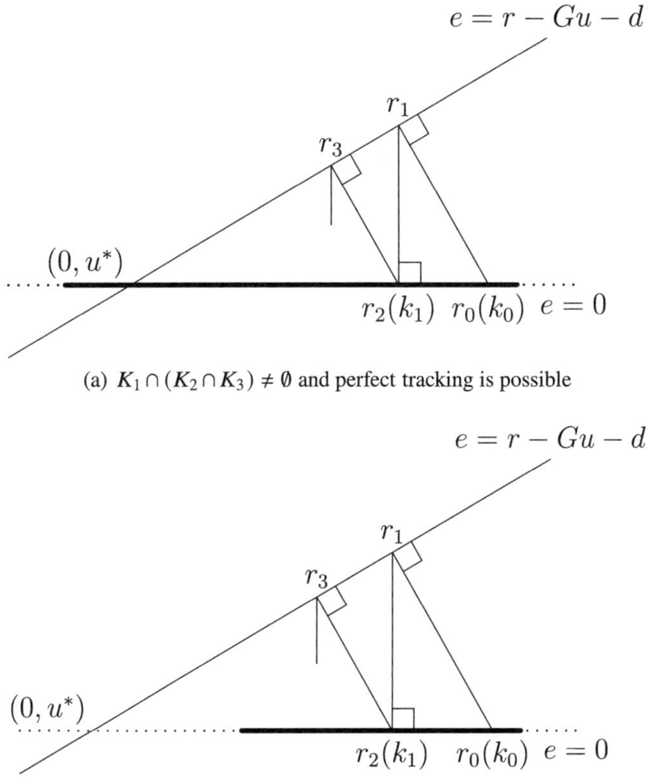

(a) $K_1 \cap (K_2 \cap K_3) \neq \emptyset$ and perfect tracking is possible

(b) $K_1 \cap (K_2 \cap K_3) = \emptyset$ and possible tracking is not possible

Fig. 6.5 Illustration of Algorithm 6.2 (Constrained ILC Algorithm 2) using successive projection

The derivation details follow similarly as that for Algorithm 6.1 (Constrained ILC Algorithm 1) and are omitted here for brevity. Note that the first step of Algorithm 6.2 (Constrained ILC Algorithm 2) is a standard NOILC step which can be implemented easily. The second step requires the solution of the problem (6.40). In practice the input constraint Ω is often a point-wise constraint and the solution of (6.40) can be computed in a straightforward manner. For example, when $\Omega = \{u \in \mathcal{U} : |u(t)| \leq M(t)\}$, the solution is simply as follows:

$$u_{k+1}(t) = \begin{cases} M(t) & : \tilde{u}_k(t) > M(t) \\ \tilde{u}_k(t) & : |\tilde{u}_k(t)| \leq M(t) \\ -M(t) & : \tilde{u}_k(t) < -M(t) \end{cases}, \tag{6.41}$$

for $t = 0, \cdots, N-1$.

6.2.3.1 Convergence Properties

This section discusses the convergence properties of Algorithm 6.2 (Constrained ILC Algorithm 2). As for Algorithm 6.1 (Constrained ILC Algorithm 1), the results are presented in two parts: $(K_1 \cap K_3) \cap K_2 \neq \emptyset$ and $(K_1 \cap K_3) \cap K_2 = \emptyset$.

When $(K_1 \cap K_3) \cap K_2 \neq \emptyset$, perfect tracking of the reference signal is possible with an input u^*. The theorem below directly follows from Theorem 6.1 (in a similar manner as that in the previous section).

Theorem 6.4 When perfect tracking is possible, Algorithm 6.2 (Constrained ILC Algorithm 2) solves the ILC problem in the sense that

$$\lim_{k\to\infty} e_k = 0, \lim_{k\to\infty} u_k = u^*. \tag{6.42}$$

Moreover, this convergence is monotonic with respect to the following performance index:

$$J_k = \|Ee_k\|_Q^2 + \epsilon^2\|Fe_k\|_R^2 \tag{6.43}$$

where

$$\begin{aligned} e_k &= r - Gu_k - d \\ E &= I - G\left(G^T QG + \epsilon^2 R\right)^{-1} G^T Q \, . \\ F &= \left(G^T QG + \epsilon^2 R\right)^{-1} G^T Q \end{aligned} \tag{6.44}$$

Furthermore,

$$\|u_{k+1} - u^*\|_R \leq \|u_k - u^*\|_R, \tag{6.45}$$

i.e., the input iterates approach the solution monotonically in norm.

When $(K_1 \cap K_3) \cap K_2 = \emptyset$, perfect tracking is not possible. The following theorem describes the algorithm's behaviour:

Theorem 6.5 When perfect tracking is not possible, Algorithm 6.2 (Constrained ILC Algorithm 2) converges to a point u_s^* which is defined by the following optimization problem:

$$u_s^* \in \arg\min_{u\in\Omega} \left\{\|Ee\|_Q^2 + \epsilon^2\|Fe\|_R^2\right\}, \tag{6.46}$$

where

$$\begin{aligned} e &= r - Gu - d \\ E &= I - G\left(G^T QG + \epsilon^2 R\right)^{-1} G^T Q \, . \\ F &= \left(G^T QG + \epsilon^2 R\right)^{-1} G^T Q \end{aligned} \tag{6.47}$$

Moreover, this convergence is monotonic with respect to the performance index $J_k = \|Ee_k\|_Q^2 + \epsilon^2\|Fe_k\|_R^2$, i.e.,

$$J_{k+1} \leq J_k, \; \forall k \geq 0.$$

Note the following:

1. Algorithm 6.2 (Constrained ILC Algorithm 2) achieves monotonic convergence in the *Weighted Tracking Error*—this is a clear contrast with Algorithm 6.1 (Constrained ILC Algorithm 1) for which monotonic tracking error norm reduction is produced.
2. For the constrained ILC problem, the best result we can achieve in terms of tracking error is defined by the following QP problem:

$$u^* \in \arg\min_{u\in\Omega} \left\{ \|e\|_Q^2 : e = r - Gu - d \right\}. \tag{6.48}$$

   This is achieved by Algorithm 6.1 (Constrained ILC Algorithm 1) (see Theorem 6.3). Compared to Theorem 6.5, it can be found that Algorithm 6.2 (Constrained ILC Algorithm 2) actually minimizes weighted norm of tracking error. In this case, only nearly optimal performance can be achieved.

6.2.3.2 A Note on the Nature of the Convergence

In this section, the nature of the convergence and the effect of the input weighting ϵ^2 on the convergence properties of Algorithm 6.2 (Constrained ILC Algorithm 2) are discussed.

Algorithm 6.2 (Constrained ILC Algorithm 2) first computes the NOILC solution and then projects this solution onto the constraint set. This approach is much simpler than the previously described algorithm in the sense that the computational load is much less and hence is a simpler way to implement successive projection in practice. Not surprisingly, this strategy may, however, lead to other problems. As will be seen next, the monotonic convergence in the tracking error norm may not be preserved anymore, as seen in the following example. As a consequence, the users may have to make a compromise based on the available computational resources and the desired convergence properties.

Example 6.2 Consider the following system:

$$G(s) = \frac{4s - 2}{s^2 - 0.1s + 3}, \tag{6.49}$$

which is sampled using a zero-order hold and a sampling time of 0.1s. The trial length is 20s, zero initial conditions are assumed, and the reference signal is generated by the square-wave input shown in Fig. 6.6. The constraint is $|u(t)| \leq 1, t = 0, 1, \cdots,$ which is satisfied by the input u^* so that perfect tracking is possible. The initial input is chosen to be $u_0 = 0$. The simulation evaluates the performance of Algorithm 6.2 (Constrained ILC Algorithm 2) over 100 iterations. The weighting matrices are chosen to be $Q = R = I$ and $\epsilon^2 = 1$ for simplicity. The tracking error norm evolution over the first 10 iterations is plotted and shown in Fig. 6.7.

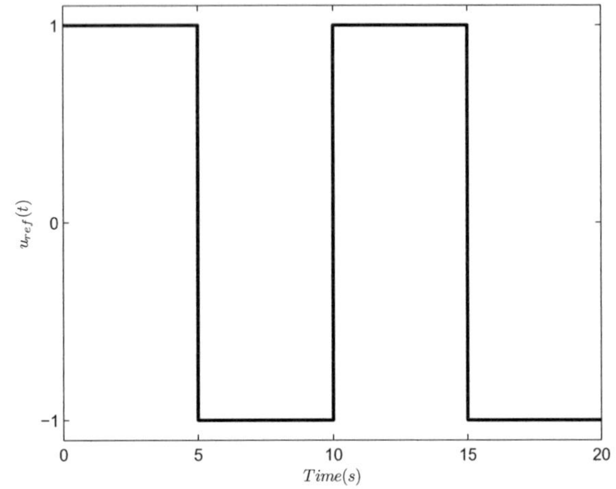

Fig. 6.6 The input signal $u_{ref}(t)$

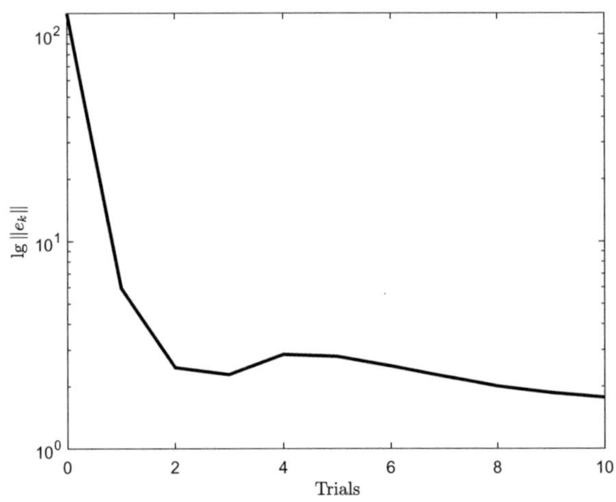

Fig. 6.7 The tracking performance of Algorithm 6.2 (Constrained ILC Algorithm 2)

From the figure, it is clear that Algorithm 6.2 (Constrained ILC Algorithm 2) may not produce monotonic convergence in the tracking error norm.

The input weighting ϵ^2 also affects the convergence. This is illustrated as follows using SISO systems (assuming G is invertible) with scalar weighing $Q = q \times I$ and $R = r \times I$. Choose $q = r = 1$ and consider the effect of variation of ϵ^2. When perfect tracking is possible, zero tracking error can be achieved and smaller ϵ^2 will result in faster convergence. When perfect tracking is not possible, reducing ϵ^2 will again result in faster convergence rate but the asymptotic error changes (in contrast to Algorithm 6.1 (Constrained ILC Algorithm 1). This can be explained as follows. According to Theorem 6.5, Algorithm 6.2 (Constrained ILC Algorithm 2) converges to the solution of the following problem:

$$u_s^* = \arg\min_{u\in\Omega} \Big\{ \|(I - G\left(G^TQG + \epsilon^2 R\right)^{-1} G^T Q)e\|_Q^2 \\ + \epsilon^2 \| \left(G^TQG + \epsilon^2 R\right)^{-1} G^T Q e\|_R^2 \Big\}$$

which for the considered scalar weighting becomes

$$u_s^* = \arg\min_{u\in\Omega} \Big\{ \|(I - G\left(G^TG + \epsilon^2 \times I\right)^{-1} G^T)e\|_Q^2 \\ + \epsilon^2 \| \left(G^TG + \epsilon^2 \times I\right)^{-1} G^T e\|_R^2 \Big\}.$$

When $\epsilon^2 \to \infty$, the first term of the last equation becomes $\|e\|_Q^2$ and the second term becomes zero. Hence, the optimization problem becomes

$$u_s^* = \arg\min_{u\in\Omega} \|e\|_Q^2. \tag{6.50}$$

This is the constrained optimal solution and is the best result that can be achieved with constrained control. However, in this case, since the weighting of input change ϵ^2 is very large, the convergence rate is expected to be very slow. On the other hand, when $\epsilon^2 \to 0$, it can be seen whilst both terms become zero the second term dominates putting emphasis on minimizing $\epsilon^2\|G^{-1}e\|_R^2$, which can be further written as $\epsilon^2\|u - u^*\|_R^2$, where u^* is the unique input generating the reference signal. Hence, the optimization problem becomes

$$u_s^* = \arg\min_{u\in\Omega} \|u - u^*\|_R^2. \tag{6.51}$$

This is just the projection of u^* onto the constraint set Ω. Clearly the tracking error may be larger than that of the constrained optimal solution. However, in this case, the convergence rate is fast.

From the discussion above, it can be seen that when perfect tracking is not possible, the input weighting parameter ϵ^2 provides a compromise between the convergence rate and the tracking performance, which is very different from that of Algorithm 6.1 (Constrained ILC Algorithm 1). This is illustrated in the following example.

Example 6.3 Consider the following system:

$$G(s) = \frac{s+6}{s^2+5s+6}, \tag{6.52}$$

which is sampled using a zero-order hold and a sampling time of 0.1 s. The trial length is 20 s, zero initial conditions are assumed, and the reference signal is generated by the sine-wave input shown in Fig. 6.3. The constraint set is defined by $|u(t)| \leq 0.8, t = 0, 1, \cdots$, which doesn't contain the input u^*. The initial input is chosen to be $u_0 = 0$. The simulation aims to investigate the effect of weighting matrices on the performance of Algorithm 6.2 (Constrained ILC Algorithm 2) over 100 iterations. Six simulations are shown with the weighting matrices

Fig. 6.8 Effect of different weighting matrices on convergence performance

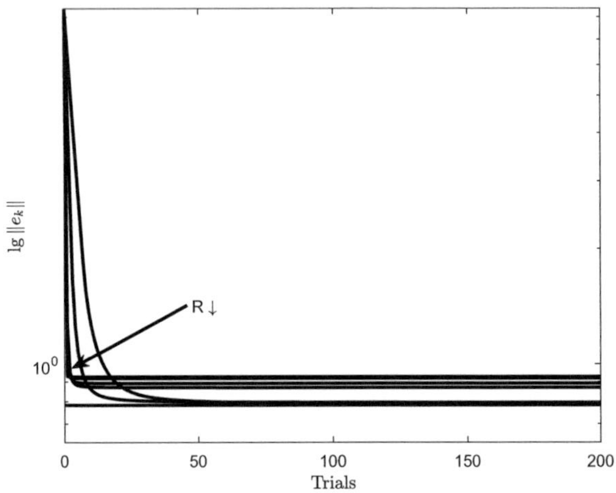

(a) Original figure

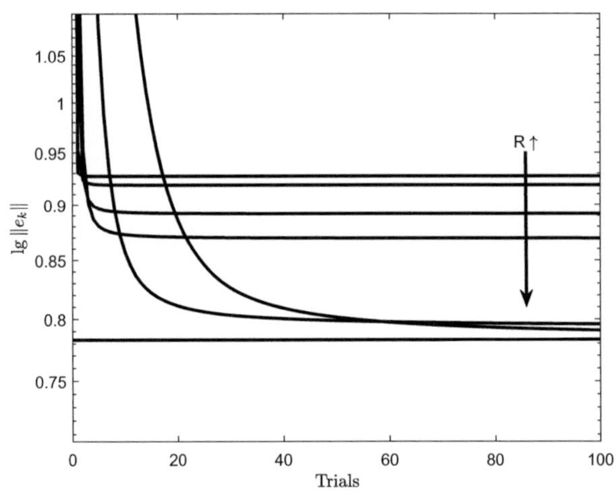

(b) Magnified figure

$Q = R = I$ and $\epsilon^2 = 3, 1, 0.1, 0.05, 0.01, 0.001$, respectively. This is equivalent to using $Q = I$ and $R = 3, 1, 0.1, 0.05, 0.01, 0.001 \times I$ and $\epsilon^2 = 1$. The results are shown in Figs. 6.8 and 6.9. Note that the best result that can be achieved, i.e., the minimum tracking error norm given by (6.50), is shown as horizontal lines in the figures.

From the figure, it can be seen that smaller ϵ^2 (or R) results in faster convergence and the weighting ϵ^2 (or R) does have an effect on the asymptotic performance/accuracy with larger values of ϵ^2 (or R) giving smaller asymptotic error

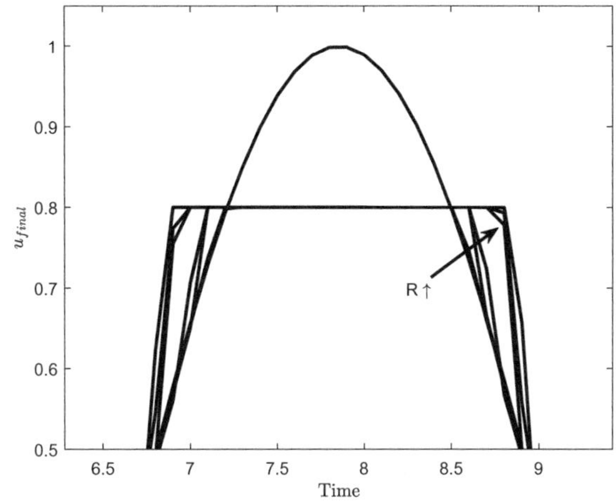

Fig. 6.9 Part of the resulting input for Example 6.3

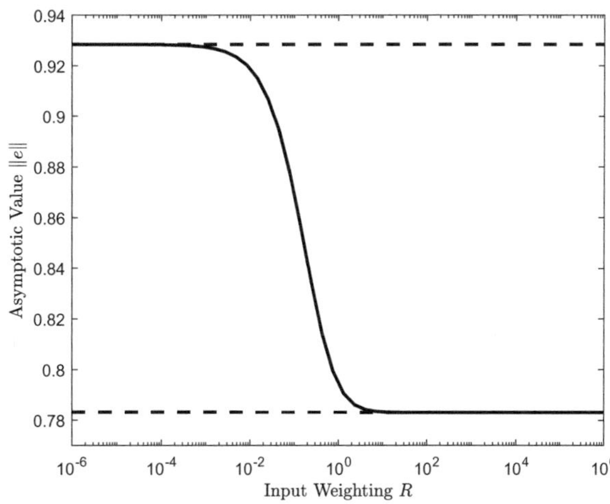

Fig. 6.10 Effect of different weighting ϵ^2 (or R) on asymptotic performance

norms. The asymptotic tracking error norm of Algorithm 6.2 (Constrained ILC Algorithm 2) against different weighting ϵ^2 (or equivalently R) is also plotted and shown in Fig. 6.10. Note that the lower horizontal line is the tracking error norm with the input (6.50) and the upper one is (6.51).

Note that when the system is MIMO or the weighting matrices are not scalars, the effect of weighting ϵ^2 would not be so easy to analyse but a similar pattern could be expected.

6.2.4 Numerical Examples

In this section, three examples are given to demonstrate and compare the effectiveness of the proposed methods. First, consider the following example where perfect tracking is achievable.

Example 6.4 Consider the following non-minimum phase system:

$$G(s) = \frac{s-5}{s^2+5s+6}, \tag{6.53}$$

which is sampled using a zero-order hold and a sampling time of 0.1s. The trial length is 20s, zero initial conditions are assumed, and the reference signal is generated by the square-wave input shown in Fig. 6.11. The constraint is $|u(t)| \leq 1, t = 0, 1, \cdots,$ which just contains the input u^*. The initial input is chosen to be $u_0 = 0$. The simulation compares the NOILC, Algorithm 6.1 (Constrained ILC Algorithm 1), and Algorithm 6.2 (Constrained ILC Algorithm 2) with over 1000 iterations. For simplicity, the weighting matrices are chosen to be $Q = R = I$ and $\epsilon^2 = 1$. The results are shown in Figs. 6.12 and 6.13.

Note that in this example, perfect tracking is possible. According to Theorems 6.2 and 6.4, perfect tracking can be achieved by both algorithms. However, it is expected that the constraint will be active during the iterations, which means the resulting input of NOILC may violate the constraint.

From Fig. 6.12, it can be seen that both Algorithm 6.1 (Constrained ILC Algorithm 1) and Algorithm 6.2 (Constrained ILC Algorithm 2) are approaching perfect tracking, which verifies the previous expectations. During the first few iterations, as $u_0 = 0$, u_k increases in point-wise magnitude gradually and doesn't violate the constraint in any of the three algorithms. In subsequent iterations, the input computed

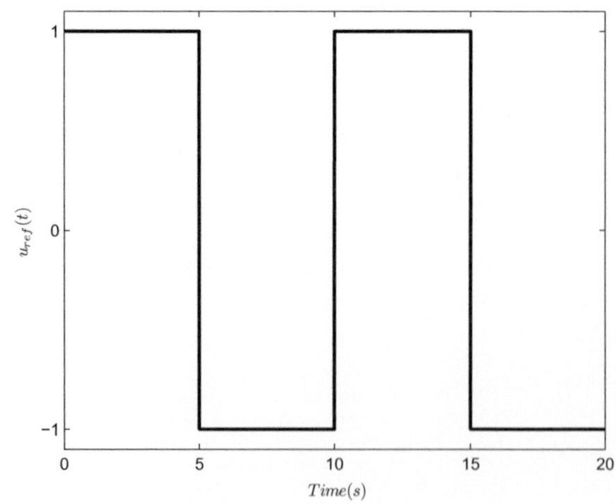

Fig. 6.11 The input signal $u_{ref}(t)$ for Example 6.4

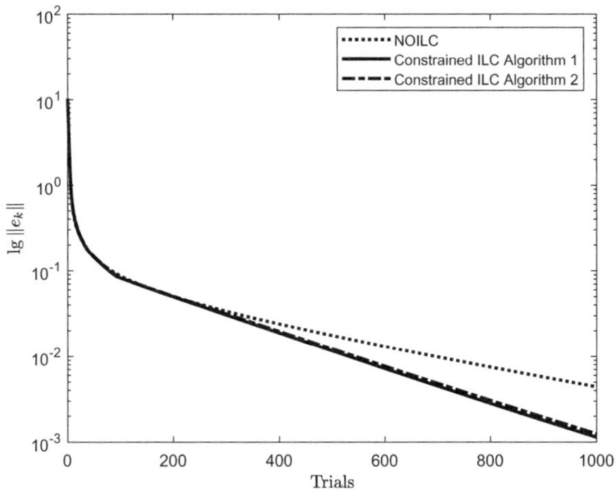

Fig. 6.12 Comparison of convergence for Example 6.4

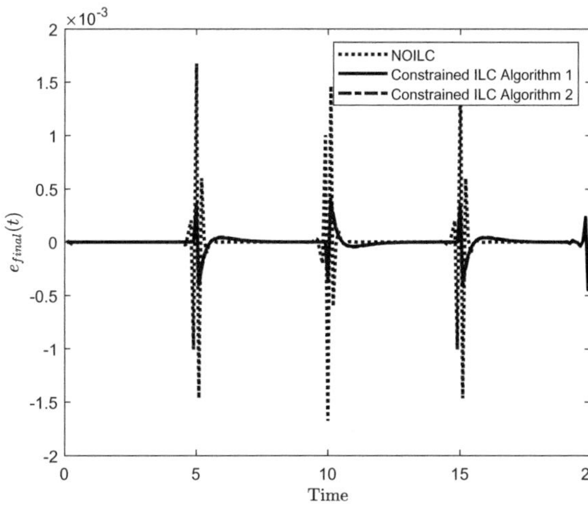

Fig. 6.13 The tracking error at 1000th iteration for Example 6.4

using NOILC then begins to violate the constraint and differences begin to emerge. It is interesting to see that Algorithm 6.1 (Constrained ILC Algorithm 1) and Algorithm 6.2 (Constrained ILC Algorithm 2) outperform NOILC at this stage. It can also be seen that Algorithm 6.1 (Constrained ILC Algorithm 1) performs a little better than Algorithm 6.2 (Constrained ILC Algorithm 2)—this is not surprising as the former is more computationally intensive than the latter and consequently improved convergence is expected.

The second example is to illustrate what will happen if perfect tracking is not possible.

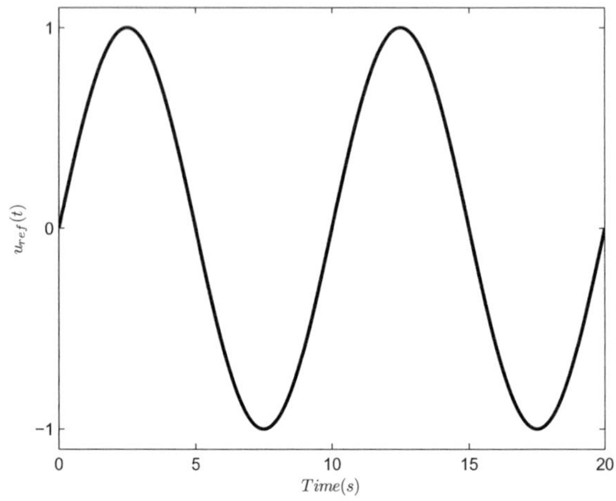

Fig. 6.14 The input signal $u_{ref}(t)$ for Examples 6.5 and 6.6

Example 6.5 Consider the same non-minimum phase system as in the previous example. The reference signal is generated by the sine-wave input as shown in Fig. 6.14. The constraint is replaced by $|u(t)| \leq 0.8, t = 0, 1, \cdots$, so that perfect tracking is not possible. The simulation compares Algorithm 6.1 (Constrained ILC Algorithm 1), Algorithm 6.2 (Constrained ILC Algorithm 2), and the constrained optimal (6.48) over 200 iterations. All other settings remain the same. In Algorithm 6.1 (Constrained ILC Algorithm 1), the constrained QP problem is solved by the MATLAB® optimization toolbox. The results are shown in Figs. 6.15, 6.16, and 6.17.

From Fig. 6.15, it can be seen that Algorithm 6.1 (Constrained ILC Algorithm 1) does converge to the constrained optimal solution, which verifies Theorem 6.3. Inspection of the data shows that Algorithm 6.2 (Constrained ILC Algorithm 2) converges to the solution of (6.46), which verifies Theorem 6.5. It can be seen that Algorithm 6.1 (Constrained ILC Algorithm 1) converges faster than Algorithm 6.2 (Constrained ILC Algorithm 2), which is not difficult to understand. Figure 6.17 shows the original input and resulting input of the three algorithms at the 200th iteration. It can be seen that the resulting input of Algorithm 6.1 (Constrained ILC Algorithm 1) converges to the constrained optimal solution, whilst Constrained ILC Algorithm 2 (Algorithm 6.2) doesn't. It is also noticed that the resulting final input of Algorithm 6.2 (Constrained ILC Algorithm 2) is not just putting saturation on the original input, instead, it adds some compensation. It should be kept in mind that although Algorithm 6.1 (Constrained ILC Algorithm 1) gives better performance, this is achieved at the expense of a larger computation load, which may not be acceptable in the real application, whereas Algorithm 6.2 (Constrained ILC Algorithm 2) achieves nearly optimal performance using quite simple computations.

The third example is to illustrate alternative solution methods of Algorithm 6.1 (Constrained ILC Algorithm 1) when perfect tracking is not possible.

Fig. 6.15 Comparison of convergence for Example 6.5

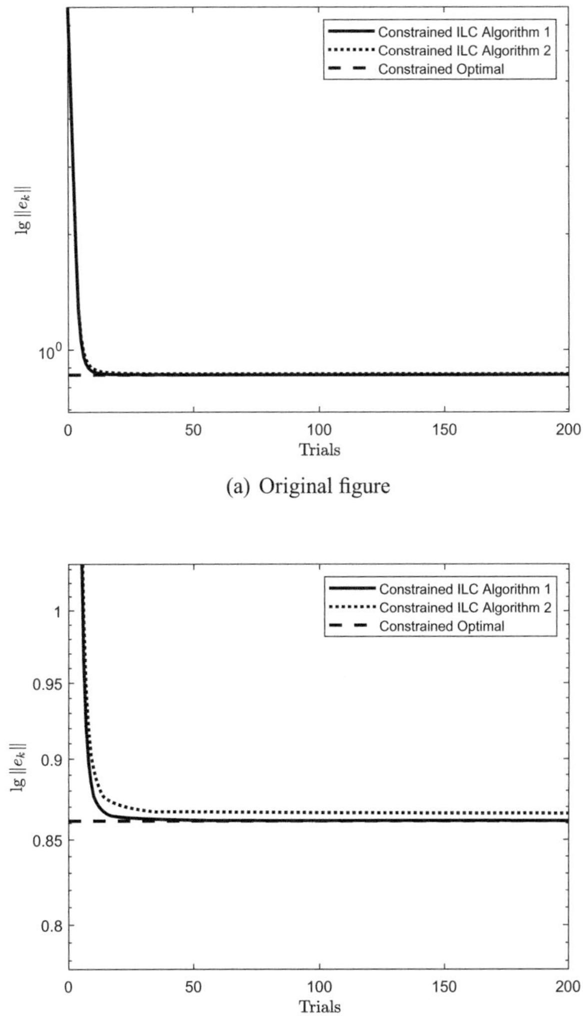

(a) Original figure

(b) Magnified figure

Example 6.6 Consider the following non-minimum phase system:

$$G(s) = \frac{s-4}{s^2+5s+6}. \tag{6.54}$$

The reference and other settings are the same as in the previous example. The simulation compares the iterative GLP algorithm, receding horizon method, and exact solution of the constrained QP problem of Algorithm 6.1 (Constrained ILC Algorithm 1) over 200 iterations. The exact solution is solved by MATLAB® optimization toolbox. In the iterative GLP algorithm, the algorithm is stopped after 10 iterations.

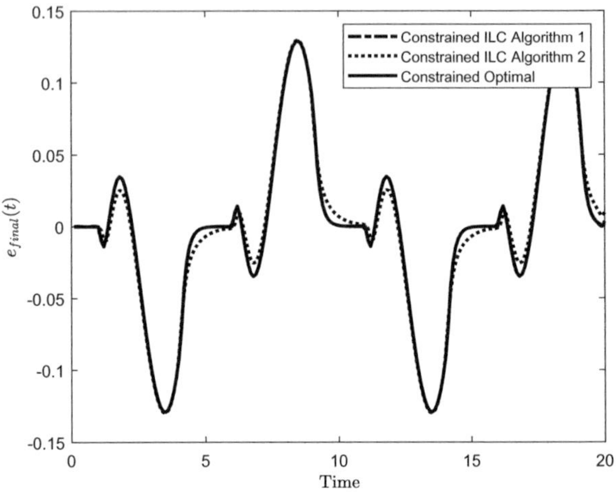

Fig. 6.16 The tracking error at 200th iteration for Example 6.5

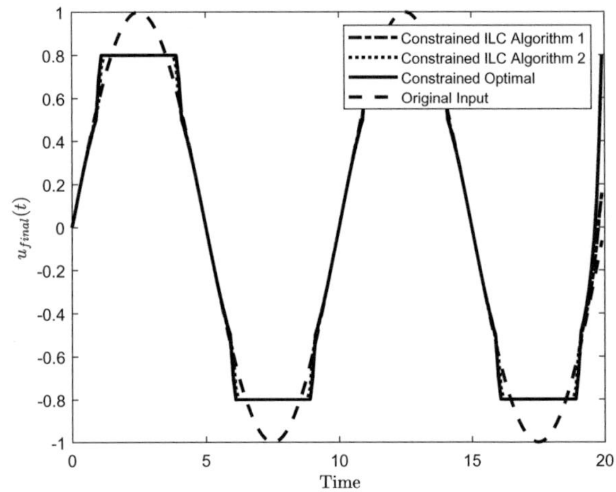

Fig. 6.17 The resulting input at 200th iteration for Example 6.5

In the receding horizon control method, the horizon is chosen to be $N_u = 10$. The results are shown in Figs. 6.18, 6.19, and 6.20.

From Fig. 6.18, it can be seen that the iterative GLP solution method converges to the constrained optimal solution as $k \to \infty$, whilst the receding horizon method doesn't. This is due to the solution accuracy of the receding horizon method. Further simulation shows that when improving the accuracy of the solution by increasing the horizon in the receding horizon method, the limiting point becomes closer to the constrained optimal solution. It can also be noticed that the convergence rates of iterative solution method and receding horizon control method are lower than that of the exact solution of constrained QP problem. The convergence rate can be further improved by increasing the iteration number in the iterative solution or by enlarging the horizon in the receding horizon method.

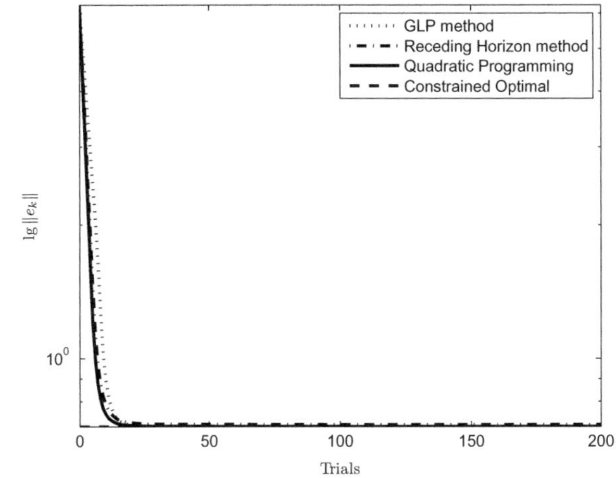

(a) Original figure

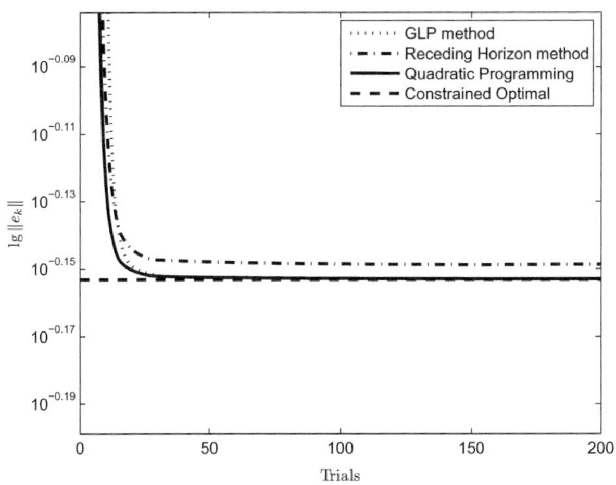

(b) Magnified figure

Fig. 6.18 Comparison of convergence for Example 6.6

6.3 Soft Constraints: An Auxiliary Optimization Problem

The section considers the soft constrained optimization problem (6.4), i.e.,

$$u_\infty = argmin\{J(z,u) : \ u \in \mathcal{U}, \ r = Gu + d, \ z = G_1 u + d_1\}$$

in a manner that meets the needs of ILC. The structure of the proposed algorithms is designed to address the two objectives of eliminating the tracking errors $e = r - y$

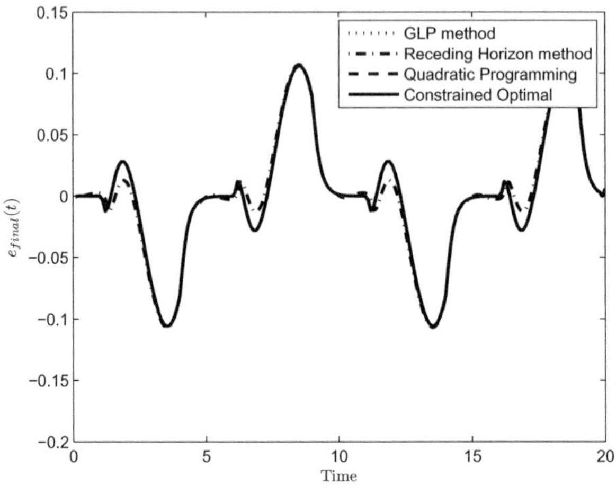

Fig. 6.19 The tracking error at 200th iteration for Example 6.6

Fig. 6.20 The resulting input at 200th iteration for Example 6.6

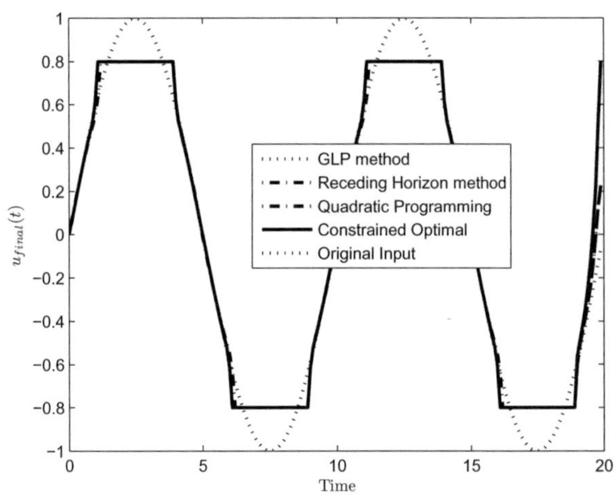

whilst also achieving the desired auxiliary minimization of $J(z, u)$. In addition, the aim is to design algorithms that use simple computational tools from linear quadratic optimization which, for state-space systems, can be realized as familiar linear quadratic optimal control methodologies. This construction permits both online and off-line computational implementations.

The following assumptions are also used to ensure invertibility of the relevant "matrix" operators GG^* used later in the development of this section.

Important Dimensionality and Invertibility Assumptions: *It is assumed that*

1. *the Hilbert space $\mathcal{Y}$ is finite dimensional and also that*
2. *the kernel (null space) of the adjoint operator satisfies $ker[G^*] = \{0\}$ (equivalently, the range of G, denoted $\mathcal{R}[G]$, is exactly $\mathcal{Y}$).*

Note that the condition on the range of G is necessary to ensure that all signals in $\mathcal{Y}$ can be tracked. In practice, if $\mathcal{R}[G]$ is a proper subspace of $\mathcal{Y}$, the problem can be removed in a formal way by regarding $\mathcal{R}[G]$ as a Hilbert space in its own right.

The two algorithms now presented are based on the idea of switching between constraints in the sense that each iteration consists of two components. The two components of each iteration for the first algorithm, Algorithm 6.3 (Switching Algorithm 1), respectively, solve the tracking problem exactly and then minimize a modified auxiliary objective function. The second algorithm, Algorithm 6.4 (Switching Algorithm 2), is a relaxation of the first that allows approximate satisfaction of the tracking constraint at each iteration whilst retaining the asymptotic convergence properties of the first algorithm. This approximation takes the form of replacing the exact solution by a series of $k_0 \geq 1$ iterations of a NOILC problem based on G and a start condition obtained from the auxiliary system iterations.

Remark 6.4 The aforementioned assumptions can be further relaxed to include some infinite dimensional systems by requiring the operator GG^* to have a bounded inverse, for which the algorithms introduced in this section can be applied directly without any problem. For problems where this is not the case, Algorithm 6.4 (Switching Algorithm 2) introduced in this section can still be applied (with some minor modifications to the algorithm properties)—for more details please refer to [21].

6.3.1 Switching Algorithm 1 *(Algorithm 6.3)*

The algorithm is given as follows and presupposes that the two factors u_0, z_0 appearing in objective function (6.3) have been chosen and that G^* and GG^* have been characterized:

Algorithm 6.3 (Switching Algorithm 1) The algorithm has the following steps:

Step 1 Set the iteration index $k = 0$ and choose $u_k = u_0$.
Step 2 Find the plant output $y_k = Gu_k + d$ and compute $e_k = r - y_k$.
Step 3 Solve the optimization problem

$$\tilde{u}_{k+1} = argmin\{\|u - u_k\|^2 : \ r = Gu + d\}$$

using the off-line formula $\tilde{u}_{k+1} = u_k + G^*(GG^*)^{-1}e_k$. (6.55)

Note the need for invertibility of the matrix GG^*.

Step 4 Solve the optimization problem

$$(u_{k+1}, z_{k+1}) = argmin\{\|z - z_k\|^2_{\mathcal{Z}} + \|u - \tilde{u}_{k+1}\|^2_{\mathcal{U}} : \ z = G_1 u + d_1\}. \quad (6.56)$$

Step 5 Replace k by $k+1$ and go to Step 2.

It is worth pointing out that the problem (6.56) is computationally identical to the first iteration of a NOILC control problem with "output" equal to the auxiliary signal z, a "reference" signal "z_k", and first input choice of "$\tilde{u}_{k+1}$". It can be computed off-line or online (as described in Sect. 5.4), and in the latter case a combined feedback and (predictive) feedforward structure has been found to be more robust to model uncertainty [21, 23].

Clearly there is an option to terminate the algorithm at any point if required but, for analysis purposes, it is assumed that the iterations are not terminated. For simplicity, the algorithm is written assuming that the control input u_{k+1} is the new input applied to the plant on iteration $k+1$ but the input $\tilde{u}_{k+1}$ could be used as an alternative.

6.3.1.1 Convergence Properties

To analyse the behaviour of the algorithm, note that steps 3 and 4 give rise to the characterization

$$\begin{aligned} u_{k+1} =& \tilde{u}_{k+1} + G_1^*(z_k - z_{k+1}) \\ =& u_k + G_1^*(z_k - z_{k+1}) + G^*(GG^*)^{-1} e_k \\ =& u_k + G_1^* G_1 (u_k - u_{k+1}) + G^*(GG^*)^{-1} e_k \end{aligned} \quad (6.57)$$

which gives

$$u_{k+1} = u_k + (I + G_1^* G_1)^{-1} G^*(GG^*)^{-1} e_k \quad (6.58)$$

and hence

$$e_{k+1} = W_\infty e_k \ \text{ where } \ W_\infty = (I - G(I + G_1^* G_1)^{-1} G^*(GG^*)^{-1}) \quad (6.59)$$

which is a matrix iteration in the finite dimensional space $\mathcal{Y}$.

Define a new equivalent topology for $\mathcal{Y}$ by the inner product $\langle u, v \rangle_0 = \langle u, (GG^*)^{-1} v \rangle_{\mathcal{Y}}$ and associated induced norm. Then with respect to this new topology W_∞ is self-adjoint and

$$0 \le W_\infty \le \frac{\|G_1\|^2}{1 + \|G_1\|^2} I < I.$$

To see $0 \le W_\infty$ note that

$$(GG^*)^{-1} W_\infty = (GG^*)^{-1} - (GG^*)^{-1} G (I + G_1^* G_1)^{-1} G^*(GG^*)^{-1} \ge 0$$

as $(I + G_1^* G_1)^{-1} < I$. To show $W_\infty \leq \frac{\|G_1\|^2}{1+\|G_1\|^2} I < I$ note

$$(GG^*)^{-1}(W_\infty - I) = -(GG^*)^{-1}G(I + G_1^* G_1)^{-1}G^*(GG^*)^{-1}$$
$$\leq -\frac{1}{1 + \|G_1\|^2}(GG^*)^{-1}.$$

As a consequence, we have the following results characterizing the tracking error convergence properties of the algorithm:

Theorem 6.6 Using the above construction, W_∞ has a real spectrum lying in the half-open interval $[0, 1)$ and the tracking error sequence $\{e_k\}_{k \geq 0}$ satisfies the condition

$$e_\infty = \lim_{k \to \infty} e_k = 0 \quad (\text{i.e., } y_\infty = \lim_{k \to \infty} y_k = r) \tag{6.60}$$

in any norm topology of $\mathcal{Y}$. In particular,

$$\|e_{k+1}\|_0 \leq \lambda \|e_k\|_0 \ \forall k \geq 0 \quad (\text{where } \lambda = \|W_\infty\|_0 < 1) \tag{6.61}$$

and e_k converges to zero as $\|e_k\|_0 \leq \lambda^k \|e_0\|_0 \to 0$ as $k \to \infty$. Finally,

$$\lim_{k \to \infty} \|\tilde{u}_k - u_k\|_{\mathcal{U}} = 0. \tag{6.62}$$

It is worth noting the following:

1. The tracking error is monotonic in the error norm $\|e\|_0$ which depends upon the original inner product chosen for $\mathcal{Y}$ modified by $(GG^*)^{-1}$. It is therefore not necessarily monotonic in the commonly used Euclidean norm.
2. Also, as each $\tilde{u}_k$ solves the tracking problem, the main interpretation of this result is that of consistency, i.e., either of the sequences $\{u_k\}_{k \geq 0}$ or $\{\tilde{u}_k\}_{k \geq 0}$ asymptotically solve the problem.

To derive the properties of the algorithm with respect to the auxiliary performance optimization, write the input iteration in the form

$$u_{k+1} = u_k + (I + G_1^* G_1)^{-1} G^*(GG^*)^{-1} W_\infty^k e_0 \tag{6.63}$$

which, as W_∞ has spectral radius < 1, proves the existence of the limit $u_\infty = \lim_{k \to \infty} u_k$ generated from the formula

$$u_\infty - u_0 = (I + G_1^* G_1)^{-1} G^*(GG^*)^{-1} \sum_{k=0}^{\infty} W_\infty^k e_0$$
$$= (I + G_1^* G_1)^{-1} G^*(GG^*)^{-1}(I - W_\infty)^{-1} e_0 \tag{6.64}$$

where u_∞ hence generates the auxiliary signal $z_\infty = G_1 u_\infty + d_1$ where

$$\lim_{k\to\infty} z_k = z_\infty = z_0 + G_1(I + G_1^* G_1)^{-1} G^*(GG^*)^{-1}(I - W_\infty)^{-1} e_0. \tag{6.65}$$

Writing the above as

$$(I + G_1^* G_1)(u_\infty - u_0) = G^* \lambda_0 \quad \text{where } \lambda_0 = (GG^*)^{-1}(I - W_\infty)^{-1} e_0 \in \mathcal{Y} \tag{6.66}$$

or using $z = G_1 u + d_1$

$$u_\infty - u_0 = G^* \lambda_0 + G_1^* \lambda_1, \quad \text{where } z_\infty - z_0 = -\lambda_1, \tag{6.67}$$

it is immediately concluded the following.

Theorem 6.7 Under the condition defined above, the pair (u_∞, z_∞) is a stationary point of the Lagrangian

$$\mathcal{L}(u, \lambda_0) = J(z, u) + 2\langle \lambda_0, r - Gu - d\rangle_{\mathcal{Y}} + 2\langle \lambda_1, z - G_1 u - d_1\rangle_{\mathcal{Z}} \tag{6.68}$$

(with multipliers $\lambda_0 \in \mathcal{Y}$, $\lambda_1 \in \mathcal{Z}$) for the linearly constrained optimization problem with

$$\text{Objective Function} \quad J(z, u) = \|z - z_0\|_{\mathcal{Z}}^2 + \|u - u_0\|_{\mathcal{U}}^2 \tag{6.69}$$

$$\text{and Constraints} \quad \{u \in \mathcal{U},\ r = Gu + d,\ z = G_1 u + d_1\}. \tag{6.70}$$

Furthermore, the sequence $\{u_k\}_{k\ge 0}$ (and hence $\{\tilde{u}_k\}_{k\ge 0}$) asymptotically solves the associated ILC tracking problem with the auxiliary optimization criterion.

Proof The linearity of the constraints and the assumption that $\mathcal{R}[G] = \mathcal{Y}$ indicate that all points of $\mathcal{Z} \times \mathcal{U}$ are *regular points* (see, for example, [19]) of the constraints. This condition, together with the observation that the objective function is convex and the constraint set is both closed and convex in $\mathcal{Z} \times \mathcal{U}$, then proves that (u_∞, z_∞) solves the required, constrained optimization problem. ■

6.3.1.2 A Note on the Convergence Rates

The norms in $\mathcal{U}$ and $\mathcal{Z}$ are assumed to be chosen by the control design engineer to reflect the desired relative weighting between $\|z - z_0\|_{\mathcal{Z}}^2$ and $\|u - u_0\|_{\mathcal{U}}^2$ in the objective function (6.3). To illustrate the influence of this relative weighting replace (6.3) by the following:

$$J_\epsilon(z, u) = \|z - z_0\|_{\mathcal{Z}}^2 + \epsilon_1^2 \|u - u_0\|_{\mathcal{U}}^2 \tag{6.71}$$

where $\epsilon_1^2 > 0$ is a parameter introduced to reveal the effects of a simple change in the relative weights of the two terms in the objective function.

The effect on the previous theory is simply that the operator G_1^* is replaced by $\epsilon_1^{-2} G_1^*$ and, in particular, W_∞ is replaced by

$$W_\infty(\epsilon_1) = (I - G(I + \epsilon_1^{-2} G_1^* G_1)^{-1} G^* (GG^*)^{-1}). \tag{6.72}$$

$W_\infty(\epsilon_1)$ has all the properties of W_∞, in particular (following a similar proof)

$$0 \le W_\infty(\epsilon_1) \le \frac{\epsilon_1^{-2}\|G_1\|^2}{1 + \epsilon_1^{-2}\|G_1\|^2} I < I.$$

From that it can be seen that

$$\lim_{\epsilon_1^2 \to \infty} W_\infty(\epsilon_1) = 0,$$

which implies that a larger ϵ_1^2 will lead to faster convergence.

6.3.2 Switching Algorithm 2 (Algorithm 6.4)

Algorithm 6.3 (Switching Algorithm 1) assumes that the optimization in Step 3 will be solved using the given formula (6.55) for $\tilde{u}_{k+1}$ (or an equivalent computation). This requirement can be relaxed by replacing this step by a finite iteration process to yield the modified algorithm written in the form of ILC "outer" iterations with index k for the sequence $\{u_k\}_{k \ge 0}$, etc., but where each such iteration consists of k_0 "inner iterations" (with index j) used in the computation of $\tilde{u}_{k+1}$. This is still an ILC process but the use of two iteration indices simplifies the presentation of the methodology. In addition to providing control over the input change, the use of inner iterations means that the tracking problem can be addressed online over several trials, e.g., by exploiting a feedback plus (predictive) feedforward implementation. Intuitively, this could lead to greater robustness in practice in dealing with plant-model mismatch. The algorithm proposed is as follows.

Algorithm 6.4 (Switching Algorithm 2)
The "Outer Iteration" Process:

Step 1 Set the iteration index $k = 0$ and choose $u_k = u_0$ and an integer $k_0 \ge 1$.
Step 2 Undertake k_0 "inner iterations" using the NOILC problem for the system $y = Gu + d$ with reference signal r and initial input choice u_k (a more detailed description is given following this algorithm statement). These iterations can be undertaken online.
Step 3 Set $\tilde{u}_{k+1}$ to be equal to the final control signal $\tilde{u}_{k+1}^{(k_0)}$ generated by this process.
Step 4 As in Algorithm 6.3 (Switching Algorithm 1), solve the optimization problem

$$(u_{k+1}, z_{k+1}) = argmin\{\|z - z_k\|_{\mathcal{Z}}^2 + \|u - \tilde{u}_{k+1}\|_{\mathcal{U}}^2 : \ z = G_1 u + d_1\}. \tag{6.73}$$

Step 5 Replace k by $k + 1$ and go to Step 2.

The "Inner Iterations": Step 2 in the above can be described in more detail as follows:

Step i Set an inner iteration index $j = 0$ and initiate inner iterations from the starting condition $\tilde{u}^{(j)}_{k+1} = u_k$ and $e_{(k+j)} = e_k$.

Step ii Compute the new input signal

$$\tilde{u}^{(j+1)}_{k+1} = \tilde{u}^{(j)}_{k+1} + \epsilon^{-2} G^*(I + \epsilon^{-2} GG^*)^{-1} e_{(k+j)} \tag{6.74}$$

obtained by minimizing

$$\|e_{(k+j+1)}\|^2_{\mathcal{Y}} + \epsilon^2 \|\tilde{u}^{(j+1)}_{k+1} - \tilde{u}^{(j)}_{k+1}\|^2_{\mathcal{U}} \tag{6.75}$$

where the new output $y^e_{(k+j+1)} = G\tilde{u}^{(j+1)}_{k+1} + d$ and the resultant tracking error $e_{(k+j+1)} = r - y_{(k+j+1)}$. This can be done either off-line by simulation or online using experimentation. In the latter case a combined feedback and (predictive) feedforward structure embeds the potential for greater robustness to model uncertainty.

Step iii Replace j by $j + 1$.

Step iv If $j > k_0$, end the inner iteration process. Otherwise return to Step ii.

The analysis is similar to that of Algorithm 6.3 (Switching Algorithm 1). The crucial difference is the inner iteration process. It is easily seen from NOILC that, $\forall\, j \geq 0$,

$$\begin{aligned} \tilde{u}^{(j+1)}_{k+1} &= \tilde{u}^{(j)}_{k+1} + \epsilon^{-2} G^* e_{(k+j+1)} \quad \text{and} \quad \text{hence} \\ e_{(k+j+1)} &= (I + \epsilon^{-2} GG^*)^{-1} e_{(k+j)}, \quad \text{i.e.,} \; e_{(k+j)} = (I + \epsilon^{-2} GG^*)^{-j} e_k. \end{aligned} \tag{6.76}$$

Hence, using induction, and $\tilde{u}^{(0)}_{k+1} = u_k$

$$\tilde{u}_{k+1} = \tilde{u}^{(k_0)}_{k+1} = u_k + \epsilon^{-2} G^* \sum_{j=0}^{k_0 - 1} (I + \epsilon^{-2} GG^*)^{-(j+1)} e_k. \tag{6.77}$$

Whenever the matrix GG^* is positive definite, all eigenvalues of $(I + \epsilon^{-2} GG^*)^{-1}$ are strictly less than unity and hence, summing the geometric series,

$$\tilde{u}_{k+1} = u_k + \epsilon^{-2} G^*(I + \epsilon^{-2} GG^*)^{-1}(I - (I + \epsilon^{-2} GG^*)^{-1})^{-1}(I - (I + \epsilon^{-2} GG^*)^{-k_0}) e_k$$

which yields

$$\tilde{u}_{k+1} = u_k + G^*(GG^*)^{-1}(I - (I + \epsilon^{-2} GG^*)^{-k_0}) e_k. \tag{6.78}$$

As $\lim_{k_0\to\infty}(I+\epsilon^{-2}GG^*)^{-k_0}=0$, it is seen that Switching Algorithm 2 (Algorithm 6.4) is equivalent to Algorithm 6.3 (Switching Algorithm 1) when $k_0=\infty$.

The use of finite values of k_0 can however affect performance. More precisely, using the above input signal in Step 4 of Algorithm 6.4 (Switching Algorithm 2) and using similar algebra to that used in Algorithm 6.3 (Switching Algorithm 1) yield the results that, $\forall\, k\geq 0$,

$$e_{k+1}=W_{k_0}e_k \quad \text{and hence} \quad e_k=W_{k_0}^k e_0 \tag{6.79}$$

where the matrix

$$\begin{aligned} W_{k_0}&=I-G(I+G_1^*G_1)^{-1}G^*(GG^*)^{-1}(I-(I+\epsilon^{-2}GG^*)^{-k_0}),\\ \text{i.e., } W_{k_0}&=W_\infty+(I-W_\infty)(I+\epsilon^{-2}GG^*)^{-k_0}. \end{aligned} \tag{6.80}$$

The rest of the analysis is identical to that of Switching Algorithm 1 (Algorithm 6.3) with W_∞ replaced by W_{k_0} with identical results and properties provided that all eigenvalues of W_{k_0} have modulus strictly less than unity. To show this, note that for any eigenvalue λ, using $\|K\|_0$ to denote the induced norm of an operator $K:\mathcal{Y}\to\mathcal{Y}$ in the topology induced by $\langle\cdot,\cdot\rangle_0$

$$\begin{aligned} |\lambda| \;\leq\; \|W_{k_0}\|_0 \;&\leq\; \|W_\infty\|_0+\|I-W_\infty\|_0\|(I+\epsilon^{-2}GG^*)^{-1}\|_0^{k_0}\\ &=\; r(W_\infty)+(1-r(W_\infty))\|(I+\epsilon^{-2}GG^*)^{-1}\|_0^{k_0}. \end{aligned}$$

Note that

$$(GG^*)^{-1}(I+\epsilon^{-2}GG^*)^{-1}=(\epsilon^{-2}I+(GG^*)^{-1})^{-1}>0$$

and

$$(GG^*)^{-1}((I+\epsilon^{-2}GG^*)^{-1}-I)=-\epsilon^{-2}I(I+\epsilon^{-2}GG^*)^{-1}<0.$$

We thus have $\|(I+\epsilon^{-2}GG^*)^{-1}\|_0<1$. Also note that $r(W_\infty)<1$. The above leads to

$$|\lambda|<1$$

as desired. Therefore Algorithm 6.4 (Switching Algorithm 2) converges for all choices of $k_0>0$.

Also note that the bound on $r(W_{k_0})$ is monotonically decreasing to the value $r(W_\infty)=\|W_\infty\|_0$ suggesting that, although k_0 has no effect on final convergence, increasing k_0 will tend to increase convergence rate.

Finally, the discussion on the convergence of the input (to the solution of the auxiliary optimization problem) as well as the effect on convergence rate by introducing extra parameter ϵ_1^2 (in the auxiliary performance optimization) also remains valid.

6.3.3 A Discussion of Algorithm Robustness

A characterization of robustness is possible in the special case where all computations are undertaken off-line but using observed plant tracking error data, i.e., a specific feedforward implementation.

Feedforward Assumption for Robustness Analysis: *Suppose that both the computation of $\tilde{u}_{k+1}$ from u_k and e_k and the auxiliary optimization problem generating u_{k+1} from z_k and $\tilde{u}_{k+1}$ are undertaken off-line but that each u_k is used online to generate the tracking error data e_k from the plant.*

It follows that differences between the plant and the model G will affect each signal y_k, and hence e_k, in an "unpredictable" way. In contrast, although there may be modelling errors in G_1 (as a representation of the behaviour of the actual values of z generated by the plant), these errors have no impact on the algorithm as the "real" z measurements are not made and are not required for the computations. The algorithm may still converge to the desired tracking signal, but, if it does indeed converge, it is natural to ask the question whether or not it converges to a solution of the defined ILC problem with auxiliary optimization defined in terms of the *model-based* values of the auxiliary variable z.

Using the feedforward assumption and, for simplicity, examining the case of Algorithm 6.3 (Switching Algorithm 1), the main change to the theoretical analysis is that the data e_k used is generated from the plant. Let G now denote the operator deduced from a model of the plant and denote the actual value of this operator by $G + \Delta G$ where, again, ΔG maps $\mathcal{U}$ into $\mathcal{Y}$ and is bounded. In a similar manner, suppose that the plant replaces the model value d by $d + \Delta d$. As a consequence, the plant output $y_k = (G + \Delta G)u_k + (d + \Delta d)$ and the observed *online* plant error is now

$$e_k = r - y_k = r - (G + \Delta G)u_k - (d + \Delta d) \ \ \forall\, k \geq 0. \tag{6.81}$$

The relevant formulae used for analysis are as follows, namely, if u_k generates the error e_k from the actual plant *online*, then the next input u_{k+1} can be characterized via *off-line* computation of $\tilde{u}_{k+1}$

$$\tilde{u}_{k+1} = u_k + G^*(GG^*)^{-1}e_k$$

using the model G. This is then followed by the model-based *off-line* NOILC minimization problem which has the formal solution

$$\begin{aligned} u_{k+1} =& \tilde{u}_{k+1} + G_1^*(z_k - z_{k+1}) \\ =& u_k + G_1^*G_1(u_k - u_{k+1}) + G^*(GG^*)^{-1}e_k, \\ \text{i.e., } u_{k+1} =& u_k + (I + G_1^*G_1)^{-1}G^*(GG^*)^{-1}e_k. \end{aligned}$$

It follows trivially, by operating on this with the real plant operator $G + \Delta G$, that

$$e^e_{k+1} = (W_\infty + \Delta W_\infty)e_k, \quad \forall\, k \geq 0, \tag{6.82}$$

where ΔW_∞ is the change induced by the plant-model mismatch. Note that the change to d has no effect on the general form of the error evolution and hence the results obtained.

Analysis of the algorithm has already proved that the spectral radius $r(W_\infty)$ of the matrix W_∞ is strictly less than unity and considerably less than unity if the weighting on control energy in the auxiliary optimization problem is large.

The matrix perturbation ΔW_∞ takes the form

$$\Delta W_\infty = -\Delta G(I + G_1^* G_1)^{-1} G^*(GG^*)^{-1}. \tag{6.83}$$

Again, using $\|K\|_0$ to denote the induced norm of $K : \mathcal{Y} \to \mathcal{Y}$ in the topology induced by the inner product $\langle \cdot, \cdot \rangle_0$, it is deduced the following.

Theorem 6.8 Under the defined conditions above, the feedforward implementation of the Algorithm 6.3 (Switching Algorithm 1) converges to a zero tracking error if the induced norm

$$\|\Delta W_\infty\|_0 = \|\Delta G(I + G_1^* G_1)^{-1} G^*(GG^*)^{-1}\|_0 \; < \; 1 - r(W_\infty). \tag{6.84}$$

In particular, it is sufficient that

$$\begin{aligned} \|\Delta G\| &= \; (sup\{\, \langle \Delta G v, \Delta G v \rangle_0 \; : \; v \in \mathcal{U} \;\&\; \langle v, v \rangle_{\mathcal{U}} \; = 1 \,\})^{1/2} \\ &< \; (1 - r(W_\infty))(1 + \delta^2) \end{aligned} \tag{6.85}$$

where δ^2 is any positive scalar in the range $0 \leq \delta^2 \leq \underline{\delta}^2$ where

$$\underline{\delta}^2 = inf\{\, \langle G_1 v, G_1 v \rangle_{\mathcal{Z}} \; : \; v \in \mathcal{U} \;\&\; \langle v, v \rangle_{\mathcal{U}} \; = 1 \,\}. \tag{6.86}$$

Proof The relation $e^e_{k+1} = (W_\infty + \Delta W_\infty)e_k$ indicates that it is only necessary to prove that, under the defined conditions, all eigenvalues of $W_\infty + \Delta W_\infty$ have modulus strictly less than unity. Note that all eigenvalues satisfy the inequality

$$\begin{aligned} |\lambda| &\leq \|W_\infty + \Delta W_\infty\|_0 \\ &\leq \|W_\infty\|_0 + \|\Delta W_\infty\|_0\,, \\ &= r(W_\infty) + \|\Delta W_\infty\|_0 \end{aligned} \tag{6.87}$$

i.e., the first inequality is sufficient to ensure that $|\lambda| < 1$. Next, note that a sufficient condition for this to be true is that

$$\|\Delta G\| \; < \; \frac{1 - r(W_\infty)}{\|(I + G_1^* G_1)^{-1}\| \|G^*(GG^*)^{-1}\|}. \tag{6.88}$$

From the definitions

$$\begin{aligned} \|G^*(GG^*)^{-1}\|^2 &= sup\{ \langle G^*(GG^*)^{-1}v, G^*(GG^*)^{-1}v\rangle_{\mathcal{U}} \ : \ \langle v, v\rangle_0 = 1 \} \\ &= sup\{ \langle v, (GG^*)^{-1}v\rangle_{\mathcal{R}} \ : \ \langle v, v\rangle_0 = 1 \} \\ &= 1 \,, \end{aligned}$$

and $\|(I + G_1^* G_1)^{-1}\| \leq (1 + \underline{\delta}^2)^{-1}$

which proves the result. ■

Two useful observations can easily be made, namely:

1. It is always possible to choose the value $\delta = 0$. This choice removes the need to consider the auxiliary system dynamics in the robustness analysis.
2. Choosing $\delta^2 = 0$ and using the lower bound

$$\begin{aligned} 1 - r(W_\infty) &= inf\{ \langle v, (I - W_\infty)v\rangle_0 \ : \ v \in \mathcal{R}, \ \langle v, v\rangle_0 = 1 \} \\ &\geq \tfrac{1}{1+\overline{\delta}^2} \ \text{ for any} \\ \overline{\delta}^2 &\geq \|G_1\|^2 = sup\{ \langle G_1 v, G_1 v\rangle_{\mathcal{Z}} \ : \ v \in \mathcal{U} \ \& \ \langle v, v\rangle_{\mathcal{U}} = 1 \}, \end{aligned} \tag{6.89}$$

the robustness condition can be replaced by the simple inequality

$$\|\Delta G\| < \frac{1}{1 + \overline{\delta}^2} \tag{6.90}$$

which bounds permissible modelling errors simply in terms of a bound on the operator norm of G_1 in $\mathcal{U}$. G and the weightings defining the inner products (and associated norms) in $\mathcal{Y}$ and $\mathcal{U}$ play a role in the numerical values appearing in the inequality as the defined measure of magnitude of ΔG is defined in terms of the $\langle \cdot, \cdot \rangle_0$ topology. Also the weightings defining the inner product in $\mathcal{Z}$ affect the numerical value of $\overline{\delta}^2$.
3. It is noted that these formulae may make it possible to provide more explicit numerical bounds on robustness. For example, if G_1 is an asymptotically stable ℓ-input, m_z-output continuous state-space system, then causality on $[0, T]$, monotonicity of norms with respect to T, and Parseval's theorem on $[0, \infty)$ suggest the choice of $\overline{\delta}^2$ as

$$\|G_1\|^2 \leq \overline{\delta}^2 = \sup_{\omega \geq 0} r(R^{-1} G_1^T(-i\omega) Q G_1(i\omega)) \tag{6.91}$$

where $G_1(s)$ is used to denote the transfer function matrix of the state-space model of the operator G_1. The general effects of Q (weighting defining the topology in $\mathcal{Z}$) and R (weighting defining the topology in $\mathcal{U}$) on robustness are now revealed, i.e., robustness tends to reduce as the input weighting R reduces (or Q increases). It also suggests that, for a given choice of weights, the form of the frequency response $G_1(i\omega)$ will have an effect, e.g., significant resonances will reduce robustness by increasing $\overline{\delta}^2$.

The final question to be answered is "what is the nature of the limit of the algorithm in the presence of such modelling errors?". This is answered by the following theorem:

Theorem 6.9 Under the conditions of the previous theorem, the feedforward implementation of Algorithm 6.3 (Switching Algorithm 1) in the presence of the defined modelling error converges to a solution $(u_\infty, z_\infty, \lambda_0, \lambda_1)$ of the equations

$$\begin{aligned} u_\infty - u_0 &= G^*\lambda_0 + G_1^*\lambda_1 \ , \\ z_\infty - z_0 = -\lambda_1 \ , \ \ & z_\infty = G_1 u_\infty + d_1 \end{aligned} \tag{6.92}$$

and

$$r - (G + \Delta G)u_\infty - (d + \Delta d) = 0\,. \tag{6.93}$$

Proof The proof follows in a similar manner to Theorem 6.7, replacing W_∞ by $W_\infty + \Delta W_\infty$ to prove the existence of the limits to the input sequence and auxiliary variable sequence. The relevant value of λ_0 is just

$$\lambda_0 = (GG^*)^{-1}(I - W_\infty - \Delta W_\infty)^{-1} e_0$$

which completes the outline proof. ■

These equations are just perturbations of the stationary point conditions for the Lagrangian associated with the optimization problem

$$\min_{u \in \mathcal{U}}\{J(z, u) : u \in \mathcal{U},\ r = Gu + d,\ z = G_1 u + d_1\} \tag{6.94}$$

suggesting that the converged solution no longer minimizes the auxiliary objective function but, if modelling errors are small, will be a good approximation.

6.3.4 Numerical Examples

Consider the same system and the setting as in the previous chapter, i.e.,

$$G(s) = \frac{s + 1}{s^2 + 5s + 6}.$$

The system is required to track a reference at intermediate time instants at $t = 1s, 2s, 3s, 4s, 5s, 6s$. Zero initial input is chosen.

To produce a smooth output response at all points $t \in [0, 6]$ (e.g., in order to reduce vibration and potential payload damage/spillage between the specified intermediate points if the output is the position of a robotic arm end effector), some degree of minimization of the mean square output acceleration is proposed. The auxiliary variable is hence $z(t) = \ddot{y}(t)$. The above systems are sampled at intervals of $T_s =$

0.1 using zero-order hold (resulting in a discrete-time plant model $S(A, B, C)$ and $S(A_z, B_z, C_z)$, respectively).

The solutions u_∞ and z_∞ of the objective function (6.3) can be calculated directly using the system model and are used to compare the converged performance of switching ILC algorithms. In what follows, we use fix weightings of $Q = I$ in both $\mathcal{Y}$ and $\mathcal{Z}$, and $R = I$ in $\mathcal{Z}$. We choose $\epsilon^2 = \epsilon_1^2$ (with different values) to study their effect on the convergence.

6.3.4.1 Switching Algorithm 1 (Algorithm 6.3) Results

Algorithm 6.3 (Switching Algorithm 1) is first implemented. 100 trials have been performed and the tracking error norm evolution as well as the change of performance index $J(z, u)$ over trials are shown in Figs. 6.21 and 6.22. The output, input, and auxiliary signals on the 100th trial, together with the optimal values (by directly solving the optimization problem), are given in Figs. 6.23, 6.24, and 6.25, respectively.

As can be seen from the figures, for all choices of the weighting parameters the algorithm achieves perfect reference tracking at the intermediate points. At the same time, the auxiliary performance index is minimized. This can be seen from the Figs. 6.23, 6.24, and 6.25 that the final converged input and auxiliary signal are the same as that the optimal values u_∞ and z_∞ (in the figure they are on top of each other and not distinguishable). Furthermore, it is seen that a larger ϵ_1^2 leads to faster convergence and a smaller ϵ_1^2 produces smoother output (as it puts more weight on the auxiliary signal)—which is just as expected.

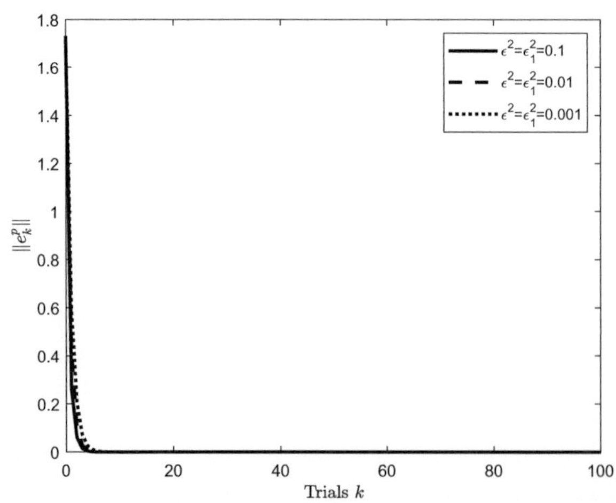

Fig. 6.21 Algorithm 6.3 (Switching Algorithm 1)—tracking error norm evolution for a range of ϵ^2, ϵ_1^2

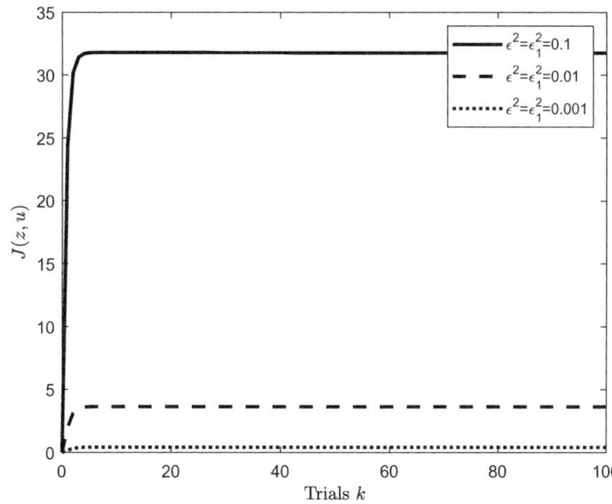

Fig. 6.22 Algorithm 6.3 (Switching Algorithm 1)—performance index evolution for a range of ϵ^2, ϵ_1^2

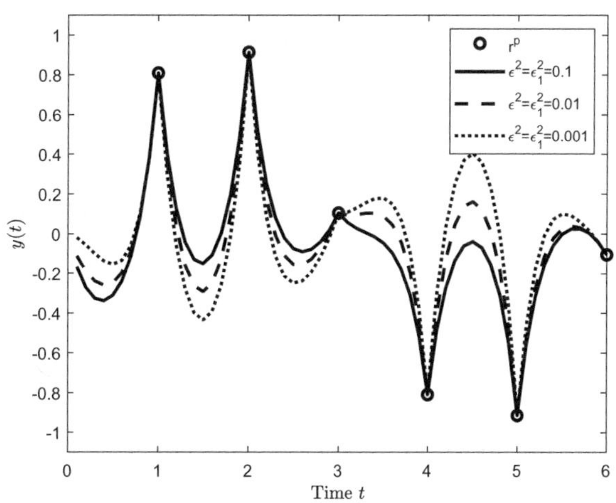

Fig. 6.23 Algorithm 6.3 (Switching Algorithm 1)—output signal on 100th trial for a range of ϵ^2, ϵ_1^2

6.3.4.2 Switching Algorithm 2 (Algorithm 6.4) Results

Algorithm 6.4 (Switching Algorithm 2) is now implemented. We first consider using a single step in the inner loop, i.e., $k_0 = 1$. The results are shown in Figs. 6.26, 6.27, 6.28, 6.29, and 6.30. As can be seen, even with $k_0 = 1$ the algorithm converges (to the solution of the auxiliary optimization problem) as the theory predicts, though the convergence rate is slower than that in Algorithm 6.3 (Switching Algorithm 1)—again not surprised.

We then investigate the effect of varying k_0. For a fair comparison we keep the weighting $\epsilon^2 = \epsilon_1^2 = 0.1$ fixed. The results are shown in Figs. 6.31, 6.32, 6.33, 6.34, and 6.35. As can be seen, increasing k_0 does not affect the final converged values, but improves the convergence speed, again, consistent with the theoretical predictions.

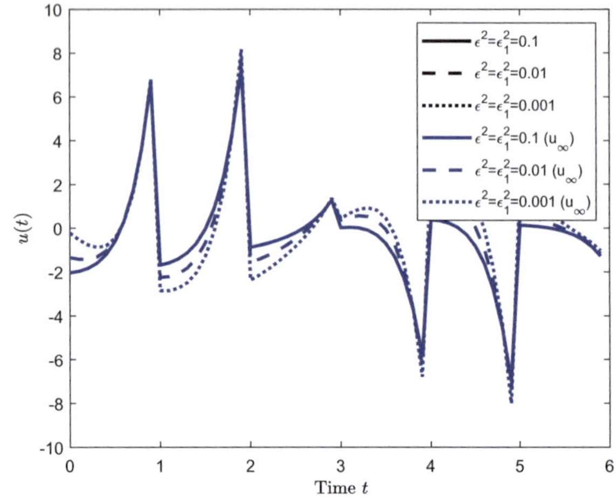

Fig. 6.24 Algorithm 6.3 (Switching Algorithm 1)—input signal on 100th trial for a range of ϵ^2, ϵ_1^2

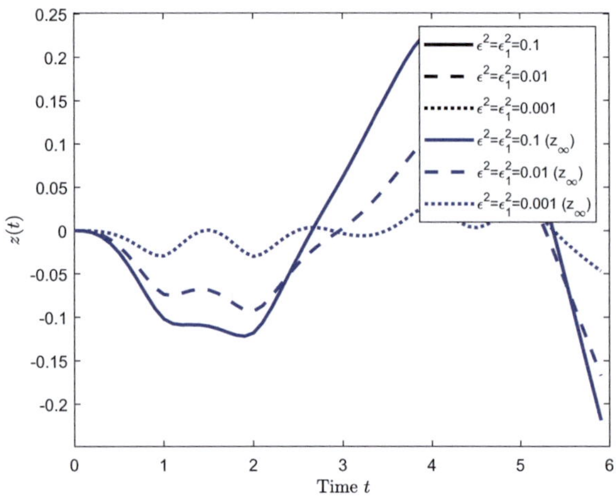

Fig. 6.25 Algorithm 6.3 (Switching Algorithm 1)—auxiliary signal on 100th trial for a range of ϵ^2, ϵ_1^2

6.4 Discussion and Further Reading

This chapter has examined norm optimal iterative learning control for linear systems with constraints, a situation that approximates to scenarios met often in practice. Both hard constraints that the control design must satisfy and soft constraints taking into account of some additional performance of interest are considered.

First, the constrained ILC problem with hard constraints is formulated into a successive projection framework. Then, based on this projection method, two algorithms are proposed to solve the constrained ILC problem. It has been shown that, when perfect tracking is possible, both algorithms can achieve perfect tracking whereas one algorithm needs much less computational effort (with some sacrifice in the convergence performance). When perfect tracking is not possible, both algorithms have

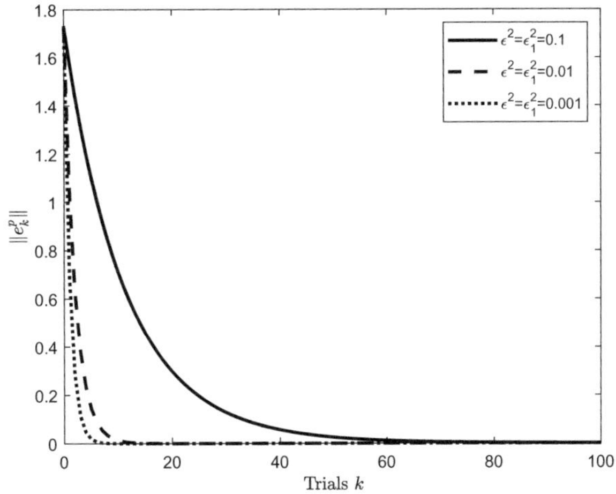

Fig. 6.26 Algorithm 6.4 (Switching Algorithm 2)—tracking error norm evolution for a range of ϵ^2, ϵ_1^2

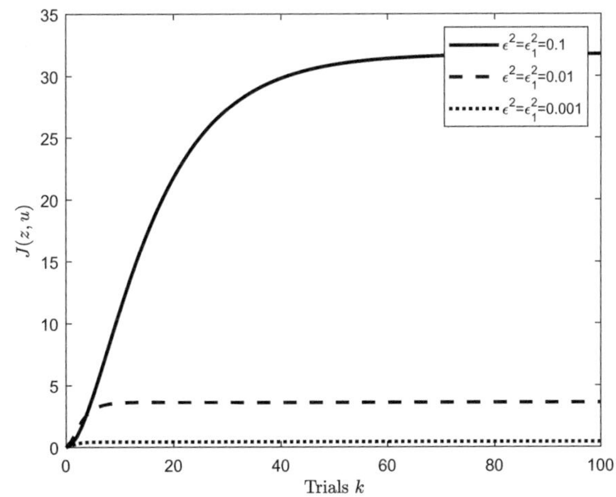

Fig. 6.27 Algorithm 6.4 (Switching Algorithm 2)—performance index evolution for a range of ϵ^2, ϵ_1^2

been shown to provide useful approximate solutions to the constrained ILC problem but that (1) the asymptotic error will be non-zero and (2) the computational complexity and convergence properties of the algorithms do differ. These observations should be taken into account when choosing the algorithm, which requires a compromise between the performance/accuracy and the computational cost. The effect of weightings on the performance of the algorithms has also been discussed and numerical simulations have been given to demonstrate their effectiveness.

After that, ILC design with soft constraints is considered. This is formulated into an auxiliary optimization problem that requires tracking of a specified finite dimensional reference whilst minimizing a quadratic objective function of input and a defined auxiliary variable. It is shown that this formulation can be used to solve the intermediate point tracking problem whilst simultaneously minimizing a general

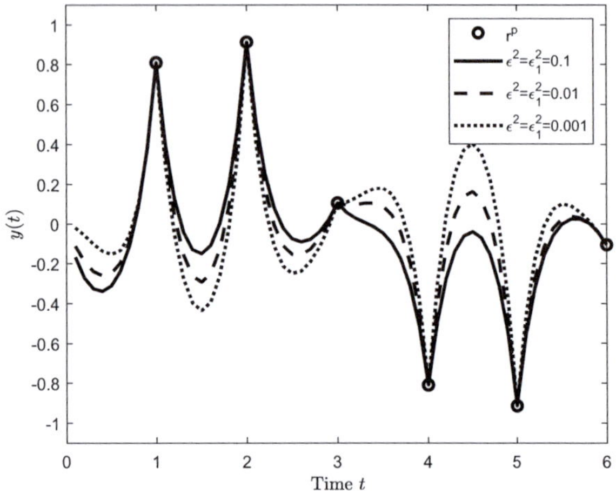

Fig. 6.28 Algorithm 6.4 (Switching Algorithm 2)—output signal on 100th trial for a range of ϵ^2, ϵ_1^2

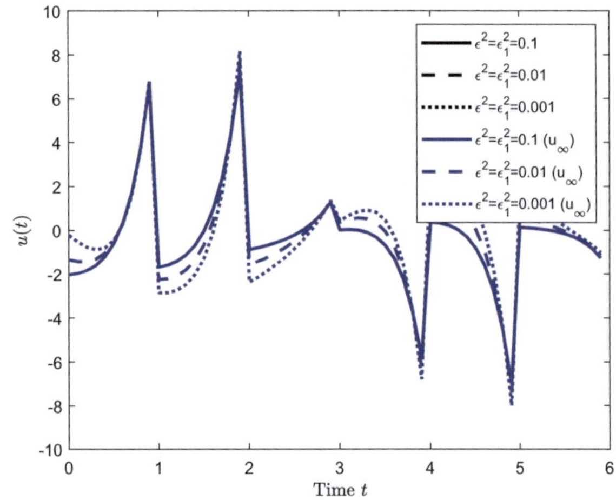

Fig. 6.29 Algorithm 6.4 (Switching Algorithm 2)—input signal on 100th trial for a range of ϵ^2, ϵ_1^2

class of objective function. A broad range of operations within fields such as industrial automation and robotics require such a framework to, for example, move payloads to preset positions during manufacturing tasks whilst reducing spillage, vibration, contact forces, and actuator wear. We derive solutions employing the norm optimal ILC framework, with implementation achieved using either a feedforward update structure or a combined feedback plus feedforward form which has the practically confirmed benefit of additional robustness to modelling uncertainty. The general analysis and algorithms proposed expressed in terms of operator notation have the advantage of efficiency and generality of proofs plus the great benefit that the results on performance and robustness are directly applicable to many system types including linear discrete-time, continuous-time, and time-varying systems. The algorithm structure is based on systematic switching between objectives and permits design

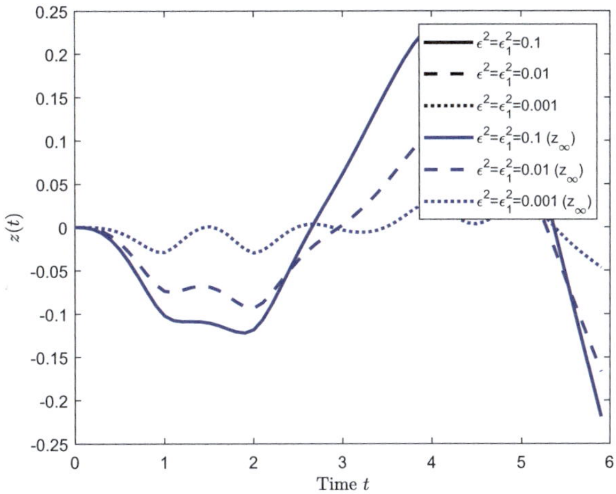

Fig. 6.30 Algorithm 6.4 (Switching Algorithm 2)—auxiliary signal on 100th trial for a range of ϵ^2, ϵ_1^2

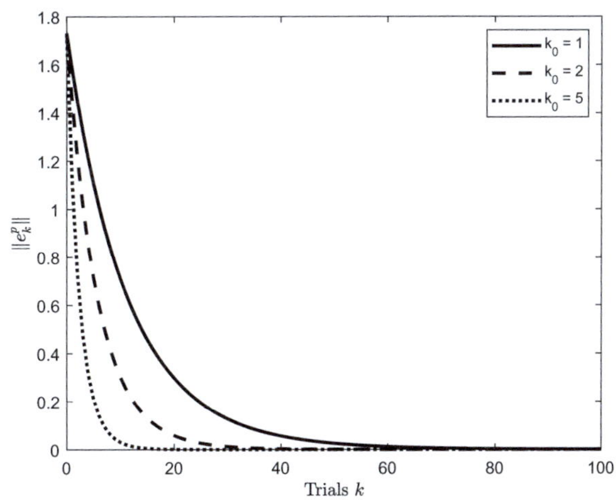

Fig. 6.31 Algorithm 6.4 (Switching Algorithm 2)—tracking error norm evolution for a range of k_0

flexibility in terms of choice of a number of parameters (e.g., Q, R, k_0) and the choice of whether off-line updates are applied to the experimental plant or updates are generated online. A rigorous robustness analysis has been performed which provides explicit uncertainty bounds which ensure convergence to a solution which precisely achieves the tracking objective despite the presence of plant-model mismatch. The various options have been illustrated using numerical simulations.

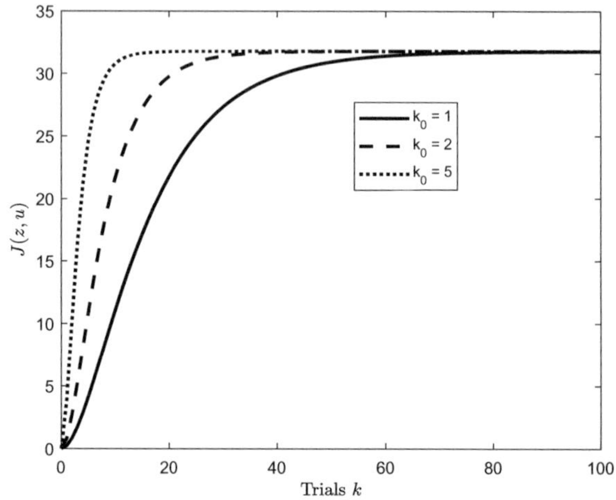

Fig. 6.32 Algorithm 6.4 (Switching Algorithm 2)—performance index evolution for a range of k_0

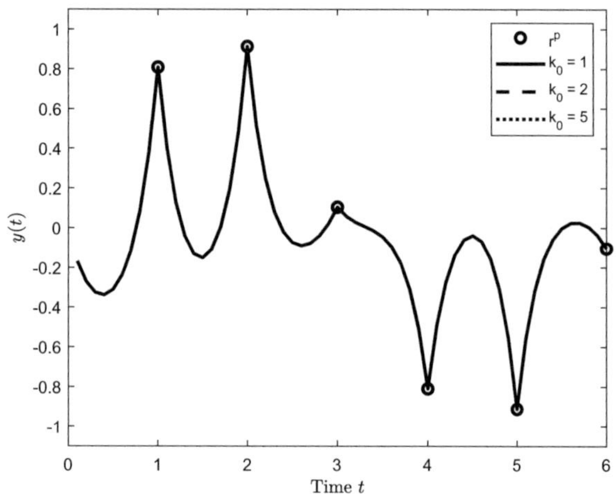

Fig. 6.33 Algorithm 6.4 (Switching Algorithm 2)—output signal on 100th trial for a range of k_0

The design methods introduced in this chapter are built on the NOILC framework that comes with appealing and well-defined properties (see, e.g., Amann et al. [1, 2], Owens et al. [24], and Owens [21]). It is worth mentioning that other methods, e.g., inverse-based ILC, can also be extended to include constraints (e.g., [25]). There are also other constrained design methods available. The readers are referred to the work, e.g., Chen and Peng [7], Gunnarsson and Norrlof [15], and Lee et al. [18] for more information.

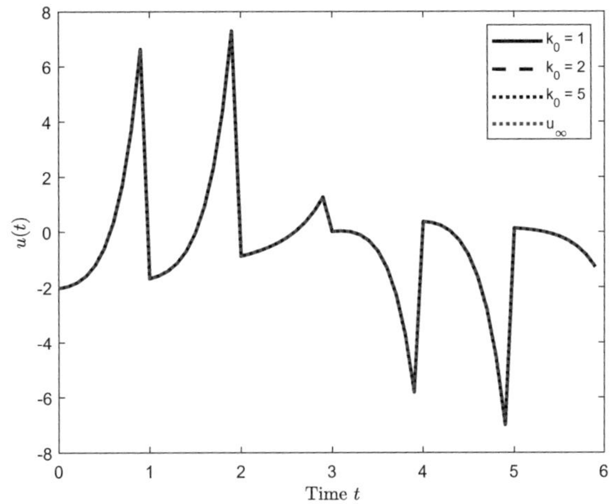

Fig. 6.34 Algorithm 6.4 (Switching Algorithm 2)—input signal on 100th trial for a range of k_0

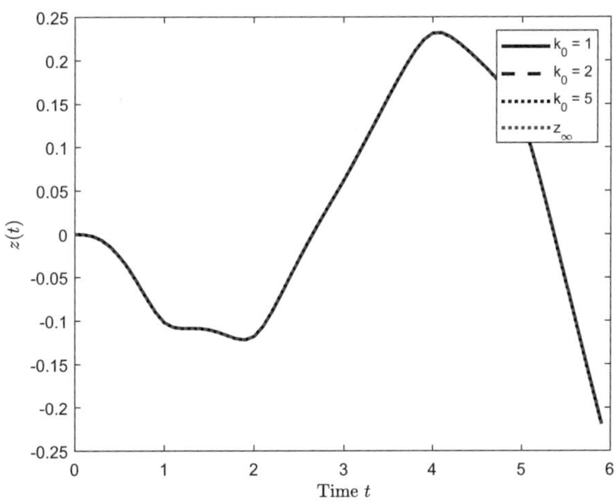

Fig. 6.35 Algorithm 6.4 (Switching Algorithm 2)—auxiliary signal on 100th trial for a range of k_0

References

1. Amann N, Owens DH, Rogers E (1996) Iterative learning control for discrete-time systems with exponential rate of convergence. IEE Proc Control Theory Appl 143(2):217–224
2. Amann N, Owens DH, Rogers E (1996) Iterative learning control using optimal feedback and feedforward actions. Int J Control 65(2):277–293
3. Bartlett RA, Biegler LT, Backstrom J, Gopal V (2002) Quadratic programming algorithms for large-scale model predictive control. J Process Control 12(7):775–795
4. Bauschke H, Borwein J (1996) On projection algorithms for solving convex feasibility problems. SIAM Rev 38(3):367–426
5. Bertsekas D (1976) On the Goldstein-Levitin-Polyak gradient projection method. IEEE Trans Autom Control 21(2):174–183
6. Bierlaire M, Toint P, Tuyttens D (1991) On iterative algorithms for linear least-squares problems with bound constraints. Linear Algebra Appl 143(1):111–143

7. Chen C, Peng S (1999) Learning control of process systems with hard input constraints. J Process Control 9(2):151–160
8. Chu B, Owens DH (2009) Accelerated norm-optimal iterative learning control algorithms using successive projection. Int J Control 82(8):1469–1484
9. Chu B, Owens DH (2010) Iterative learning control for constrained linear systems. Int J Control 83(7):1397–1413
10. Chu B, Freeman CT, Owens DH (2015) A novel design framework for point-to-point ILC using successive projection. IEEE Trans Control Syst Technol 23(3):1156–1163
11. Chu B, Rauh A, Aschemann H, Rogers E, Owens DH (2021) Constrained iterative learning control for linear time-varying systems with experimental validation on a high-speed rack feeder. IEEE Trans Control Syst Technol 30(5):1834–1846
12. Gill PE, Murray W, Saunders MA (2002) SNOPT: An SQP algorithm for large-scale constrained optimization. SIAM J Optim 12(4):979–1006
13. Goldstein A (1967) Constructive real analysis. Harper & Row, New York
14. Gubin L, Polyak B, Raik E (1967) The method of projections for finding the common point of convex sets. USSR Comput Math Math Phys 7:1–24
15. Gunnarsson S, Norrlof M (2001) On the design of ILC algorithms using optimization. Automatica 37(12):2011–2016
16. Halperin I (1962) The product of projection operators. Acta Scientiarum Mathematicarum 23:96–99
17. He BS (1992) A projection and contraction method for a class of linear complementarity-problems and its application in convex quadratic-programming. Appl Math Optim 25(3):247–262
18. Lee J, Lee K, Kim W (2000) Model-based iterative learning control with a quadratic criterion for time-varying linear systems. Automatica 36(5):641–657
19. Luenberger DG (1969) Optimization by vector space methods. Wiley, New York, Chichester
20. von Neumann J (1950) Functional operators, vol II. Princeton University Press
21. Owens DH (2015) Iterative learning control—an optimization paradigm. Springer
22. Owens DH, Jones RP (1978) Iterative solution of constrained differential/algebraic systems. Int J Control 27(6):957–974
23. Owens DH, Freeman CT, Van Dinh T (2012) Norm-optimal iterative learning control with intermediate point weighting: theory, algorithms, and experimental evaluation. IEEE Trans Control Syst Technol 21(3):999–1007
24. Owens DH, Freeman CT, Chu B (2013) Multivariable norm optimal iterative learning control with auxiliary optimization. Int J Control 86(6):1026–1045
25. Owens DH, Freeman CT, Chu B (2014) An inverse approach to intermediate point norm optimal iterative learning control with auxiliary optimization. Int J Control 87(8):1646–1671
26. Rossiter J (2003) Model-based predictive control: a practical approach. CRC Press

Chapter 7
Accelerating NOILC Convergence

This chapter will use a similar operator notation to previous chapters. In particular, the plant is described by a linear operator $G : \mathcal{U} \to \mathcal{Y}$ mapping a (Hilbert) space of inputs $\mathcal{U}$ into a (Hilbert) space $\mathcal{Y}$ of outputs with dynamics represented by

$$y = Gu + d \ . \tag{7.1}$$

For a given choice of parameters (more precisely, the parameters in the inner products in input and output spaces and the weight ϵ^2), Norm-Optimal Iterative Learning control (NOILC) has guaranteed convergence in the absence of modelling error. In the presence of a wide class of modelling errors, the algorithm is robust, producing a monotonically decreasing sequence of tracking error norms. As each iteration, notionally, needs a real-time experiment on the real system, the rate at which the measured error norm sequence $\{\|e_k\|_{\mathcal{Y}}\}_{k\geq 0}$ converges to the desired tracking accuracy are important practical concerns. In this chapter, algorithm development will assume that there is no modelling error, robustness being discussed after algorithms have been defined. It will also be assumed that:

1. The iteration index k will denote the index of the kth *online experiment/trial/iteration performed on the plant*. This is important as plant experiments/trials require much more investment of resources (financial, human, time, etc.) and may be subject to safety constraints. In contrast, offline, model-based calculations are much cheaper.
2. In addition, it is assumed that iterations will be terminated when the measured, online, error norm $\|e_k\|_{\mathcal{Y}}$ satisfies some measure of a desired tracking accuracy. This could take the form of a specified threshold value, iterations being terminated when the measured, online, error norm drops below this value. Convergence rates will then have been improved, in a practical sense, if the number of iterations needed before termination is reduced.

The number of variations implicit in the ideas presented in the chapter are numerous and it is not feasible to address every variant of every idea. This will become self-

B. Chu and D. H. Owens, *Optimal Iterative Learning Control*, Advances in Industrial Control, https://doi.org/10.1007/978-3-031-80236-2_7

evident to the reader who is invited to explore variants of his or her choice/creation as an exercise or research program.

Previous chapters have already noted that convergence rates of NOILC increase as the weight parameter ϵ^2 is reduced. However, the freedom to vary this parameter is, typically, limited. This is because it has already been seen that the robustness of the algorithm decreases as ϵ^2 is reduced and that, for systems described by state-space models, small values of this parameter will tend to generate state feedback realizations with high feedback gains. High gains have a number of other disadvantages amongst which is an increased sensitivity to noise. For these reasons, decreasing the value of ϵ^2 is not, in practice, necessarily a satisfactory approach to achieving significantly improved convergence rates. In what follows, it is assumed that a value is chosen, and further improvements will be sought by other methods of accelerating convergence.

Given the above, alternative approaches must be sought. This chapter describes and explores a number of options including:

1. the incorporation of pre-compensators into the system to produce conditioned dynamics that improve the ability of NOILC to converge more rapidly and/or
2. the use of extrapolation methods that, given a computed new input u_{k+1} on iteration $k+1$, construct a replacement that reduces the tracking error norm on implementation.
3. Starting with initial plant input and output data u_0, y_0, a simple approach to acceleration, following any plant experiment k, is to perform a number, $M_{k+1} - 1 \geq 0$ say, of offline, model-based iterations followed by a plant iteration, or
4. modifying that idea to reduce the offline computations by generating a changed $(k+1)$th NOILC iteration on the plant where that reference signal r is replaced, temporarily, by a modified version. It will be shown that suitable choices of an additional parameter introduce the flexibility to eliminate, or substantially reduce, selected eigenvector components in the error signal.

Remark 7.1 Finally, as improving convergence rates by reducing ϵ^2 tends to reduce the robustness of the NOILC algorithm, it is anticipated that all methods of improving convergence will have the same effect.

7.1 Feedback, Feedforward Compensation and Acceleration

A necessary practical condition for safe iteration is the stability of the plant G. If a process G_0, with output y and input v, is unstable or has difficult, undesirable dynamics (for example, oscillation) then automatic control systems will need to be added for stabilization and to satisfy other classical performance objectives. The resultant composite system of G_0 with its controller becomes the plant G responding to a new input u. It is the process "seen" by the Iterative Learning Control (ILC) algorithm. By design, G will be an acceptably "well-behaved" dynamical system from the user's standpoint.

The embedded control system could take any chosen form, including state feedback and/or output feedback and feedforward control. For example, if the controller for G_0 is a unity negative feedback plus feedforward controller (as shown in Fig. 7.1), then the closed-loop system has the operator form,

$$\begin{aligned} v &= Ku + K_0(u - y)\,, \quad \textit{so that} \\ G &= (I + G_0 K_0)^{-1} G_0 (K + K_0) \\ &= I - (I + G_0 K_0)^{-1}(I - G_0 K)\,, \end{aligned} \tag{7.2}$$

where K_0 is the feedback compensator and K is the feedforward compensator. Note that the new input u to G is just the reference signal for this closed-loop system.

Classical design considerations then suggest the following observations that influence the Iterative Control problem for G:

1. If $K = 0$, the design of K_0 will tend to ensure that the "error" $y - u$ is small in some defined low frequency band and it might be expected that G (and hence G^*) is, in some sense, close to the identity.
2. Suppose that $K_0 = 0$, then the remaining feedforward controller K will provide good tracking if
    a. it is close to the inverse system G_0^{-1} over a wide frequency band. Again, G is then close to the identity.
    b. Alternatively, K could be chosen to approximately cancel or compensate for some of the dynamics of G_0. For example, reducing phase lag effects in G_0.
3. If both K and K_0 are non-zero, the benefits described above are combined.

The general conclusion is intuitive but compelling, namely that Iterative Control of a system G incorporating embedded controllers could be much more effective than Iterative Control of the original system G_0. The benefits should manifest themselves in ease of operation, increased safety and more rapid convergence.

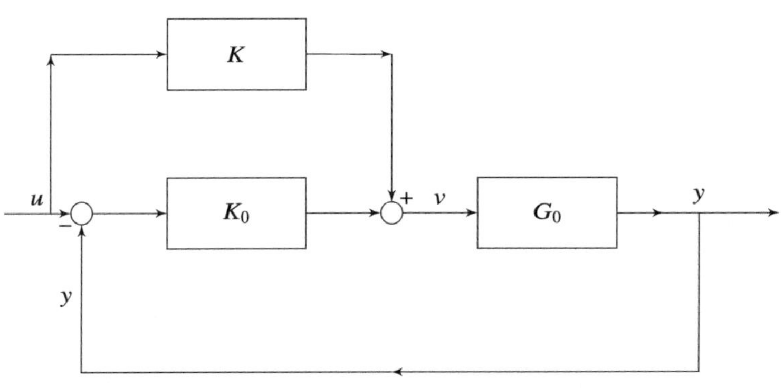

Fig. 7.1 An embedded control system structure

7.2 Extrapolation Methods

Embedding control systems within G is always a good design decision but the effect on Iterative Control convergence rates may be limited by the need for the embedded control systems to be acceptable when the plant is operating in a normal rather than an iterative mode. More direct changes to the convergence rates can be achieved using the details of the iterative algorithm used.

Consider the feedforward implementation of the NOILC algorithm and the consequent input update equation

$$u_{k+1} = u_k + \epsilon^{-2} G^* L e_k \ , \quad L = (I + \epsilon^{-2} G G^*)^{-1} \ . \tag{7.3}$$

This input minimizes $J(u)$ but not the error norm $\|e_{k+1}\|^2_{\mathcal{Y}}$. Additional flexibility in input choice can be achieved by replacing this expression by the extrapolated form,

$$u_{k+1} = u_k + \lambda_{k+1} \epsilon^{-2} G^* L e_k \ , \tag{7.4}$$

where the scalar *extrapolation factor* $\lambda_{k+1} > 0$ is to be chosen below.

7.2.1 Extrapolation and Feedforward NOILC

Error reductions on iteration $k+1$ will be increased if $\|e_{k+1}\|^2_{\mathcal{Y}}$ is minimized. Write, after a little algebra, the error evolution in the form

$$\underbrace{e_{k+1} = (I - \lambda_{k+1} \epsilon^{-2} G G^* L) e_k}_{error\ evolution} = (I - \lambda_{k+1}(I - L))\, e_k \ , \tag{7.5}$$

which is identical to that obtained for feedforward NOILC except that the multiplier ϵ^{-2} is replaced by $\lambda_{k+1}\epsilon^{-2}$.

Note that:

1. $(I - L)e_k = e_k - \tilde{e}_{k+1}$ where $\tilde{e}_{k+1} = L e_k$ is the predicted error, computed offline, using the un-extrapolated input (7.3). The only data needed for this computation is the measured error e_k.
2. Note also that, when $\lambda_{k+1} = 1$, the extrapolated algorithm is identical to NOILC.

A simple calculation then gives

$$\|e_{k+1}\|_{\mathcal{Y}}^2 = \|e_k\|_{\mathcal{Y}}^2 - 2\lambda_{k+1}\langle e_k, e_k - \tilde{e}_{k+1}\rangle_{\mathcal{Y}} + \lambda_{k+1}^2\|e_k - \tilde{e}_{k+1}\|_{\mathcal{Y}}^2 . \tag{7.6}$$

which is a quadratic relationship in λ_{k+1}. Minimizing yields the value

$$\underbrace{\lambda_{k+1,best} = \frac{\langle e_k, e_k - \tilde{e}_{k+1}\rangle_{\mathcal{Y}}}{\|e_k - \tilde{e}_{k+1}\|_{\mathcal{Y}}^2} \geq 1}_{optimal\ extrapolation} , \tag{7.7}$$

as $L = L^*$ and $0 \leq (I - L)^2 \leq I - L < I$.

Using $\lambda_{k+1,best}$ produces the best reduction in error norm $\|\cdot\|_{\mathcal{Y}}$ and certainly improves on the error that would be achieved using the un-extrapolated algorithm. It is not essential that this value is used however, as any value of λ_{k+1} in the range

$$1 < \lambda_{k+1} \leq \lambda_{k+1,best} , \tag{7.8}$$

will also reduce the error.

7.2.2 Extrapolation and Robustness

Finally, extrapolation does, indeed, influence the robustness of the algorithm. The case of right multiplicative errors U and $\lambda_{k+1} = 1$ was considered in Chap. 5 describing NOILC and its properties. The error dynamics now becomes

$$\underbrace{e_{k+1} = (I - \lambda_{k+1}\epsilon^{-2}GUG^*L)e_k}_{error\ evolution\ seen\ on-line} . \tag{7.9}$$

An examination of the error evolution shows that the factor GUG^*L remains unchanged but that the multiplier ϵ^{-2} is replaced by $\lambda_{k+1}\epsilon^{-2}$. The robustness analysis described in the NOILC chapter requires guaranteed reductions in a modified error norm $\|e_k\|_0$ at every iteration. Using the techniques of that chapter then yields the robustness criterion expressed as the operator inequality

$$U + U^* > \lambda_{k+1}\beta U^*U , \quad \beta = \frac{\epsilon^{-2}\|G\|^2}{1 + \epsilon^{-2}\|G\|^2} , \tag{7.10}$$

which is the condition derived for NOILC with the multiplier β replaced by $\lambda_{k+1}\beta$.

The robustness condition (7.10) suggests that any increase in $\lambda_{k+1}\beta$ will reduce the robustness of the algorithm in the sense that the set of U satisfying the condition will become smaller. Although only a sufficient condition, just as increasing β (i.e., reducing ϵ^{-2}) reduces the predicted robustness of NOILC, (7.10) suggests that using $\lambda_{k+1} > 1$ will also reduce the robustness of the algorithm. There is therefore a compromise to be reached between retaining the basic robustness of NOILC and achieving increased algorithm convergence rates using extrapolation.

Finally, the theory predicts a degree of robustness and describes a class of plant dynamics GU for which robust monotonicity is retained. This robustness depends very clearly on parameters, particularly the choices made for λ_{k+1} and ϵ^2. Insight can be obtained into the issues by considering parameter values which the theory predicts will ensure that the both the plant GU and the model G (i.e., $U = I$) possesses the robust monotonicity property. This is not a necessary condition arising from the theory but it will reveal a need to limit the range of λ_{k+1} in practice.

Setting $U = I$ in (7.10) requires that $\lambda_{k+1}\beta < 2$. That is,

$$\lambda_{k+1}\frac{\epsilon^{-2}\|G\|^2}{1+\epsilon^{-2}\|G\|^2} < 2\,. \tag{7.11}$$

This is satisfied for all values of ϵ^2 if, and only if, $\lambda_{k+1} \leq 2$. For values of $\lambda_{k+1} > 2$, ϵ^2 will be limited in magnitude. For example, for, $\lambda_{k+1} > 2$, the condition reduces to $\epsilon^2 > \frac{1}{2}(\lambda_{k+1} - 2)\|G\|^2$. For a given value of ϵ^2, this limits the range of λ_{k+1}. However, it is satisfied for all ϵ^2 if $\lambda_{k+1} \leq 2$. Intuitively, it may be advisable to limit the choice of all λ_{k+1} by adding the requirement that it lies in a user specified range $1 \leq \lambda_{k+1} \leq \lambda_{max}$, with $\lambda_{max} \leq 2$?

Example 7.1 Consider the following system

$$G(s) = \frac{2s+3}{s^2+3s+3}, \tag{7.12}$$

which is sampled using zero-order hold and a sampling time of 0.1 s. The trial length is 20 s and the reference signal is generated by the square-wave input shown in Fig. 7.2. The initial input is chosen to be $u_0 = 0$. The simulation compares the performance of NOILC and the accelerated algorithm using extrapolation over 50 iterations. The weighting matrices are again chosen to be $Q = R = I$, $\epsilon^2 = 1$ for simplicity. In this simulation, the optimal extrapolation parameter $\lambda_k = \lambda_{k,best}$ is used. The results are summarized in Figs. 7.3, 7.4 and 7.5.

From Figs. 7.3a and 7.5, it can be seen that the accelerated algorithm outperforms NOILC algorithm over 50 iterations with $\lambda_k = \lambda_{k,best}$. The improvement is very significant. Firstly, from the figure, it can be seen that the convergence rate of the accelerated algorithm (error norm reduction) in the first few iterations is much faster than NOILC, which is of great value in practice. The norm of the tracking error in the first eight iterations is shown in Table 7.1 to add a more detailed picture. Secondly, note that the initial benefits of acceleration are maintained with the final error at 50th iteration with the choice of $\lambda_k = \lambda_{k,best}$ being almost 10% that achieved by NOILC. The fact that this benefit continues to accrue is illustrated in Fig. 7.3b where results up to 200 iterations are given. The accelerated algorithm outperforms NOILC at 200th iteration by a factor of almost 10^3.

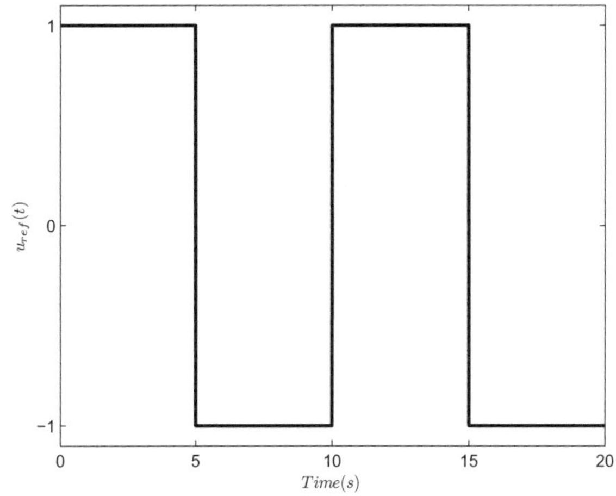

Fig. 7.2 The input signal $u^*(t)$

Finally, it is worth noting that convergence is excellent despite the discontinuous limit $u^*(t)$. The discontinuous limit explains the form of $e_{50}(t)$ which is very small away from the discontinuity but has localized "activity" in the vicinity of the "jumps" in u^*.

7.3 Offline Plus Online Iterations

This section recognizes the different status of online experiments and offline, model-based iterations in providing acceleration in the sense of reducing the number of online iterations required to achieve a desired accuracy. It does this by exploring the idea of reducing the number of online, experimental iterations by adding virtual (inner) iterations that are undertaken by offline, model-based computational methods. Any error norm reducing algorithm could be used for these offline iterations provided that they are designed to reduce the error norm. In contrast, the experimental, online (outer) iterations on the plant are assumed to be based on NOILC.

7.3.1 Formal Statement of the Approach

Although, in principle, the offline, "virtual" iterations could use any ILC algorithm, in what follows NOILC will be used for both online and offline iterations. More precisely, suppose that, following iteration k on the plant:

Fig. 7.3 Comparison of convergence for Example 7.1

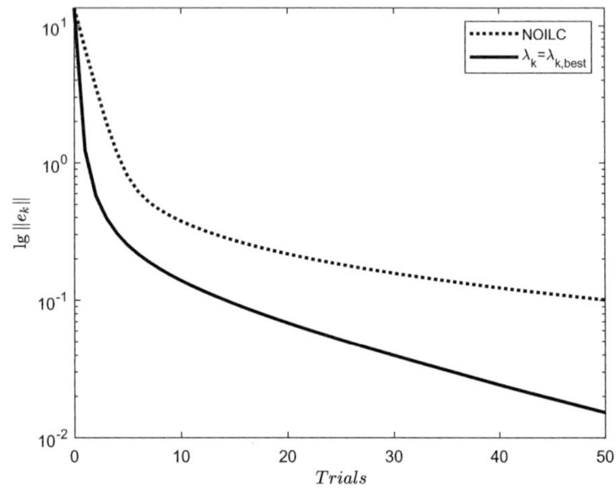

(a) Simulation Results over 50 Iterations

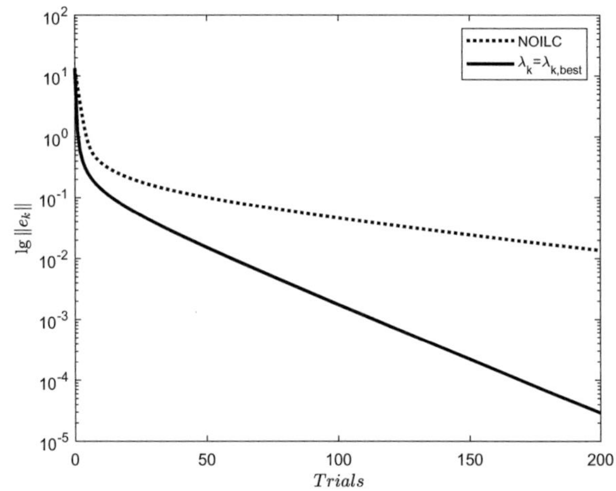

(b) Simulation Results over 200 Iterations

1. The experimental data u_k, y_k are available from an experimental iteration undertaken on the plant.
2. Choose a number, $M_{k+1} - 1 \geq 0$, of offline, model-based, computational iterations, and set $u_{k+1}^{(0)} = u_k$. Then use offline, NOILC computations to construct a sequence of inputs $u_{k+1}^{(j)}$, $1 \leq j \leq M_{k+1} - 1$ as the inputs minimizing the objective function

$$J(u, u_{k+1}^{(j-1)}) = \|e\|_{\mathcal{Y}}^2 + \epsilon^2 \|u - u_{k+1}^{(j-1)}\|_{\mathcal{U}}^2 , \quad 1 \leq j \leq M_{k+1} - 1 ,$$
$$e = r - y \quad , \quad \& \ y = Gu + d \quad .$$

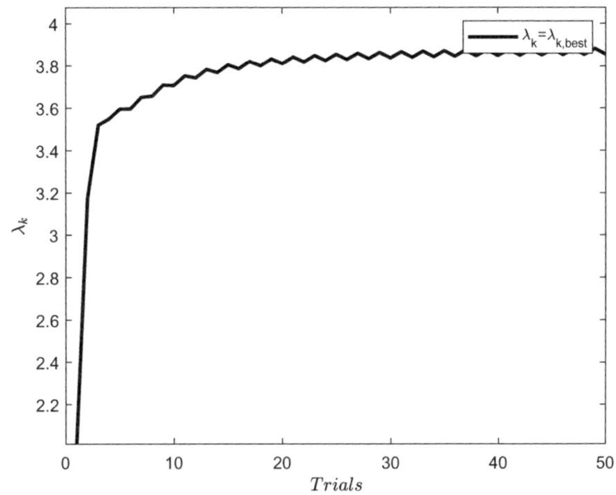

Fig. 7.4 Acceleration gain values for Example 7.1

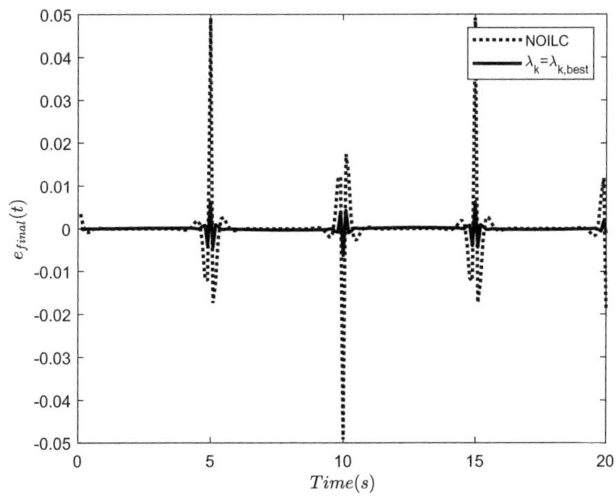

Fig. 7.5 The final tracking error $e_{50}(t)$ for Example 7.1

This yields the data $e_{k+1}^{(j)} = r - y_{k+1}^{(j)}$ and $y_{k+1}^{(j)} = Gu_{k+1}^{(j)} + d$ and, setting $e_{k+1}^{(0)} = e_k$, the familiar relationship

$$e_{k+1}^{(j)} = Le_{k+1}^{(j-1)}, \quad j \geq 1. \tag{7.13}$$

Note also that $e_{k+1}^{(M_{k+1}-1)} = L^{(M_{k+1}-1)} e_k$.

3. Given the computed input $u_{k+1}^{(M_{k+1}-1)}$, undertake an online experimental iteration using a controller $u_{k+1}^{(M_{k+1})}$ that minimizes the objective function $J(u, u_{k+1}^{(M_{k+1}-1)})$. Note that $e_{k+1}^{(M_{k+1})} = Le_{k+1}^{(M_{k+1}-1)}$.
4. Set $u_{k+1} = u_{k+1}^{(M_{k+1})}$ and $e_{k+1} = e_{k+1}^{(M_{k+1})}$, increase k by one and go to the next offline and online iteration sequence.

Table 7.1 Tracking error norms for Example 7.1

Iteration (k)	NOILC	$\lambda_k = \lambda_{k,best}$
0	13.5265	13.5265
1	6.8927	1.2319
2	3.5956	0.5794
3	1.9689	0.3938
4	1.1797	0.3050
5	0.8031	0.2521
6	0.6188	0.2162
7	0.5186	0.1898
8	0.4554	0.1693

Clearly,

$$\|e_{k+1}\|_{\mathcal{Y}} \leq \underbrace{\cdots \leq \|e_{k+1}^{(j+1)}\|_{\mathcal{Y}} \leq \|e_{k+1}^{(j)}\|_{\mathcal{Y}} \leq \cdots}_{off-line\ computational\ iterations} \leq \|e_k\|_{\mathcal{Y}} \tag{7.14}$$

so that the experimentally observed, online error norm sequence $\{\|e_k\|_{\mathcal{Y}}\}_{k\geq 0}$ converges to zero more rapidly than the case of $M_k = 1\ ,\ k \geq 1$, i.e., the normal NOILC algorithm.

7.3.2 The Case of $M_k = M,\ k \geq 0$

Assuming that the value of ϵ^2 used throughout the iteration sequence remains unchanged then the error evolution is described by

$$e_{k+1} = L^{M_{k+1}} e_k\ , \quad i.e., \quad e_k = L^{\sum_{j=1}^{k} M_j} e_0\ , \qquad k \geq 1\ , \tag{7.15}$$

or, if $M_k = M \geq 1$, for all iteration indices k, then

$$e_{k+1} = L^M e_k\ , \quad i.e., \quad e_k = L^{kM} e_0\ , \qquad k \geq 0\ . \tag{7.16}$$

This expression has the same form as the error evolution for NOILC with L replaced by L^M. The consequences of this observation and the facts that

$$L^M \leq L \qquad \& \qquad \mathcal{R}(I - L^M) = \mathcal{R}(I - L) \tag{7.17}$$

are that the analysis of the error and input convergence of the algorithm is identical to that of NOILC and has the same conclusions.

7.3.3 Robustness Characterization

This section assumes that $M_{k+1} = M$ is iteration dependent. The analysis of the robustness of the approach using a feedforward implementation of NOILC in all experimental plant iterations can be approached using similar techniques to those used for NOILC. The feedforward description of the offline computed inputs is

$$u_{k+1}^{(j)} = u_{k+1}^{(j-1)} + \epsilon^{-2} G^* e_{k+1}^{(j)} = u_{k+1}^{(j-1)} + \epsilon^{-2} G^* L^j e_k \ , \quad 1 \le j \le M \tag{7.18}$$

so that

$$u_{k+1} = u_k + \epsilon^{-2} G^* (L + L^2 + \cdots L^M) e_k \ . \tag{7.19}$$

In the presence of a right multiplicative modelling error U, $G \mapsto GU$ and the experimental errors observed in online experiments are related by the expression

$$\begin{aligned} e_{k+1} = e_k - \epsilon^{-2} G U G^* \Phi(L) e_k = (I - \epsilon^{-2} G U G^* \Phi(L)) e_k \ , \\ where \quad \Phi(L) = L + L^2 + \cdots + L^M \ , \end{aligned} \tag{7.20}$$

which is identical to that derived for the NOILC algorithm with the replacement $L \mapsto \Phi(L)$.

As in the analysis of the robustness of NOILC, the norm used must be changed. This is done by using the equivalent topology in $\mathcal{Y}$ defined by the inner product $\langle y, z \rangle_0 = \langle y, \Phi(L) z \rangle_{\mathcal{Y}}$ and the associated norm $\|e\|_0 = \sqrt{\langle e, e \rangle_0}$. The robust monotonic reduction of the norm sequence $\{\|e_k\|_0\}_{k \ge 0}$ is then guaranteed if (a sufficient condition)

$$\underbrace{U + U^* > \epsilon^{-2} U^* G^* \Phi(L) G U \quad on\ the\ range\ space\ \mathcal{R}[G^* \Phi(L)]}_{A\ Condition\ for\ Robust\ Monotonicity} \ . \tag{7.21}$$

Two further simplifications can now be introduced:

1. The range $\mathcal{R}[G^* \Phi(L)]$ is not, necessarily, an easy object to characterize so a more conservative but more useful result replaces it by the whole space $\mathcal{U}$.
2. Also the presence of $\Phi(L)$ in (7.21) makes checking a complex problem, but, as in the NOILC analysis, it can be removed. This can be done by noting that, $LG = GL_0$ where $L_0 = (I + \epsilon^{-2} G^* G)^{-1}$ so that, for any $j \ge 1$,

$$\begin{aligned} \epsilon^{-2} G^* L^j G = \epsilon^{-2} G^* G L_0^j = (I - L_0) L_0^{j-1} \ , \quad and\ hence \ , \\ \epsilon^{-2} G^* \Phi(L) G = (I - L_0) \left(I + L_0 + \cdots + L_0^{M-1} \right) = \left(I - L_0^M \right) \ . \end{aligned}$$

   Using the inequality

$$L_0^M \ge \left(\frac{1}{1 + \epsilon^{-2} \|G\|^2)} \right)^M I \ , \tag{7.22}$$

then implies that robustness is achieved for all U satisfying

$$\underbrace{U + U^* > \beta(M) U^* U \quad on \ \mathcal{U}}_{A \ Simplified \ Robustness \ Criterion} , \quad where$$
$$0 < \beta(M) = 1 - \left(\frac{1}{1+\epsilon^{-2}\|G\|^2} \right)^M < 1 . \tag{7.23}$$

In simplified terms, analysis of robustness can hence be approached using identical techniques to those developed for NOILC but with the parameter β replaced by $\beta(M)$. In particular:

a. A sufficient condition for robust monotonicity of $\{\|e_k\|_0\}_{k \geq 0}$ is that the operator norm condition
$$\left\| \frac{1}{\beta(M)} I - U \right\| < \frac{1}{\beta(M)} \tag{7.24}$$
is satisfied.
b. Also, as M increases and/or ϵ^2 decreases, the parameter $\beta(M) \to 1-$ and the robustness of the algorithm is predicted to reduce, reinforcing the speculation that all mechanisms for accelerating convergence will reduce the robustness of the NOILC algorithms.
c. However, any error U satisfying $\|I - U\| < 1$ satisfies the robustness condition for any $\beta(M)$, and hence any M and ϵ^2. Robust monotonic convergence is therefore tolerant to all such modelling errors.

In summary, the use of offline iterations does, indeed, accelerate the algorithm but, for a given value of ϵ^2, reduced robustness can be expected as M increases. Despite this, there is a substantial set of modelling errors U that can always be tolerated.

7.4 Error Corrected NOILC Algorithm(s)

Using the concepts introduced in Sect. 7.3, this section explores a special case of using different algorithms for offline virtual iterations whilst retaining NOILC for the online, experimental plant iteration. It then combines the two steps into a single, online algorithm that eliminates the need for the offline computations.

More precisely, it considers the case of accelerating the iterative algorithm using a combination of offline Gradient and online NOILC approaches by (the main outcome) *combining the two steps into a single, equivalent, online NOILC iteration that uses a modified, error data dependent, reference signal,*

$$\underbrace{r_{k+1} = r + \epsilon^2 \beta_{k+1} e_k}_{Error \ Corrected \ Reference} . \tag{7.25}$$

The approach is based on the underlying algorithm where given experimental data u_k, e_k from the plant on iteration k, a two step iteration $k+1$ has the form:

1. an *offline* model-based, (inner) gradient iteration using the data u_k and e_k and a "gain" β_{k+1} to generate an input signal $u_{k+1}^{(1)}$ and an error signal $e_{k+1}^{(1)}$, followed by
2. the use of $u_{k+1}^{(1)}$ in a feedback or feedforward Norm Optimal (outer) Iterative Learning Control, *online* and application of the resultant control input u_{k+1} to the plant, recording the observed tracking error e_{k+1}.

As both gradient ILC and NOILC generate error norm reductions, the resultant two-step algorithm will produce an online, experimental error norm sequence that converges more rapidly than the application of NOILC alone. The following sections add more detail to this idea and observation, and demonstrate that the two steps can be combined into one online, NOILC iteration.

7.4.1 The Underpinning Algorithm

The algorithm is described assuming that a model is available for the plant and that there is no modelling error.

7.4.1.1 The Offline Gradient Calculations

Expressed in terms of the notation used above, an offline gradient calculation (Chap. 4) would use the adjoint G^* of the operator G in the system model $y = Gu + d$ to compute

$$u_{k+1}^{(1)} = u_k + \beta_{k+1} G^* e_k \,, \quad \textit{so that}$$

$$e_{k+1}^{(1)} = (I - \beta_{k+1} GG^*)\, e_k \; = L_G(\beta_{k+1}) e_k,$$

$$\textit{where the operator} \quad L_G(\beta) : \mathcal{Y} \to \mathcal{Y}\,, \quad L_G(\beta) = I - \beta GG^* \,.$$

For state-space systems, both G and G^* can be represented by state-space models so these calculations are simulations with appropriate end and/or intermediate point conditions.

The gradient step is predicted to produce a reduction in the error norm $\|e_{k+1}^{(1)}\|_{\mathcal{Y}} \leq \|e_k\|_{\mathcal{Y}}$ for any choice of β_{k+1} in the range

$$0 < \beta_{k+1} < 2\|G\|^{-2} \,, \tag{7.26}$$

with no guaranteed reduction (but certainly no increase) in norm for the choices $\beta_{k+1} = 0$ or $\beta_{k+1} = 2\|G\|^{-2}$.

7.4.1.2 The NOILC Calculations for Online Use

The online NOILC step (Chap. 5) can either be in a feedback or feedforward form. In both cases the input u_{k+1} is obtained, using $u_{k+1}^{(1)}$ and the reference signal r, as the input minimizing the objective function

$$J(u, u_{k+1}^{(1)}) = \|e\|_{\mathcal{Y}}^2 + \epsilon^2 \|u - u_{k+1}^{(1)}\|_{\mathcal{U}}^2 , \quad e = r - y , \tag{7.27}$$

subject to the plant dynamics $y = Gu + d$. The solution has the formal representation

$$\begin{aligned} u_{k+1} &= u_{k+1}^{(1)} + \epsilon^{-2} G^* e_{k+1} \quad \textit{so that} \\ e_{k+1} &= L e_{k+1}^{(1)} , \quad L = (I + \epsilon^{-2} G G^*)^{-1} . \end{aligned} \tag{7.28}$$

The known properties of the NOILC algorithm ensure the further reduction $\|e_{k+1}\|_{\mathcal{Y}} < \|e_{k+1}^{(1)}\|_{\mathcal{Y}}$.

7.4.2 *Error Corrected Reference Algorithm*

The *Error Corrected Algorithm* is deduced from the above.

7.4.2.1 Algorithm Structure

The gradient step described could be undertaken offline in the manner assumed above. Alternatively, it can be eliminated by incorporating it into the NOILC step using the following observation. More precisely, combining the two input update equations,

$$\underbrace{u_{k+1}^{(1)} = u_k + \beta_{k+1} G^* e_k}_{\textit{Gradient Input Update}} \quad \& \quad \underbrace{u_{k+1} = u_{k+1}^{(1)} + \epsilon^{-2} G^* e_{k+1}}_{\textit{NOILC Input Update}} \tag{7.29}$$

gives the single expression

$$\underbrace{u_{k+1} = u_k + \epsilon^{-2} G^* \left(\underbrace{\epsilon^2 \beta_{k+1} e_k + e_{k+1}}_{} \right)}_{\textit{Gradient and NOILC update combined}} . \tag{7.30}$$

Note that the underbraced modified error signal $\epsilon^2 \beta_{k+1} e_k + e_{k+1}$ is the tracking error that would be computed from a NOILC problem where the output y_{k+1} is attempting to track the modified reference $r + \epsilon^2 \beta_{k+1} e_k$. It must be concluded that the offline, gradient calculation is unnecessary as it can be incorporated into an online, NOILC iteration by replacing the reference r by the "*Error Corrected Reference*",

$$\underbrace{r_{k+1} = r + \epsilon^2 \beta_{k+1} e_k}_{Error\ Corrected\ Reference} \ , \quad k \geq 0 \ , \tag{7.31}$$

and using the input u_{k+1} that minimizes the modified objective function

$$\underbrace{J_{k+1}(u, u_k) = \|r_{k+1} - y\|^2_{\mathcal{Y}} + \epsilon^2 \|u - u_k\|^2_{\mathcal{U}}}_{Modified\ NOILC\ Objective\ Function} \ , \tag{7.32}$$

subject to the plant dynamics $y = Gu + d$. That is,

$$u_{k+1} = argmin\{\ J_{k+1}(u, u_k)\ :\ y = Gu + d\ \} \ , \tag{7.33}$$

the minimization of which gives rise to the desired control signal (7.30).

In effect, the acceleration benefits of the combined gradient plus NOILC sequence of iterations will be obtained from a sequence of (gradient-free) NOILC iterations with error corrected references computed using (7.31). The replacement of r by r_{k+1} in the iteration does not change the structure of the computational procedure. It simply changes the data used in that computation. As the change is simply to a signal r, many calculations, such as the solution of Riccati equations are unaffected.

7.4.2.2 Error Convergence Analysis

The expressions defining the underlying two-step algorithm are $e_{k+1} = Le^{(1)}_{k+1}$ and $e^{(1)}_{k+1} = L_G(\beta_{k+1}) e_k$ so that the evolution errors in the *Error Corrected NOILC Algorithm* are described by

$$\underbrace{e_{k+1} = L_e(\beta_{k+1}) e_k, \ \ k \geq 0}_{Error Evolution} \ , \ \ with$$
$$L_e(\beta) = L L_G(\beta) = (I + \epsilon^{-2} G G^*)^{-1} \left(I - \beta G G^* \right) \ . \tag{7.34}$$

Error convergence analysis starts by noting that $L_e = L_e^*$ is self-adjoint and, if each of the parameters $\{\beta_k\}_{k \geq 0}$ are either zero or satisfy (7.26), error convergence can be anticipated. The iteration dependence on β_k complicates the analysis so, for simplicity, two cases are considered here, based on the properties of NOILC and the observations that,

$$\|L_G(\beta)\| \le 1 \,, \quad LL_G(\beta) = L_G(\beta)L$$
$$\& \quad (-1 + \mu(\beta))I < L_e(\beta) \le I, \tag{7.35}$$

for some $\mu(\beta) > 0$. The two cases are as follows:

1. Note that $e_k = L_e(\beta_k) \cdots L_e(\beta_1)e_0$. Suppose that $ker[G^*] = \{0\}$, and write

$$e_k = L_G(\beta_k) \cdots L_G(\beta_1)L^k e_0 \,, \qquad so\ that$$
$$\|e_k\|_{\mathcal{Y}} \le \|L^k e_0\|_{\mathcal{Y}}\| \prod_{j=1}^k \|L_G(\beta_j)\| \le \|L^k e_0\|_{\mathcal{Y}} \,.$$

   The sequence $\{L^k e_0\}_{k \ge 0}$ is just the error sequence of a NOILC algorithm without error correction. This converges to zero which implies that the error corrected algorithm will also converge to zero. In addition, the error norm in the Error Corrected Algorithm is less, iteration by iteration, than that that would be seen with the application of NOILC alone.
2. Next, consider the simplest case of constant $\beta_k = \beta$ after a finite number of iterations, N, say. More precisely, convergence of the error is guaranteed by the relation $e_k = L_e^{k-N}(\beta)e_N, \; k \ge N$, and the properties of $L_e(\beta)$. The error converges to the orthogonal projection of e_N onto the kernel $ker[I - L_e(\beta)] = ker[G^*]$, which is independent of β, and hence identical to the projection of e_0 onto that subspace.

The next section considers the choice of the, as yet unspecified, parameters β_{k+1}. They can be regarded simply as empirical, gradient parameters chosen by trial and error. Alternatively, they could be chosen in a more systematic manner, guided by rules derived from an interpretation of theoretical principles.

7.4.3 Spectral Analysis and Parameter Choice

7.4.3.1 Spectral Analysis of GG^*

The easiest way to obtain a useful insight into parameter choice is to use matrix-like methodologies. More precisely, using the same methodologies as those used in spectral analysis of NOILC:

1. Assume that GG^* has a spectrum that consists only of eigenvalues of $\sigma_0^2 = 0$ (if $ker[G^*] \ne \{0\}$) and positive values $\sigma_j^2 > 0$, ordered in the natural ascending way,
$$0 < \cdots \le \sigma_{j+1}^2 \le \sigma_j^2 \le \cdots \le \sigma_2^2 \le \sigma_1^2 \,. \tag{7.36}$$
2. The operator norm has an exact relationship to the largest eigenvalue,

$$\sigma_1^2 = \|GG^*\| = \|G\|^2 = \|G^*\|^2. \quad (Operator\ Norm\ and\ Eigenvalues) \tag{7.37}$$

3. The span of the eigenvectors of GG^* corresponding to zero eigenvalues is precisely $ker[GG^*] = ker[G^*]$.
4. Suppose also that GG^* has a complete (finite or infinite) set of eigenvectors $\{w_j\}_{j\geq 1}$ spanning $\overline{\mathcal{R}[GG^*]}$. As GG^* is self-adjoint, they can be assumed to be orthonormal, namely, they satisfy,

$$GG^* w_j = \sigma_j^2 w_j \,, \quad j \geq 1 \,, \quad \& \quad \langle w_j, w_p \rangle_{\mathcal{Y}} = \delta_{j,p} \\ (Eigenvector\ Equation\ for\ GG^*) \tag{7.38}$$

where $\delta_{j,p}$ is the Kronecker delta. In addition, $GG^* w_0 = 0$ for all $w_0 \in ker[GG^*]$. All problems with finite dimensional signal space $\mathcal{Y}$ (e.g., those associated with discrete-time state-space models) automatically have this property.
5. The set of eigenvectors form a basis for $\overline{\mathcal{R}[GG^*]}$. In this circumstance, any error, for example, can be expanded in terms of the eigenvectors

$$e_k = a_{0,k} w_0 + \sum_{j\geq 1} a_{j,k} w_j \,, \ (Eigenvector\ Expansion) \\ w_0 \in ker[G^*] \,, \qquad \|w_0\|_{\mathcal{Y}} = 1 \,, \tag{7.39}$$

by suitable choice of the coefficients $a_{j,k}$. The norm of e_k can then be expressed in the form

$$\|e_k\|^2_{\mathcal{Y}} = \sum_{j\geq 0} a^2_{j,k} \,, \tag{7.40}$$

so the coefficients are square summable.

Finally, the number of eigenvalues in an ILC problem can be small and finite, very large (but finite) or infinite. In all cases the eigenvalues lie in the interval $0 \leq \sigma^2 \leq \|G\|^2$. Little can be said, in general, about the values or distribution within that interval other than the intuitive observation that, strictly proper, time-invariant, state-space systems (described by a convolution operator G) have an infinite number of eigenvalues, including an infinite number of small eigenvalues associated with:

1. high frequency behaviour when output responses to high frequency inputs become extremely small,
2. small outputs for oscillatory inputs of frequency close to system zeros that are very close to the imaginary axis of the complex plane, and
3. non-minimum phase TFM characteristics.

7.4.3.2 Spectral Allocation and Choice of $\{\beta_k\}_{k\geq 1}$

The important function in what follows is

$$f_e(\sigma^2, \beta) = \frac{1 - \beta\sigma^2}{1 + \epsilon^{-2}\sigma^2} \,, \quad 0 \leq \sigma^2 \leq \|G\|^2 \,, \tag{7.41}$$

which has the properties that:

1. $f_e(\sigma_j^2, \beta)$, $j \geq 0$, are the eigenvalues of $L_e(\beta)$ with orthonormal eigenvectors $\{w_j\}_{j\geq0}$,
2. $f_e(\sigma_j^2, 0)$ are the eigenvalues of the NOILC algorithm,
3. $f_e(0, \beta) = 1$ for all β,
4. $f_e(\sigma^2, \sigma^{-2}) = 0$ for all σ^2, including the eigenvalues of GG^*, and
5. in the range (7.26) of β specified, $-f_e(\sigma_j^2, 0) < f_e(\sigma_j^2, \beta) < f_e(\sigma_j^2, 0)$, i.e., the use of non-zero values of β reduces the magnitude of the equivalent NOILC eigenvalue.

This function has other useful properties:

1. in the parameter range,
$$0 < \beta < \|G\|^{-2}, \tag{7.42}$$
the eigenvalues satisfy $0 < f_e(\sigma_j^2, \beta) < f_e(\sigma_j^2, 0)$ and
2. if
$$\|G\|^{-2} \leq \beta < 2\|G\|^{-2}, \tag{7.43}$$
then $f_e(\sigma^2, \beta)$ takes the value of zero at some point in the interval $\frac{1}{2}\|G\|^2 < \sigma^2 \leq \|G\|^2$.

The plot of $f_e(\sigma^2, \beta)$ against σ^2 in the range $0 \leq \sigma^2 \leq \|G\|^2$ shows, therefore, the impact of the using the value β in error correction on eigenvectors corresponding to eigenvalues equal to or close to σ^2.

Note that, eigenvalues and eigenvectors are not, normally, known but this does not prevent a useful interpretation of the plot when answering questions of the form:

Given ϵ^2 and a chosen value of β, what will be the effect of one iteration of the Error Corrected NOILC Algorithm on the eigenvectors corresponding to an eigenvalue of GG^* equal, or close to, a particular value σ^2 of interest?

Insight into the answer is obtained by reading off the value $f_e(\sigma^2, \beta)$ and interpreting this as the reduction in the contribution of eigenvectors of interest. Manipulation of this attenuation in a desired manner can achieved by suitable parameter choice.

The form of the graph of f_e and its variation with β are illustrated in the following example.

Example 7.2 Consider the discrete-time SISO system

$$G(z) = \frac{0.64}{(z-0.2)^2}, \qquad \sup_{|z|=1} |G(z)| = 1, \tag{7.44}$$

with zero initial conditions and inner products

$$\langle u, v\rangle_{\mathcal{U}} = \sum_{t=0}^{N} u(t)R(t)v(t), \quad \langle y, w\rangle_{\mathcal{Y}} = \sum_{t=0}^{N} y(t)Q(t)w(t).$$

Take time independent values $Q = R = 1$ and $N = 200$. It follows that $\|G^*\|^2 \approx 1$. Figure 7.6 shows the plot of $f_e(\sigma^2, \beta)$ for $\epsilon^2 = 1$ with $\beta = 0\ ,\ 0.5\ ,\ 1\ ,\ 1.5\ ,\ 2$. The first represents NOILC, the third a midpoint of interest and the fifth the largest value of β that could be considered (i.e., $2\|G^*\|^{-2}$). It is clearly seen that for any choice of β, the spectrum of the learning operator is less than that of NOILC and thus faster convergence can be expected. For example, if $\beta = 1$, the eigenvectors corresponding to eigenvalues close to unity will be almost eliminated in one iteration.

The effect of f_e on one iteration of the error sequence is revealed by analysis of the eigenvector expansions. Write,

$$e_k = a_{0,k}w_0 + \sum_{j\geq 1} a_{j,k}w_j\ , \quad so\ that$$

$$e_{k+1} = L_e(\beta_{k+1})e_k = a_{0,k}w_0 + \sum_{j\geq 1} f_e(\sigma_j^2, \beta_{k+1})a_{j,k}w_j\ , \tag{7.45}$$

$$i.e.\ \ a_{0,k+1} = a_{0,k} \quad and \quad a_{j,k+1} = f_e(\sigma_j^2, \beta_{k+1})a_{j,k}\ , \quad k \geq 1\ .$$

This shows that the component of the error in $ker[G^*]$ remains unchanged, whilst the contribution of other eigenvectors w_j corresponding to the eigenvalues σ_j^2 are reduced by a factor $f_e(\sigma_j^2, \beta_{k+1})$. In more detail:

1. For gradient gains, β, in the range (7.26), the contribution of *every* eigenvector is reduced in magnitude.
2. For gains in the range (7.42), the contribution of every eigenvector is reduced but is not eliminated.
3. For gains in the range (7.43), the contribution of eigenvectors corresponding to eigenvalues that are close to β_{k+1}^{-1} are almost eliminated.

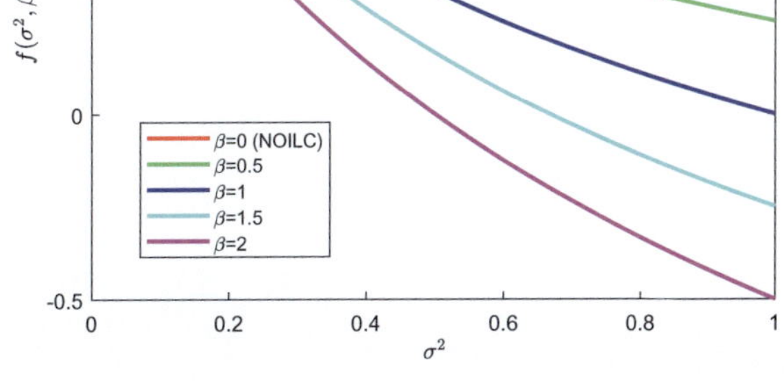

Fig. 7.6 The accelerating effect of β on the spectrum of L_e for Example 7.2. Reprinted with permission from [10], ©2024 IEEE

The effect of the choice of the sequence $\{\beta_k\}_{k \ge 1}$ on the evolution of the sequence can also be revealed by analysis of the contribution of each eigenvector w_j. More precisely, using (7.45), the contribution on the kth iteration has magnitude, for all $k \ge 1$,

$$a_{0,k} = a_{0,0} \quad and \quad a_{j,k} = \left(\prod_{p=1}^{k} f_e(\sigma_j^2, \beta_p) \right) a_{j,0}, \quad j \ge 1. \tag{7.46}$$

Three cases are considered here:

1. The sequence $\{\beta_k\}_{k \ge 1}$ could be taken to take only constant values $\beta_k = \beta$. The improvement in algorithm convergence over the first *two* iterations is revealed in Fig. 7.7 (for the system in Example 7.2). This represents the attenuation over one iteration by $f_e(\sigma^2, \beta)$ and, over two iterations, by $f_e^2(\sigma^2, \beta)$ for the illustrative choice of $\beta = \|G\|^{-2}$ and $\epsilon^2 = \|G\|^2$. These plots are compared with the equivalent plots of $f_e(\sigma^2, 0)$ and $f_e^2(\sigma^2, 0)$ representing the convergence of NOILC for this data. The plot of $f_e^2(\sigma^2, \beta)$ is substantially smaller than that for NOILC, particularly over the range $\frac{1}{2}\|G\|^2 \le \sigma^2 \le \|G\|^2$ with the eigenvectors corresponding to eigenvalues in the vicinity of $\sigma^2 = \|G\|^2$ being annihilated or almost annihilated in the error signal.
2. Alternatively, the gain β_k can be varied. For simplicity, consider the first two iterations, $k = 1, 2$ using values β_1 and β_2. The improved reduction in error norm over the first *two* iterations is revealed in Fig. 7.8 which shows $f_e(\sigma^2, \beta_1) f_e(\sigma^2, \beta_2)$ for the illustrative choice of $\beta_1 = \|G\|^{-2}$, $\beta_2 = 2\|G\|^{-2}$ and $\epsilon^2 = \|G\|^2$ and compared with the plot of $f_e^2(\sigma^2, 0)$ of NOILC for this data. Again, the graph represents the relative convergence rates. The plot of $f_e(\sigma^2, \beta_1) f_e(\sigma^2, \beta_2)$ is substantially smaller than that for NOILC, particularly over the range $\frac{1}{2}\|G\|^2 \le \sigma^2 \le \|G\|^2$, and is an improvement over the use of the constant gain case. This can be attributed, intuitively as being due to the almost annihilation of eigenvector contributions to the error associated with eigenvalues near to the *two* points $\sigma^2 = \|G\|^2$ and $\sigma^2 = \frac{1}{2}\|G\|^2$. Their performance over four iterations are also shown, again, confirming the significant improvement.
3. If the output space is finite dimensional (of dimension N) such as the cases where G represents a discrete-time state-space system or the tracking of end-point and intermediate points in either discrete or continuous state-space systems, then there are only a finite number, N, of eigenvalues in the spectrum of L_e. What can be achieved in practice will depend on the form of the operator GG^*, the dimension N and the interval containing the eigenvalues.
   Consider the special case when GG^* only has eigenvalues σ_j^2, $1 \le j \le N$ in the interval
   $$\frac{1}{2}\|G\|^2 \le \sigma^2 \le \|G\|^2. \tag{7.47}$$

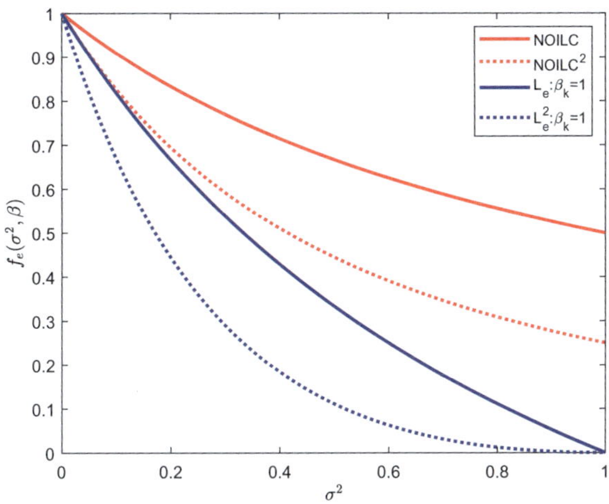

Fig. 7.7 Comparison of the spectrum of L, L^2, L_e, L_e^2 with constant βs

Choosing $\beta_k = \sigma_k^{-2}$, $k = 1, \ldots, N$ (or any other reordering) yields $f_e(\sigma_k^2, \beta_k) = 0$, $k \geq 1$. As a consequence, for all initial errors e_0, $e_N = 0$, i.e., the algorithm converges to zero in a finite number of iterations (less than or equal to N)! This may only be useful if N is small, but it relies on a knowledge of the eigenvalues. A more useful approach may be to choose β_k using a more empirical approach aimed at fast convergence by trying to get close to each and every eigenvalue on some iteration. One such choice is a sequence that "covers" the interval in a uniform way. There is no unique approach to this but the idea can be illustrated by an "interval halving" choice illustrated by

$$\begin{aligned} \beta_1 &= \|G\|^{-2}, & \beta_2 &= 2\|G\|^{-2}, & \beta_3 &= \tfrac{1}{2}(\beta_1 + \beta_2), \\ \beta_4 &= \tfrac{1}{2}(\beta_1 + \beta_3), & \beta_5 &= \tfrac{1}{2}(\beta_2 + \beta_3), & \beta_6 &= \tfrac{1}{2}(\beta_1 + \beta_4), \\ \beta_7 &= \tfrac{1}{2}(\beta_3 + \beta_4), & & \cdots & & . \end{aligned} \tag{7.48}$$

After a small number of such variations, the error will have been reduced considerably, the residual error being further reduced by, for example, continuing with a constant, chosen value of β_k.

In general, variation of β_k seems to be an effective accelerator of the convergence as it can systematically annihilate eigenvector contributions to the error corresponding to eigenvalues in the range $\frac{1}{2}\|G\|^2 \leq \sigma^2 \leq \|G\|^2$. This statement is certainly true in the absence of modelling errors but it must be assumed that modelling errors may modify this conclusion or limit its applicability. Robustness is considered in a later section.

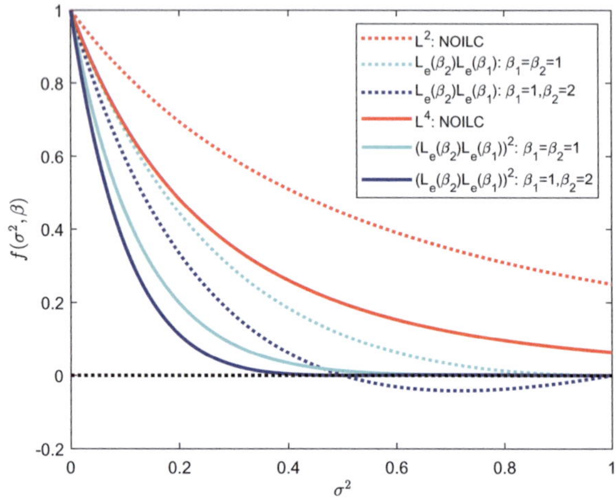

Fig. 7.8 Comparison of the spectrum of L^2, L^4, $L_e(\beta_2)L_e(\beta_1)$, $(L_e(\beta_2)L_e(\beta_1))^2$ with constant and varying βs. Reprinted with permission from [10], ©2024 IEEE

7.4.4 Simulation Examples

The examples in this section are based on iterative control of a linear, time-invariant, ℓ-input, m-output, state dimension n, discrete-time state-space system with sample interval h on a time interval $[0, t_f]$ with sample points $t = 0, h, 2h, \ldots, Nh = t_f$. Simplifying the notation by assuming that $h = 1$, the model has the classical form,

$$x(t+1) = Ax(t) + Bu(t), \ x(0) = x_0,$$
$$y(t) = Cx(t), \ t = 0, 1, 2, \ldots, N. \tag{7.49}$$

The input time series $\{u(0), u(1), \ldots, u(N-1)\}$ can be regarded as generating the output time series $\{y(1), y(2), \ldots, y(N)\}$.

7.4.4.1 End Point Control

Consider the case of end point control where it is desired that the control sequence is chosen to reach a desired end point $y(N) = y_f$. In ILC terms, the aim is to find the input time series column vector $u = [u^T(0), \ldots, u^T(N-1)]^T$ that takes the output $y = y(N)$ to the final value $r = y_f$. The error on the kth iteration will simply be

$$e_k = y_f - y_k(N)\,. \tag{7.50}$$

The dynamic equations have the alternative supervector form,

$$y = Gu + d$$

where

$$G = [CA^{N-1}B, \ldots, CAB, CB], \quad d = CA^N x_0.$$

In terms of the notation of previous sections, use the inner product $\langle y, w \rangle = y^T Q w$ in the output Hilbert space $\mathcal{R}^m$ of $m \times 1$ columns of real numbers. The $m \times m$ matrix Q is symmetric and positive definite. The Hilbert space of inputs will be taken to be of the form defined by the inner product $\langle u, v \rangle = u^T R v$ in the Hilbert space $\mathcal{R}^{\ell(N)}$. Typically R is block diagonal of the form $R = blockdiag[R_1, R_2, \ldots, R_N]$, where each $\ell \times \ell$ matrix R_j is symmetric and positive definite. The adjoint operator

$$G^* = R^{-1}G^T Q \quad \Rightarrow \quad GG^* = GR^{-1}G^T Q \tag{7.51}$$

which is an $m \times m$ matrix with m positive, real eigenvalues $\eta_1 \geq \eta_2 \geq \cdots, \eta_m \geq 0$ with $\eta_m > 0$ if and only if G has full rank. In principle, this matrix (and its eigenstructure) can be estimated using a model and the quantity $\|GG^*\| = \|G^*G\| = \|G\|^2 = \|G^*\|^2$ set equal to the largest eigenvalue. The simplest case is for single-input–single output systems with $m = \ell = 1$ and Q and R multiples of the identity, when

$$\|GG^*\| = \|GR^{-1}G^T Q\| = R^{-1}Q\|H\|^2 , \tag{7.52}$$

where $\|H\|^2 = \sum_{t=1}^{N}(CA^{t-1}B)^2$ is the squared Euclidean norm of the impulse response sequence for the system truncated to the interval of interest.

The operators and norms associated with error evolution for NOILC and the error corrected algorithm are $m \times m$ matrices,

$$L = (I_m + \epsilon^{-2}GR^{-1}G^T Q)^{-1}, \ L_G(\beta) = I - \beta GR^{-1}G^T Q ,$$
$$\|G^*\|^2 = \eta_1, \ 0 \leq \beta \leq 2\eta_1^{-1} .$$

The range of GG^* is the span of eigenvectors of $GR^{-1}G^T Q$ associated with non-zero eigenvalues. The kernel of GG^* is the span of eigenvectors of $GR^{-1}G^T Q$ having zero eigenvalues. In order that all possible output final values y_f can be reached it is essential that all eigenvalues are non-zero.

Starting with an initial control sequence, u_0, the problem to be solved for the $(k+1)$th iteration is, given the error from the kth iteration and a choice of β_{k+1}, to construct the error $e_k = y_f - y_k$ and the corrected reference $r_{k+1} = y_f + \epsilon^2 \beta_{k+1} e_k$ and then to find the input time series vector u_{k+1} minimizing

$$J = (r_{k+1} - y(N))^T Q (r_{k+1} - y(N)) + \epsilon^2 \sum_{t=0}^{N-1} (u(t) - u_k(t))^T R_{t+1} (u(t) - u_k(t)) . \tag{7.53}$$

This is a well-defined linear quadratic optimal control problem that can be computed offline (a feedforward solution) or solved online (a state feedback solution) using Riccati matrix methods.

For end point problems with only non-zero eigenvalues, the first thing to note is that the choice of the parameters $\beta_p = \eta_p^{-1}$, $p = 1, \ldots, m$ will, in theory reduce the end-point error to zero on the mth iteration $k = m$. More generally, choosing β_p, sequentially, to be equal to the distinct eigenvalues will reduce the end-point error to zero in $\leq m$ iterations. In the process, the algorithm finds the control that transfers the system from its initial state to the desired final output, using a control time series that minimizes

$$\|u - u_0\|_{\mathcal{U}}^2 = \sum_{t=0}^{N-1} (u(t) - u_0(t))^T R_{t+1}(u(t) - u_0(t)). \tag{7.54}$$

Remark 7.2 The $SISO$ case will achieve the objective in one iteration using the error corrected reference $r_1 = y_f + \epsilon^2 R^{-1} Q \|H\|^2 e_0$. More generally, the eigenvalues are not known but, if the norm $\|GG^*\|$ can be estimated, an approximation to the desired range of parameters β_{k+1} will be available.

Finally, note that:

1. u_0 is also a design variable as it will influence the final form of the computed input. It is natural to use one of a desirable or convenient form even if it fails to meet the end point condition. For example, u_0 could be generated by a one or more NOILC iterations attempting to track a nominal reference r_{nom} time series starting at Cx_0 and terminating at $y(N) = y_f$. The nominal trajectory could be chosen to represent the sort of motion that fits the characteristics/requirements of the application.
2. The intermediate point control problem also leads to relatively low order matrix descriptions of L and GG^* of dimension $\ll N$ so similar comments can be made for this case. The details are omitted for brevity but, as an exercise, the reader will be able to derive the relevant supervector description, G^* and hence GG^* in this case.

The following example illustrates the error corrected algorithm for end point control. Note that the weighting was chosen as $\epsilon^2 = 1$.

Example 7.3 Consider the state-space model (7.49) with

$$A = \begin{bmatrix} 1 & 0 & 0 & 0 & 0 & 0 & 0 & 0 \\ 0.001 & 1 & 0 & 0 & 0 & 0 & 0 & 0 \\ 0 & 0.001 & 1 & 0 & 0 & 0 & 0 & 0 \\ 0 & 0 & 0.001 & 1 & 0 & 0 & 0 & 0 \\ 0 & 0 & 0 & 0 & 1 & 0 & 0 & 0 \\ 0 & 0 & 0 & 0 & 0.001 & 1 & 0 & 0 \\ 0 & 0 & 0 & 0 & 0 & 0.001 & 1 & 0 \\ 0 & 0 & 0 & 0 & 0 & 0 & 0.001 & 1 \end{bmatrix},$$

$$B = \begin{bmatrix} 0.001 & 5\times 10^{-7} & 0 & 0 & 0 & 0 & 0 & 0 \\ 0 & 0 & 0 & 0 & 0.001 & 5\times 10^{-7} & 0 & 0 \end{bmatrix}^T$$

$$C = \begin{bmatrix} 0 & 0 & 0 & 1 & 0 & 0 & 0 & 0 \\ 0 & 0 & 0 & 0 & 0 & 0 & 0 & 1 \end{bmatrix}$$

with $x_0 = 0$ and a sampling time of $h = 0.001$s. This is a model used in functional electrical stimulation-based stroke rehabilitation after applying a feedback linearisation controller (see [5] for more detail). The input to this model is the electrical stimulation levels applied to two pairs of arm muscles, and the two outputs are angle positions of two arm joints. The design objective is to compute the input such that the output reaches the following desired position at $t_f = 6$s (or equivalently at sample number $N = 6000$),

$$y_f = \begin{bmatrix} -0.3 \\ -0.6 \end{bmatrix}.$$

With the choice of $R = I$, $Q = \begin{pmatrix} 2 & 0 \\ 0 & 1 \end{pmatrix}$, model data give

$$GG^* = \begin{bmatrix} 2.2217 & 0 \\ 0 & 1.1109 \end{bmatrix},$$

whose eigenvalues are $\eta_1 = 2.2217$, $\eta_2 = 1.1109$. From the above discussion, if we choose $\beta_1 = \eta_1^{-1} = 0.4501$, $\beta_2 = \eta_2^{-1} = 0.9002$, the tracking error will be eliminated in 2 iterations. To show this, if we choose $u_0 = 0$, the tracking error can be computed as follows:

$$e_0 = \begin{bmatrix} -0.3 \\ -0.6 \end{bmatrix}, \; e_1 = \begin{bmatrix} 0 \\ -0.1421 \end{bmatrix}, \; e_2 = \begin{bmatrix} 0 \\ 0 \end{bmatrix},$$

confirming convergence to zero error in two iterations. This is true for all choices of u_0 and almost all R and Q. Exceptions include, for example, the choice of $R = I$ and Q as a multiple of the identity matrix when convergence is achieved in *one* iteration!

7.4.4.2 Interval Tracking Problems

The objective is to find an input time series, represented by the column matrix $u = [u^T(0), \ldots, u^T(N-1)]^T$, such that the output $y = [y^T(1), \ldots, y^T(N)]^T$ tracks a reference time series $r = [r^T(1), \ldots, r^T(N)]^T$. The transfer function matrix of the system is denoted $G(z)$ and the associated operator is denoted by G. The norm $\|G^*\|^2 = \|G\|^2 = \|G^*G\| = \|GG^*\|$ can be estimated using the computations found in Chap. 2. More precisely, using the Hilbert space $\mathcal{U}$ of inputs with the inner product $\langle u, v\rangle = \sum_{t=0}^{N-1} u^T(t)Rv(t)$ where R is $\ell \times \ell$, symmetric and positive definite and the Hilbert space $\mathcal{Y}$ of outputs with the inner product $\langle y, w\rangle = \sum_{t=1}^{N} y^T(t)Qw(t)$ where Q is $m \times m$, symmetric and positive definite, the operator norm is bounded by

$$\|G\|^2 \leq \sup_{|z|=1} r\left(R^{-1}G^T(z^{-1})QG(z)\right), \tag{7.55}$$

where $r(\cdot)$ is the spectral radius of the argument. The accuracy of the upper bound increases as the length of the time interval Nh increases, becoming exact when $N \to \infty$.

Given a starting input time series u_0 and the error e_0, iterations are based, for all $k \geq 0$, on tracking the error corrected reference $r(t) + \epsilon^2\beta_{k+1}e_k(t)$, $t \geq 0$ by using optimal control theory to minimize the quadratic objective function

$$\begin{aligned} J = \sum_{t=1}^{N} &((r_{k+1}(t) - y(t))^T Q(r_{k+1}(t) - y(t)) \\ &+\epsilon^2(u(t-1) - u_k(t-1))^T Q(u(t-1) - u_k(t-1))). \end{aligned} \tag{7.56}$$

The choice of the parameters $\{\beta_{k+1}\}_{k\geq 0}$ can be based on the ideas outlined in the section on endpoint control but, for tracking problems of the type discussed here, there is a useful frequency domain interpretation.

Remark 7.3 (Frequency Domain Interpretations) Frequency domain interpretations are often of great value in control theory, giving insights into the behaviour of the controlled plant. In Error Corrected NOILC, it will provide insight into the choice of parameters and the effect on the iterations and measured data.

The general conclusion is that,

any given choice of β will almost annihilate eigencomponents of GG^* at all frequencies where the eigenvalues (which are all real and positive) of the frequency dependent matrix $G(z)R^{-1}G^T(z^{-1})Q$ are close to the value β^{-1}.

The full details are omitted here for brevity (see Sect. 2.14.5 and [9] for more details) but the simplest case of $SISO$ systems can be summarized as follows: if z_j are the distinct Nth complex roots of unity then $W(z_j) = [1, z_j, \ldots, z_j^{N-1}]^T$, $1 \leq j \leq N$, are a (complex) basis for $\mathcal{Y}$ representing the span of "sinusoidal" variation in the measured signals. The useful relationship here is as follows from consideration of the state-space solutions (in supervector form),

$$GG^*W(z_j) \simeq R^{-1}Q|G(z_j)|^2W(z_j) , \quad 1 \le j \le N , \tag{7.57}$$

the approximation being increasingly accurate as N increases. That is, $R^{-1}Q|G(z_j)|^2$ can be regarded as an "approximate eigenvalue" and the oscillatory time series $W(z_j)$ as an approximate eigenvector of GG^*. This suggests the conclusion that,

choosing $\beta = RQ^{-1}|G(z)|^{-2}$ will approximately annihilate the frequency component of the error at the frequency z.

The following examples are designed to verify the validity of the theory and its interpretation. In all cases, $\epsilon^2 = 1$.

Example 7.4 (Annihilating a Single Frequency/Spectral Value) Consider the linear SISO discrete system with transfer function

$$G(z) = \frac{0.7}{z - 0.7} , \qquad \sup_{|z|=1} |G(z)| = 2.333 . \tag{7.58}$$

Zero initial conditions are assumed and a reference signal defined by the oscillating time series

$$r = \{r(t)\}_{0 \le t \le N} , \quad r(t) = \frac{1}{c} sin(\frac{\pi}{2}t) = \frac{1}{c} Im \left[i^t \right] , \tag{7.59}$$

with a frequency $z = i$, trial length $N = 2000$, and a constant $c = 31.6228$ added to normalize the reference norm to be unity. The weights in the performance index are chosen as $Q = R = 1$.

As the reference signal has only one frequency component, it is possible to consider approximate annihilation of the single frequency reference by choosing $\beta_1 = 3 \approx RQ^{-1}|G(i)|^{-2}$.

Note 7.1 The gain used does not satisfy (7.26) and is not necessarily an acceptable choice for further iterations. It is used only to illustrate the annihilation property.

The algorithm is initiated with $u_0 = 0$. The first iteration gives the error norm values

$$\underbrace{\|e_0\| = \|r\| = 1 , \qquad and \qquad \|e_1\| = 0.0222,}_{The\ almost\ annihilation\ property} \tag{7.60}$$

i.e., the error norm after, just one iteration, is of the order of 2% of the initial error norm. This substantial reduction in one iteration provides a simple illustration of the frequency content interpretation of the approximate annihilation properties of the new algorithm.

Example 7.5 (Annihilating a Frequency/Spectral Band) Reference signals, in most cases, contain a wide band of frequencies and plant dynamics are more complex. A simple illustration of possible approaches to such cases is the choice

$$G(z) = \frac{0.64}{(z - 0.2)^2} , \quad where \quad \max_{|z|=1} |G(z)| = 1 , \tag{7.61}$$

with zero initial conditions and a reference signal $r(t) = t,\ t = 0, 1, 2, \ldots, N$. Again take $Q = R = 1$ and $N = 200$ and note that $\|G^*\|^2 \approx \max_{|z|=1} R^{-1}Q|G(z)|^2 = 1$.

Taking the initial control $u_0 = 0$, the results obtained from the use of NOILC are shown in the plot of the evolution of the error norm in Fig. 7.9 with the conclusion that the error norm reduces to approximately 10^{-2} in 13 iterations and is reduced a little more by 20 iterations.

Acceleration of the algorithm is now approached in two ways:

1. Approach One: Take $\beta_{k+1} = \beta$ to be iteration independent and let $\beta = 1$ to approximately annihilate the contribution of the eigenvectors corresponding to the largest eigenvalues of GG^*. Alternatively, using the frequency annihilation interpretation, the choice will approximately annihilate the DC ($|z| = 1$) and part of the low frequency bandwidth content of the error. The results are given in Fig. 7.9 together with the NOILC result. Note that the error norm reduces to approximately 10^{-4} in 13 iterations and is reduced a little more by 20 iterations. The initial and long term improvements in rate of convergence are clear and, given the simplicity of the error correction mechanism, well worth the use of the modification.
2. Approach Two: By letting the gains vary from iteration to iteration, different frequency contributions can be approximately annihilated. The constraint of (7.26) is taken to ensure that overall error reduction is guaranteed as the iterations progress. This means that we must focus on the frequencies where

$$0.5 \le |G(z)|^2 \le 1\,. \tag{7.62}$$

The values of these frequencies are not required for the choice of β_{k+1}. All that is needed is to choose $\beta_{k+1} \in [1, 2]$. To illustrate the possibilities, the choices are made to "cover" this interval, for example,

$$\beta_1 = 1.0,\ \beta_2 = 1.2,\ \beta_3 = 1.4,\ \beta_4 = 1.6,\ \beta_5 = 1.8 \tag{7.63}$$

and, for simplicity, $\beta_{k+1} = 2.0$ thereafter. As the gain of $G(z)$ decreases with frequency, the motivation for this choice is to sequentially and approximately annihilate the frequency content of the error in the range represented by (7.62), beginning with the lowest frequency.

The algorithm behaviour is shown in Fig. 7.9 including the basic NOILC behaviour. Although NOILC produces monotonic reductions in norm, the simple error corrected reference mechanism again accelerates the algorithm producing even more rapid reductions in error norm.

In summary, the two approaches considered for gain selection show different but beneficial behaviours. For the first iteration, the norm reduction is the same (as the gain value is the same). For Approach One, error norm reductions are two orders of magnitude better than NOILC over twenty iterations. For further iterations, varying the gains in Approach Two adds to the acceleration by a further two orders of magnitude as it is more effective in eliminating frequency content in the range (7.62).

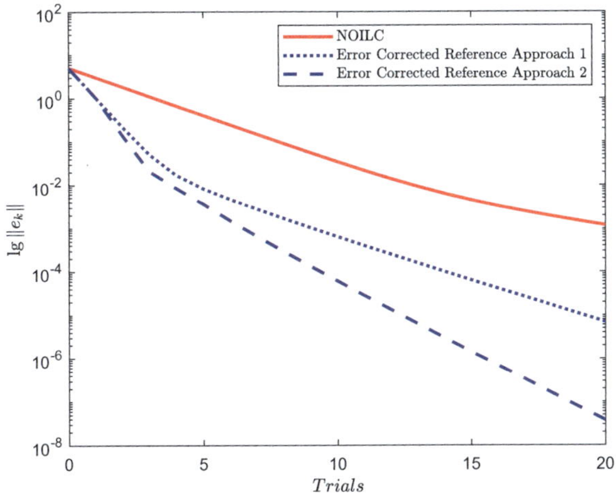

Fig. 7.9 Logarithmic error norm convergence for NOILC and error corrected algorithms showing order of magnitude increases in convergence rates for the Example. Reprinted with permission from [10], ©2024 IEEE

The examples provide evidence that the acceleration produced by the simple, easily implemented, error correction mechanism can, in the absence of modelling errors, be substantial. The next section considers algorithm robustness.

7.4.5 Robustness Analysis

Following the approach of the previous section, the robustness of the Error Corrected Algorithm is approached assuming that the modelling error is linear and can be represented by a right multiplicative representation. That is, the available model of the plant is characterized by the operator G whilst, in reality, the plant is obtained by replacing G by the operator GU where $U : \mathcal{U} \to \mathcal{U}$.

It is assumed that the input update on iteration $k+1$ is constructed offline, using the model, to generate a *feedforward* version. To do this, write the equations representing the input obtained from the Error Corrected NOILC Algorithm using the model in the form,

$$\begin{aligned} u_{k+1} &= u_k + \epsilon^{-2} G^* \left(r_{k+1} - y_{k+1} \right) \\ &= u_k + \epsilon^{-2} G^* \left((r_{k+1} - r) + e_{k+1} \right) \,. \end{aligned} \tag{7.64}$$

Replacing e_{k+1} from the error evolution (7.34) gives, after a little algebra, the feedforward form of u_{k+1}, namely,

$$\underbrace{u_{k+1} = u_k + \epsilon^{-2} \left(1 + \epsilon^2 \beta_{k+1}\right) G^* \, \underbrace{Le_k}_{}}_{Feedforward\ Representation\ of\ the\ Input\ Update} \,. \tag{7.65}$$

Le_k can be computed from the model and the *measured* plant data e_k. More precisely, the term Le_k is the tracking error $e_k - z_{k+1}$ generated by an uncorrected NOILC algorithm, namely, the offline solution of the NOILC problem,

$$\begin{aligned} v_{k+1} &= argmin\{J(v) \ : \ z = Gv\} \quad where \\ J(v) &= \|e_k - z\|^2_{\mathcal{Y}} + \epsilon^2 \|v\|^2_{\mathcal{U}} \, . \end{aligned} \tag{7.66}$$

Using (7.65), the tracking errors observed when applying the feedforward control signal u_{k+1} to the plant are therefore related by

$$e_{k+1} = \left(I - \left(\epsilon^{-2} + \beta_{k+1}\right) GUG^* L\right) e_k \tag{7.67}$$

which is almost identical to the error evolution for NOILC in the presence of a modelling error. The difference is that ϵ^{-2} is replaced by $\epsilon^{-2} + \beta_{k+1}$. As a consequence, conditions for robust monotonicity of the error norm $\|e_k\|_0 = \sqrt{\langle e_k, Le_k\rangle_{\mathcal{Y}}}$ are,

$$\begin{aligned} &U + U^* > \beta_e U^* U \, , \quad or \ \|\tfrac{1}{\beta_e} I - U\| < \tfrac{1}{\beta_e} \\ &where \ \ \beta_e = (1 + \epsilon^2 \beta_{k+1}) \left(\tfrac{\epsilon^{-2}\|G\|^2}{1+\epsilon^{-2}\|G\|^2} \right) \geq \tfrac{\epsilon^{-2}\|G\|^2}{1+\epsilon^{-2}\|G\|^2} \, . \end{aligned} \tag{7.68}$$

Note that β_e increases as β_{k+1} increases, the smallest value occurring when $\beta_{k+1} = 0$, when β_e reduces to the value deduced for NOILC in previous chapters. That is, the theory, unsurprisingly, suggests that robustness decreases as β_{k+1} increases.

In more detail:

1. The model (obtained by setting $U = I$) satisfies the robustness condition if, and only if, $\beta_e < 2$, i.e.,
$$\beta_{k+1} < \epsilon^{-2} + 2\|G\|^{-2} \, . \tag{7.69}$$
This is satisfied in the two parameter ranges (7.42) and (7.43) for all weights ϵ^2.
2. More generally, writing
$$\beta_e = 1 + \left(\frac{\beta_{k+1}\|G\|^2 - 1}{1 + \epsilon^{-2}\|G\|^2} \right) \tag{7.70}$$
reveals that $\beta_e \leq 1$ for all gradient gains in the interval $0 \leq \beta_{k+1} \leq \|G\|^{-2}$ (i.e., the closure of (7.42)). In contrast, if β_{k+1} lies in the interval (7.43), then $1 < \beta_e < 2$ for all weights ϵ^2.
3. The value of ϵ^2 affects robustness in a subtle manner. More precisely, examination of the variation of β_e as ϵ^2 increases from zero to $+\infty$, reveals that:
    a. If $0 \leq \beta_{k+1}\|G\|^2 < 1$, then β_e decreases, monotonically, from the value $\beta_e = 1$ to the value $\beta_{k+1}\|G\|^2 < 1$. Note that the predicted degree of robustness *increases* as ϵ^2 increases.

b. If $\beta_{k+1}\|G\|^2 = 1$, then β_e is constant, taking the value $\beta_e = 1$. That is, ϵ^2 has no effect on the degree of robustness.
c. If $1 < \beta_{k+1}\|G\|^2 < 2$, then β_e increases, monotonically, from the value $\beta_e = 1$ to the value $\beta_{k+1}\|G\|^2$. In this case, note that the degree of robustness *decreases* as ϵ^2 increases.

Using the first case and the robustness condition $\|\beta_e^{-1}I - U\| < \beta_e^{-1}$, shows that there is a considerable region of permissible modelling errors U, all of which lie inside the region $\|I - U\| < 1$, which is the permissible region of U obtained in case two. The third case is more complex as $\beta_e > 1$ and the set of permissible U becomes smaller as β_{k+1} increases approaching the set $\|\frac{1}{2} - U\| < \frac{1}{2}$ as $\beta_{k+1} \to 2\|G\|^{-2}$. The model (with $U = I$) lies on the boundary of this set with errors U arbitrarily close to the identity that are in the set and others that are outside.

4. The discussion above shows that the choice of ϵ^2, in cases where β_{k+1} is to vary widely over the interval $[0, 2\|G\|^{-2}]$, needs further comment. If reasonable robustness for all choices is to be achieved and high gain NOILC is to be avoided, then the choice of $\beta_{k+1} = \|G\|^{-2}$ ensures accelerated convergence and also that the robustness condition is independent of ϵ^2. An exact value is not essential but it does provide a ballpark figure to begin a design discussion. Other values of β_{k+1} will affect robustness in a manner dependent on ϵ^2.
More detail can be obtained by computing the value of β_e for the worst case value $\beta_{k+1} = 2\|G\|^{-2}$ and a variety of values of weight, and considering the condition $\|\beta^{-2} - U\| < \beta^{-1}$. Typical values are illustrated as follows

$$\begin{array}{ll} \epsilon^2 = 2\|G\|^2 \to \beta_e = \frac{5}{3}\ , & \epsilon^2 = \|G\|^2 \to \beta_e = \frac{3}{2}\ , \\ \epsilon^2 = \frac{1}{2}\|G\|^2 \to \beta_e = \frac{4}{3}\ , & \epsilon^2 = \frac{1}{3}\|G\|^2 \to \beta_e = \frac{5}{4}\ . \end{array} \tag{7.71}$$

7.4.6 *Multi-gradient Error Correction*

An obvious extension of the error correction methodology is based on the idea of using the approach of Sect. 7.3 with $M_{k+1} - 1 \geq 0$ offline, model-based gradient calculations prior to each online NOILC iteration.

7.4.6.1 Underlying Concepts

More precisely, consider the algorithm where:

1. For iteration $k + 1$, suppose that the input u_k and online measured error e_k are known and choose an integer $M_{k+1} \geq 1$. Then undertake offline gradient computations on iteration $k + 1$ to yield inputs and errors defined by the sequences $\{u_{k+1}^{(j)}\}$ and $e_{k+1}^{(j)}$, $0 \leq j \leq M_{k+1} - 1$, where

$$\begin{aligned} u_{k+1}^{(j)} &= u_{k+1}^{(j-1)} + \beta_{k+1}^{(j)} G^* e_{k+1}^{(j-1)} \ , \quad \& \\ e_{k+1}^{(j)} &= \left(I - \beta_{k+1}^{(j)} GG^*\right) e_{k+1}^{(j-1)} \ , \quad where \\ & 1 \leq j \leq M_{k+1} - 1 \ , \end{aligned} \tag{7.72}$$

starting with the plant data $u_{k+1}^{(0)} = u_k$ and $e_{k+1}^{(0)} = e_k$. Here, the gradient parameters satisfy

$$0 \leq \beta_{k+1}^{(j)} < 2\|G\|^{-2} \ . \tag{7.73}$$

The known properties of the gradient algorithm predict a sequential reduction in the error norm $\|e_{k+1}^{(j)}\|_{\mathcal{Y}}$ sequence, $0 \leq j \leq M_{k+1} - 1$.

2. Next, use the final input data $u_{k+1}^{(M_{k+1}-1)}$ to compute the next input, u_{k+1} to be used in an online plant iteration by minimizing the NOILC objective function

$$J(u, u_{k+1}^{(M_{k+1}-1)}) = \|r - y\|_{\mathcal{Y}}^2 + \epsilon^2 \|u - u_{k+1}^{(M_{k+1}-1)}\|_{\mathcal{U}}^2 \ . \tag{7.74}$$

That is, u_{k+1} is the solution of the optimization problem,

$$u_{k+1} = argmin\{ \ J(u, u_{k+1}^{(M_{k+1}-1)}) \ ; \ y = Gu + d \ \} \ . \tag{7.75}$$

The input and the measured tracking error e_{k+1} are then stored for use in the next iteration

7.4.6.2 Error Correction Equivalent

The NOILC algorithm component of the procedure described above can be described by the error evolution

$$e_{k+1} = L e_{k+1}^{(M_{k+1}-1)} \ , \qquad L = (I + \epsilon^{-2} GG^*)^{-1} \ . \tag{7.76}$$

Bringing the gradient calculations together gives $e_{k+1}^{(M_{k+1}-1)} = e_k$ if $M_{k+1} = 1$ and, otherwise,

$$\begin{aligned} e_{k+1}^{(M_{k+1}-1)} &= \left(\prod_{j=1}^{M_{k+1}-1} (I - \beta_{k+1}^{(j)} GG^*)\right) e_k \\ &= \underbrace{\left(I - GG^* \Phi_{k+1}(GG^*)\right)}_{An \ operator \ L_G} e_k \\ &= L_G e_k \ , \end{aligned} \tag{7.77}$$

where $\Phi_{k+1}(\lambda)$ is a polynomial of degree $M_{k+1} - 2$ with $\Phi_{k+1}(0) = \sum_{j=1}^{M_{k+1}-1} \beta_{k+1}^{(j)}$. As the gradient calculations do not increase the error norm, it must be true that the self adjoint operator L_G is bounded as follows,

$$-I \leq L_G \leq I \ , \tag{7.78}$$

which, in particular, implies that the polynomial $\Phi_{k+1}(\sigma^2)$ satisfies

$$0 \leq \Phi_{k+1}(\sigma^2) \leq 2 \,, \quad for \quad 0 \leq \sigma^2 \leq \|G\|^2 \,. \tag{7.79}$$

Combining the gradient and NOILC expressions indicates that the error evolution is hence described by

$$\underbrace{e_{k+1} = LL_G e_k = L\left(I - GG^*\Phi_{k+1}(GG^*)\right) e_k}_{Error\ Evolution} \,, \quad k \geq 0 \,, \tag{7.80}$$

and hence governed by the operator LL_G, which, given the above, must satisfy

$$-L \leq LL_G \leq L \,. \tag{7.81}$$

Next, note that the error evolution expression above suggests that the gradient calculations can be combined into one expression based on the input update formula

$$u_{k+1}^{(M_{k+1}-1)} = u_k + G^*\Phi_{k+1}(GG^*)e_k \,. \tag{7.82}$$

Combining this with the NOILC online update formula gives

$$\begin{aligned} u_{k+1} &= u_{k+1}^{(M_{k+1}-1)} + \epsilon^{-2}G^*e_{k+1} \\ &= u_k + \epsilon^{-2}G^*\left(\epsilon^2\Phi_{k+1}(GG^*)e_k + e_{k+1}\right) \end{aligned} \tag{7.83}$$

which is the input obtained from the optimization problem

$$u_{k+1} = argmin\{\ J_{k+1}(u, u_k)\ ;\ y = Gu + d\ \} \tag{7.84}$$

with

$$\underbrace{J_{k+1}(u, u_k) = \|r_{k+1} - y\|^2_{\mathcal{Y}} + \|u - u_k\|^2_{\mathcal{U}}}_{Modified\ NOILC\ Objective\ Function} \tag{7.85}$$

which is a NOILC problem with r replaced by

$$\underbrace{r_{k+1} = r + \epsilon^2\Phi_{k+1}(GG^*)e_k}_{Error\ Corrected\ Reference} \,. \tag{7.86}$$

This signal can be computed from r and the measured data e_k, although, if $M_{k+1} \geq 3$, some processing using the model operator G and its adjoint G^* is necessary.

7.4.6.3 Algorithm Properties

The error correction suggested above has a number of important properties outlined below:

1. *The inclusion of multiple gradient computations again leads to a single NOILC optimization problem with a more general form of error corrected reference* (7.86).
2. *The main additional computation is the calculation of* r_{k+1}. The simplest cases are summarized as follows

$$\begin{array}{ccc} \underline{M_{k+1}} & & \underline{\Phi_{k+1}(GG^*)} \\ 1 & \rightarrow & 0 \\ 2 & \rightarrow & \beta^{(1)}_{k+1} I \\ 3 & \rightarrow & -\beta^{(1)}_{k+1}\beta^{(2)}_{k+1} GG^* + (\beta^{(1)}_{k+1} + \beta^{(2)}_{k+1}) I \\ \cdots & & \cdots \end{array} . \tag{7.87}$$

This reveals that the evaluation is obtained by linear combinations of terms of the form $(GG^*)^{j+1}e_k$ which can be written as $(GG^*)^{j+1}e_k = GG^*\left((GG^*)^j e_k\right)$. This takes the form of a recursive calculation of the form $\psi_{p+1} = GG^*\psi_j$ with $\psi_0 = e_k$ which is evaluated using the system model in two steps, namely computing $\phi_p = G^*\psi_p$, followed by computation of $\psi_{p+1} = G\phi_p$.

Note 7.2 The evaluation of r_{k+1} involves $2(M_{k+1} - 1)$ simulations, if $M_{k+1} \geq 3$, using the system model and its adjoint. The computational load, hence increases linearly as M_{k+1} increases. This is just one factor that needs to be taken into account when considering the choice of M_{k+1}.

3. A *Feedforward Implementation* of the algorithm can be approached using very similar techniques to those used in previous sections. The relationship between the inputs u_k and u_{k+1} used on the plant is,

$$\begin{aligned} u_{k+1} &= u_k + \epsilon^{-2}G^*(r_{k+1} - y_{k+1})\,, \\ &= u_k + \epsilon^{-2}G^*(r_{k+1} - r + e_{k+1})\,, \\ &= u_k + \epsilon^{-2}G^*\left(\epsilon^2\Phi_{k+1}(GG^*) + L(I - GG^*\Phi_{k+1}(GG^*))\right)e_k \\ &= \underbrace{u_k + \epsilon^{-2}G^*\left(I + \epsilon^2\Phi_{k+1}(GG^*)\right)Le_k}_{Feedforward\ Implementation\ Formula}\,. \end{aligned} \tag{7.88}$$

Remark 7.4 Note that Le_k is the feedforward term obtained from the measured error signal e_k. The computation of this signal has been described in feedforward versions of NOILC and Error Corrected algorithms. Given this signal, $\left(I + \epsilon^2\Phi_{k+1}(GG^*)\right)Le_k$ can be computed using similar methods to those described for r_{k+1}.

4. *Spectral manipulation can be achieved by choice of* M_{k+1} *and* $\beta_{k+1}^{(j)}$, $1 \leq j \leq M_{k+1} - 1$. Following the approach of Sect. 7.4.3, the function describing the effect of the algorithm on the spectrum of the evolution operator in (7.80) is

$$f_{me}(\sigma^2) = \frac{1 - \sigma^2 \Phi_{k+1}(\sigma^2)}{1 + \epsilon^{-2}\sigma^2}\ , \quad 0 \leq \sigma^2 \leq \|G\|^2 \ . \tag{7.89}$$

For parameters in the range (7.73), the accelerating effect of the algorithm is confirmed, using (7.81), as

$$|f_{me}(\sigma^2)| \leq \underbrace{\frac{1}{1 + \epsilon^{-2}\sigma^2}\ , \quad 0 \leq \sigma^2 \leq \|G\|^2}_{NOILC\ Spectral\ Function}\ , \tag{7.90}$$

and, increasing M_{k+1} by one will further reduce the value of f_{me}.
For a given choice of parameters, convergence is accelerated as M_{k+1} increases but, more insight can be obtained using the spectral/eigenvalue and parameter choice techniques of Sect. 7.4.3. Those ideas indicate that, for those values in the range (7.43), the contribution of eigenvectors corresponding to spectral/eigenvalues equal to or near the values $(\beta_{k+1}^{(j)})^{-1}$ will be eliminated or almost eliminated.
For example, in the case of $M_{k+1} = 3$, the error evolution is associated with the function,

$$f_{me}(\sigma^2) = \frac{(1 - \beta_{k+1}^{(1)}\sigma^2)(1 - \beta_{k+1}^{(2)}\sigma^2)}{1 + \epsilon^{-2}\sigma^2}\ , \quad 0 \leq \sigma^2 \leq \|G\|^2 \ . \tag{7.91}$$

In this case, four examples illustrate the effect of parameter choice and the improvement on convergence properties of NOILC:

a. If $M_{k+1} = 1$ and $\epsilon^2 = \|G\|^2$ (the NOILC Algorithm with no gradient components), then representative values of f_{me} are

$$f_{me}(\|G\|^2) = \frac{1}{2}\ , \ f_{me}(\frac{3}{4}\|G\|^2) = \frac{4}{7}\ , \ f_{me}(\frac{1}{2}\|G\|^2) = \frac{2}{3}\ . \tag{7.92}$$

b. If $M_{k+1} = 2$ (i.e., the Error Corrected Algorithm of Sect. 7.4.1), $\epsilon^2 = \|G\|^2$ and $\beta_{k+1}^{(1)} = \|G\|^2$, then representative values of f_{me} are

$$f_{me}(\|G\|^2) = \frac{1}{2}\ , \ f_{me}(\frac{3}{4}\|G\|^2) = \frac{1}{7}\ , \ f_{me}(\frac{1}{2}\|G\|^2) = \frac{1}{3}\ . \tag{7.93}$$

c. If $M_{k+1} = 3$, $\epsilon^2 = \|G\|^2$ and $\beta_{k+1}^{(1)} = \beta_{k+1}^{(1)} = \|G\|^{-2}$, then values of f_{me} are

$$f_{me}(\|G\|^2) = 0\ , \ f_{me}(\frac{3}{4}\|G\|^2) = \frac{1}{28}\ , \ f_{me}(\frac{1}{2}\|G\|^2) = \frac{1}{6}\ . \tag{7.94}$$

d. If $M_{k+1} = 3$, $\beta_{k+1}^{(1)} = \|G\|^{-2}$, $\beta_{k+1}^{(2)} = 2\|G\|^{-2}$ and $\epsilon^2 = \|G\|^2$, then

$$f_{me}(\|G\|^2) = f_{me}(\frac{1}{2}\|G\|^2) = 0 \ \ \& \ \ f_{me}(\frac{3}{4}\|G\|^2) = -\frac{1}{14} \, . \tag{7.95}$$

These simple calculations suggest that there are theoretical benefits to be obtained in improved convergence rates by increasing M_{k+1} and by using distinct values of $\beta_{k+1}^{(j)}$, $1 \le j \le M_{k+1} - 1$. There may however be implications for the robustness of the algorithm.

7.4.6.4 Algorithm Robustness

The *robustness* of the algorithm can be approached using very similar techniques to those used in Sect. 7.4.5. Assuming a right multiplicative modelling error U, and a feedforward implementation leads to the relationship between the inputs u_k and u_{k+1} used on the plant

$$u_{k+1} = u_k + \epsilon^{-2} G^* \left(I + \epsilon^2 \Phi_{k+1}(GG^*) \right) L e_k . \tag{7.96}$$

When implemented online, the measured e_k and e_{k+1} are related by

$$e_{k+1} = \left(I - \epsilon^{-2} G U G^* \left(I + \epsilon^2 \Phi_{k+1}(GG^*) \right) L \right) e_k . \tag{7.97}$$

A form of robustness is defined by requiring the monotonicity of a norm on $\mathcal{Y}$ that is topologically equivalent to $\| \cdot \|_{\mathcal{Y}}$. The norm chosen is defined by the inner product

$$\begin{array}{c} \langle y, w \rangle_0 = \langle y, L(I + \epsilon^2 \Phi_{k+1}(GG^*))w \rangle_{\mathcal{Y}} \\ with \ \ \|e\|_0 = \sqrt{\langle e, e \rangle_0} . \end{array} \tag{7.98}$$

Note 7.3 This is a valid inner product and norm. More precisely, L and Φ_{k+1} are self-adjoint, commute and $\Phi_{k+1}(GG^*) \ge 0$. This follows from the fact that the gradient algorithm guarantees norm reduction for each parameter choice in the range of $\beta_{k+1}^{(j)}$ considered, i.e., $-I \le I - GG^*\Phi_{k+1}(GG^*) \le I$ so that $2I \ge GG^*\Phi_{k+1}(GG^*) \ge 0$, the result following.

A sufficient condition for robust monotonicity of the error sequence with respect to this norm is that

$$U + U^* > \underbrace{U^*G^*L(I + \overbrace{\epsilon^2 \Phi(GG^*)}^{\ge 0})GU \ge U^*G^*LGU}_{Robustness \ deterioration \ relative \ to \ NOILC} . \tag{7.99}$$

This criterion can be simplified by writing, after a little manipulation, with $L_0 = (I + \epsilon^2 G^* G)^{-1}$,

$$\begin{array}{c} G^* L(I + \epsilon^2 \Phi(GG^*))G = G^* G L_0 (I + \epsilon^2 \Phi(G^* G)) = \\ (I - L_0)(I + \epsilon^2 \Phi(G^* G)) \leq \beta_{me} I \ , \end{array} \tag{7.100}$$

where the scalar quantity β_{me} is defined by

$$\begin{aligned} 0 < \beta_{me} &= \max_{0 \leq \sigma^2 \leq \|G\|^2} \left(\tfrac{\sigma^2(\epsilon^{-2} + \Phi_{k+1}(\sigma^2))}{1 + \epsilon^{-2}\sigma^2} \right) \\ &= \max_{0 \leq \sigma^2 \leq \|G\|^2} \left(1 - f_{me}(\sigma^2)\right) \\ &= 1 - \min_{0 \leq \sigma^2 \leq \|G\|^2} f_{me}(\sigma^2) \\ &\leq 2 \, . \end{aligned} \tag{7.101}$$

Remark 7.5 Note that the shape of the graph of the spectral function f_{me} is not only the main influence on convergence characteristics but also has an influence on robustness.

The simplified robustness criteria hence take the familiar forms,

$$\underbrace{U + U^* > \beta_{me} U^* U \ \ or \ \ \left\| \frac{1}{\beta_{me}} I - U \right\| < \frac{1}{\beta_{me}}}_{Simplified\ Robustness\ Criteria} \, . \tag{7.102}$$

Clearly the robustness of the algorithm decreases as β_{me} increases as the set of U satisfying the conditions gets smaller.

Note that the robustness conditions reduce to the NOILC criterion if $M_{k+1} = 1$ and the derived criterion for Error Corrected Algorithm if $M_{k+1} = 2$. For $M_{k+1} \geq 3$, the interpretation is identical with value of β_{me} depending on M_{k+1}, ϵ^2 and the parameters $\{\beta_{k+1}^{(j)}\}_{k \geq 0}$.

Finally, the robustness analysis of this more general Error Corrected algorithm has led to identical robustness conditions to those of NOILC, differing in the crucial sense that the single parameter β_{me} changes, hence changing the set of uncertainties U that can be tolerated dependent on parameter choice. It is the way that the parameters influence the shape of f_{me} that provides the link to understanding the consequences of parameter choice. Two general comments can be made:

1. If $0 \leq f_{me}(\sigma^2) \leq 1, \ 0 \leq \sigma^2 \leq \|G\|^2$, then $\beta_{me} \leq 1$. This will happen if all β_{k+1} lie in the interval $0 \leq \beta_{k+1} \leq \|G\|^{-2}$.
2. If at least one of the β_{k+1} lie in the interval $\|G\|^{-2} < \beta \leq 2\|G\|^{-2}$ (as is required for application of the spectral annihilation methodology) then $\beta_{me} > 1$.

7.5 Discussion and Further Reading

The chapter indicates that there are a number of simple techniques for improving the convergence rate of the measured error norm sequence $\{\|e_k\|_{\mathcal{Y}}\}_{k\geq 0}$ and offer the possibility that the number of *online, experimental* iterations can be reduced, whilst also reducing the need for the use of higher gain (small ϵ^2) Norm Optimal Iterative Learning Control strategies. In the absence of modelling error, the potential benefits could be significant although, as NOILC calculations are model-based, the presence of modelling errors will limit what can be achieved and may limit the range of free parameters that can be used.

The approach to robustness used is useful but does not cover all forms of plant-model mismatch. In particular, neither nonlinear modelling errors nor plant parameter errors are covered by the analysis so, in practice, the results are mainly indicative, providing some confidence that modelling errors could be tolerated in accelerated NOILC but there are no guarantees.

The following comments can be made about the proposed accelerated algorithms:

1. **Avoiding High Gains:** The discussion is motivated by the observation that the main parameter used to control the convergence rate of NOILC is the weight $\epsilon^2 > 0$. If this is very small (represented by the condition $\epsilon^2 \ll \|G\|^2$), a feedback implementation would have high feedback gains. Avoiding the use of high gains can be a practical and a computational need in many control problems. Unfortunately the use of larger values of ϵ^2 can reduce the convergence speed of NOILC considerably. Iterative Control offers a number of alternatives discussed in this chapter.
2. **Precompensation:** The idea of preconditioning the underlying system using control schemes is a natural step to ensure good, general plant operation. The added benefit is that the resultant system "seen" by the Iterative Algorithm is "well behaved", a property that will both ease the use of, for example, NOILC and, also, improve convergence rates. The techniques on designing the compensators can be found in the textbooks, e.g., [6–8, 11].
3. **Extrapolation:** The use of extrapolation is well known in the more general area of Numerical Analysis. In NOILC, it can be used in feedforward implementations with an extrapolation factor computed from measured signals and signals generated by model simulations. When compared with the non-extrapolated algorithm, extrapolation reduces the robustness of the algorithm but the effects of this can be limited by imposing a maximum value on the extrapolation factor. Such a constraint will limit the benefits to convergence rate.
4. **Including Offline Iterations:** Useful reductions in the number of online experiments/iterations needed to reach a desired tracking accuracy can be obtained by undertaking a small number of offline, virtual iterations between each plant experiment. In the absence of modelling errors, this will produce an accelerated convergence but, in the presence of modelling errors, the analysis of the robustness of a feedforward implementation on the plant, suggests that robustness reduces as the number of virtual iterations is increased.

5. **Error Corrected References:** Noting that an offline, gradient iteration and an online NOILC iteration can be combined to accelerate algorithm convergence, it has been shown that the two steps can be replaced by a single online NOILC iteration with a modified reference signal computed from the previously measured tracking error. In the absence of modelling errors, this algorithm accelerates NOILC in a very useful way. The approach introduces a new parameter, the gradient gain, that can be varied on every iteration if desired. The choice of this parameter can be guided by analysis of the spectrum of the operator in the error evolution and is, theoretically, capable of systematically removing the spectral/frequency content of the error. This offers the possibility of:

   a. more rapid algorithm convergence and
   b. tailoring of the algorithm to suit the reference signal.

   The algorithm has a degree of robustness for sufficiently small values of $(1 + \epsilon^2 \beta_k)$. The analysis used in the derivation of robustness conditions is known to generate only sufficient (and not necessary) conditions. The degree of conservatism of the result is unknown and depends on the modelling error. The existence of a range of parameters that ensure robustness is however reassuring.
   Finally, the natural extension of the Error Corrected Algorithm to undertake more than one offline gradient iteration between online experiments can be analysed and produces theoretical benefits. There is an increased computational load associated with this approach which may not be acceptable and an analysis of robustness based on sufficient conditions does however show that the approach tends to reduce robustness dependent on parameter choice.

Interested readers are referred to [1–4, 10] as well as the companion text [9] for more detailed discussions.

References

1. Chu B, Owens DH (2009) Accelerated norm-optimal iterative learning control algorithms using successive projection. Int J Control 82(8):1469–1484
2. Chu B, Owens DH (2011) Accelerated predictive norm-optimal iterative learning control. Proc Inst Mech Eng Part I: J Syst Control Eng 225(6):744–759
3. Chu B, Freeman CT, Owens DH (2015) A novel design framework for point-to-point ILC using successive projection. IEEE Trans Control Syst Technol 23(3):1156–1163
4. Chu B, Owens DH, Freeman CT, Liu Y (2017) Point-to-point ILC with accelerated convergence. In: 2017 6th Data Driven Control and Learning Systems (DDCLS). IEEE, pp 530–535
5. Freeman CT (2015) Upper limb electrical stimulation using input-output linearization and iterative learning control. IEEE Trans Control Syst Technol 23(4):1546–1554
6. Maciejowski JM (1989) Multivariable feedback design. Addison-Wesley
7. Owens DH (1978) Feedback and multivariable systems. Peter Peregrinus, Stevenage

8. Owens DH (1981) Multivariable and optimal systems. Academic Press, London
9. Owens DH (2015) Iterative learning control—an optimization paradigm. Springer
10. Owens DH, Chu B (2024) Error corrected references for accelerated convergence of low gain norm optimal iterative learning control. IEEE Trans Autom Control 69(9):5836–5851
11. Skogestad S, Postlethwaite I (2005) Multivariable feedback control: analysis and design. Wiley

Chapter 8
Predictive Norm Optimal Iterative Learning Control

This chapter uses the notation and terminology of previous chapters to introduce the idea of prediction to improve the performance of Iterative Control algorithms. The underlying intuition is that, if, on the current iteration, the effect of control input decisions on future iterations can be taken into account, then it may be possible to choose the input to improve algorithm performance, particularly convergence rates. The inclusion of prediction will, inevitably, increase the computational load off-line, and possibly on-line.

The idea is applied to Norm Optimal Iterative Learning Control (NOILC) to develop the "Predictive Norm Optimal Iterative Learning Control" algorithm, abbreviated to **PNOILC**, using an optimization problem to generate Iterative Learning Control (ILC) algorithms that improve on the performance of "basic" NOILC whilst retaining the possibility of using more familiar Optimal Control methods as part of the computational structure and implementations.

The major application of the theory will be to the control of systems with linear state-space models but the algorithms will also apply to more general linear system descriptions. To incorporate this degree of generality into the development, input-output plant dynamics will, again, be described by the relationship

$$y = Gu + d \quad \textit{with input} \;\; u \in \mathcal{U} \;\; \textit{and output} \;\; y \in \mathcal{Y} \tag{8.1}$$

where $\mathcal{U}$ and $\mathcal{Y}$ are real Hilbert spaces. The bounded linear operator $G : \mathcal{U} \to \mathcal{Y}$ and the term $d \in \mathcal{Y}$ represents initial conditions and similar repeatable contributions to plant dynamics. The control objective is to track a specified reference signal $r \in \mathcal{Y}$ using an Iterative Control procedure that can be implemented using a combination of off-line calculations and on-line experiments to assess and generate new data to be used in the next iteration. The tracking error is $e = r - y$ and, as in previous chapters, signals f on iteration k are denoted by f_k. Iterations are initiated by an input signal u_0 generating an initial error $e_0 = r - y_0$ with output $y_0 = Gu_0 + d$ being obtained either from simulation or from experiments on the physical plant.

B. Chu and D. H. Owens, *Optimal Iterative Learning Control*, Advances in Industrial Control, https://doi.org/10.1007/978-3-031-80236-2_8

8.1 The Multi-model Used for Prediction

The *Multi-model* describing the prediction process consists of the plant (or its off-line model) and $M - 1$ identical models of plant behaviour. Conceptually, this offers the possibilities of running the models, in real-time, in parallel to the plant and/or using M models of the plant, off-line, for control computations such as those required for a feedforward implementation.

In the absence of modelling error, the mathematical form of the Multi-model takes the form of M identical equations

$$\underbrace{y^{(j)} = Gu^{(j)} + d \ , \quad with\ errors\ \ e^{(j)} = r - y^{(j)} \ \ , \ 1 \leq j \leq M}_{The\ Multi-model} \ , \tag{8.2}$$

where $y^{(j)}$, $u^{(j)}$ and $e^{(j)}$ are the output, input, and error, respectively, associated with the jth copy.

Note 8.1 One of these models will be regarded as the plant (or a model associated with then plant), but, for the moment, the effect of making this choice is deferred.

The first step is to put the Multi-model into a familiar form. Using a matrix style notation, set

$$\begin{gathered} y^e = \begin{bmatrix} y^{(1)} \\ y^{(2)} \\ \vdots \\ y^{(M)} \end{bmatrix} , \quad u^e = \begin{bmatrix} u^{(1)} \\ u^{(2)} \\ \vdots \\ u^{(M)} \end{bmatrix} , \quad r^e = \begin{bmatrix} r \\ r \\ \vdots \\ r \end{bmatrix} , \quad d^{(rep)} = \begin{bmatrix} d \\ d \\ \vdots \\ d \end{bmatrix} , \\ e^e = \begin{bmatrix} e^{(1)} \\ e^{(2)} \\ \vdots \\ e^{(M)} \end{bmatrix} , \quad \& \quad G_e = \underbrace{\begin{bmatrix} G & 0 & \cdots & 0 \\ 0 & G & \cdots & 0 \\ \vdots & & & \vdots \\ 0 & 0 & \cdots & G \end{bmatrix}}_{(block\ diagonal)} . \end{gathered} \tag{8.3}$$

Note 8.2 In this chapter, the superscript (or subscript) "e" denotes an "expanded" version of the vector or operator whilst the superscript "(rep)" denotes a repeated version of a vector.

Using the above, the dynamics of the Multi-model is described by

$$\begin{aligned} y^e &= G_e u^e + d^{(rep)} \ , \quad &(Multi-model\ in\ a\ compact\ form) \\ e^e &= r^e - y^e \ , \quad &(Multi-model\ error) \end{aligned} \tag{8.4}$$

which has the same structure as the plant model.

The next step is to put the approach into a Hilbert space setting. Construct the inner products from those on $\mathcal{Y}$ and $\mathcal{U}$ using the definitions,

$$\underbrace{\begin{array}{c}\langle y^e, w^e\rangle_e = \sum_{j=1}^{M} \lambda^{2(j-1)} \langle y^{(j)}, w^{(j)}\rangle_{\mathcal{Y}} \\ \& \\ \langle u^e, v^e\rangle_e = \sum_{j=1}^{M} \lambda^{2(j-1)} \langle u^{(j)}, v^{(j)}\rangle_{\mathcal{U}} \ ,\end{array}}_{Multi-Model\ Inner\ Products\ in\ \mathcal{Y}_e\ and\ \mathcal{U}_e} \tag{8.5}$$

where $\lambda > 0$ is a real number. With these definitions, signals y^e, r^e and $d^{(rep)}$ can be regarded as vectors in a Hilbert space $\mathcal{Y}_e$ with inputs u^e in a real Hilbert space $\mathcal{U}_e$. The operator G_e is then a mapping from $\mathcal{U}_e$ into $\mathcal{Y}_e$.

These definitions lead naturally to the norms defined, for example, by

$$\underbrace{\|y^e\|_e^2 = \sum_{j=1}^{M} \lambda^{2(j-1)} \|y^{(j)}\|_{\mathcal{Y}}^2 \ \& \ \|u^e\|_e^2 = \sum_{j=1}^{M} \lambda^{2(j-1)} \|u^{(j)}\|_{\mathcal{U}}^2 \ .}_{Norms\ in\ \mathcal{Y}_e\ and\ \mathcal{U}_e} \tag{8.6}$$

Finally, it follows from the definitions that G_e, regarded as an operator from $\mathcal{U}_e \to \mathcal{Y}_e$, has the adjoint

$$G_e^{\dagger} = \underbrace{\begin{bmatrix} G^* & 0 & \cdots & 0 \\ 0 & G^* & \cdots & 0 \\ \vdots & & & \vdots \\ 0 & 0 & \cdots & G^* \end{bmatrix}}_{(block\ diagonal)} \ : \ \mathcal{Y}_e \to \mathcal{U}_e \ , \tag{8.7}$$

where, to avoid any confusion, two symbols for adjoints are used, namely:

1. † denotes the adjoint with respect to the inner products in $\mathcal{U}_e$ and $\mathcal{Y}_e$ and
2. * denotes the adjoint with respect to the inner products in $\mathcal{U}$ and $\mathcal{Y}$.

8.2 The Objective Function and Choice of λ

The approach to developing a predictive iterative algorithm is to examine iteration $k+1$ with the assumption that the input u_k, output y_k and hence the error e_k from iteration k are known. The use of the Norm-Optimal Algorithm over iterations $k+1, \ldots, k+M$ in the "*Prediction Horizon*" would lead to sequential minimization of the NOILC objective functions

$$J(u, u_k)\ , \quad J(u, u_{k+1})\ , \quad \ldots \quad , \ J(u, u_{k+M})\ , \tag{8.8}$$

subject to the plant dynamics. For Predictive NOILC, this sequence is replaced by the construction of inputs $\{u_{k+1}^{(j)}\}_{1 \le j \le M}$ as those that minimize the single objective function

$$J_M(\lambda, u^{(1)}, u^{(2)}, \ldots, u^{(M)}, u_k) = \sum_{j=1}^{M} \lambda^{2(j-1)} J(u^{(j)}, u^{(j-1)})$$
$$with \quad u^{(0)} = u_k \quad \& \tag{8.9}$$
$$J(u^{(j)}, u^{(j-1)}) = \|e^{(j)}\|_{\mathcal{Y}}^2 + \epsilon^2 \|u^{(j)} - u^{(j-1)}\|_{\mathcal{U}}^2$$

subject to the multi-model dynamics (8.4).

Note that:

1. The optimization problem is a linear, quadratic problem and can be solved using standard optimization tools.
2. In general, it cannot be assumed that the solution $\{u_{k+1}^{(j)}\}_{1 \le j \le M}$ will be equal to that computed from M iterations of the NOILC Algorithm.
3. Suppose that there exists an input u_∞ that produces a zero error from the plant. As the sequence $u^{(1)} = u^{(2)} = \cdots = u^{(M)} = u_\infty$ is suboptimal, it follows that, at the minimum,

$$\begin{aligned} J_2(\lambda, u_{k+1}^{(1)}, u_{k+1}^{(2)}, \ldots, u_{k+1}^{(M)}, u_k) &\le J_M(\lambda, u_\infty, \ldots, u_\infty, u_k) \\ &= \epsilon^2 \|u_\infty - u_k\|_{\mathcal{U}}^2 \end{aligned} \tag{8.10}$$

and, in particular,

$$\begin{aligned} &\sum_{j=1}^{M} \lambda^{2(j-1)} \|e_{k+1}^{(j)}\|_{\mathcal{Y}}^2 \le \epsilon^2 \|u_\infty - u_k\|_{\mathcal{U}}^2 , \\ &so\ that \ \ \|e_{k+1}^{(j)}\|_{\mathcal{Y}} \le \epsilon \lambda^{1-j} \|u_\infty - u_k\|_{\mathcal{U}} \end{aligned} \tag{8.11}$$

so that,

a. for a fixed value of prediction horizon M,

$$\lim_{\lambda \to \infty} \|e_{k+1}^{(j)}\|_{\mathcal{Y}} = 0 , \quad 2 \le j \le M , \tag{8.12}$$

so that increasing λ and using any of the inputs $u_{k+1}^{(j)}$, $2 \le j \le M$ in a plant trial will produce arbitrarily small tracking errors. In effect, the optimization problem provides almost exact tracking in one computation as $\lambda \to \infty$.

b. A similar effect is achieved if $\lambda > 1$ is fixed and M is increased. In particular,

$$\limsup_{M \to \infty} \|e_{k+1}^{(M)}\|_{\mathcal{Y}} = 0 , \tag{8.13}$$

That is, increasing M with any fixed $\lambda > 1$ produces arbitrarily small tracking errors for all sufficiently large indices j.

From the above, it must be concluded that increases in M and/or λ create, asymptotically, inputs that produce arbitrarily small tracking errors. This suggests that an algorithm based on suitable choice of both M and λ may be capable of significantly accelerating the performance of NOILC. However, care may be needed as:

1. increasing M requires more models to be used in the multi-model with the associated increased computational load whilst,
2. increases in the value of the scalar λ are easier to implement, although it is not inconceivable that this mechanism will be limited by any ill-conditioning that may be introduced into the computations.

Remark 8.1 This suggests that the most practical way forward is to choose $M = 2$ and to then adjust λ to achieve the acceleration required. This choice will play a significant role in the rest of this chapter in investigating the properties of the algorithm and illustrating its predicted performance in practice.

8.3 A State-Space Example Formulation

Consider an ℓ-input, m-output, linear, time-invariant, discrete-time system with state-space model of the form

$$\begin{aligned} x(t+1) &= Ax(t) + Bu(t)\,, \quad x(0) = x_0\,, \\ y(t) &= Cx(t)\,, \quad 0 \le t \le N\,, \end{aligned} \tag{8.14}$$

with reference signal time series $r = \{r(0), r(1), r(2), \ldots, r(N)\}$, error sequence $e(t) = r(t) - y(t)$, $0 \le t \le N$. Given an input u_k on the kth iteration, the basic NOILC algorithm would use an objective function

$$J(u, u_k) = \sum_{t=0}^{N} \left(e^T(t)Q(t)e(t) + \epsilon^2 (u(t) - u_k(t))^T R(t)(u(t) - u_k(t)) \right) \,. \tag{8.15}$$

Converting the iteration into a Predictive Control iteration replaces $J(u, u_k)$ by the objective function

$$\begin{aligned} J_M(\lambda, u^{(1)}, u^{(2)}, \ldots, u^{(M)}, u_k) &= \textstyle\sum_{j=1}^{M} \lambda^{2(j-1)} J(u^{(j)}, u^{(j-1)}) \\ with \quad u^{(0)} &= u_k\,, \end{aligned} \tag{8.16}$$

and computes the inputs $\{u_{k+1}^{(j-1)}\}_{1 \le j \le M}$ minimizing J_M subject to the multi-model

$$\begin{aligned} x^{(j)}(t+1) &= Ax^{(j)}(t) + Bu^{(j)}(t)\,, \qquad x^{(j)}(0) = x_0\,, \\ y^{(j)} &= Cx^{(j)}(t)\,, \quad 0 \le t \le N\,, \quad 1 \le j \le M\,, \end{aligned} \tag{8.17}$$

written as

$$\underbrace{\begin{aligned} x^e(t+1) = A_e x^e(t) + B_e u^e(t)\,,\ \ x^e(0) = x_0^{(rep)}\,, \\ y^e(t) = C_e x^e(t)\,, \quad 0 \le t \le N\,, \end{aligned}}_{Multi-model\ state\ space\ form}\,, \quad x_0^{(rep)} = \begin{bmatrix} x_0 \\ \vdots \\ x_0 \end{bmatrix}, \tag{8.18}$$

where, in block matrix form,

$$\begin{aligned} y_e(t) &= \begin{bmatrix} y^{(1)}(t) \\ \vdots \\ y^{(M)}(t) \end{bmatrix}, \quad A_e = \underbrace{\begin{bmatrix} A & 0 & \cdots & 0 \\ 0 & A & \cdots & 0 \\ \vdots & & & \vdots \\ 0 & 0 & \cdots & A \end{bmatrix}}_{M\ column\ blocks}, \quad B_e = \begin{bmatrix} B & 0 & \cdots & 0 \\ 0 & B & \cdots & 0 \\ \vdots & & & \vdots \\ 0 & 0 & \cdots & B \end{bmatrix} \\ C_e &= \begin{bmatrix} C & 0 & \cdots & 0 \\ 0 & C & \cdots & 0 \\ \vdots & & & \vdots \\ 0 & 0 & \cdots & C \end{bmatrix}, \quad x_e(t) = \begin{bmatrix} x^{(1)}(t) \\ \vdots \\ x^{(M)}(t) \end{bmatrix}, \quad u_e(t) = \begin{bmatrix} u^{(1)}(t) \\ \vdots \\ u^{(M)}(t) \end{bmatrix}. \end{aligned} \tag{8.19}$$

The objective function can be put into a form compatible with NOILC by writing,

$$\begin{bmatrix} u^{(1)}(t) - u^{(0)}(t) \\ \vdots \\ u^{(M)}(t) - u^{(M-1)}(t) \end{bmatrix} = F_0\left(u^e(t) - u_k^{(rep)}(t)\right), \quad u_k^{(rep)}(t) = \begin{bmatrix} u_k(t) \\ \vdots \\ u_k(t) \end{bmatrix} \tag{8.20}$$

with

$$F_0 = \begin{bmatrix} I & 0 & 0 & \cdots & 0 \\ -I & I & 0 & \cdots & 0 \\ 0 & -I & I & \cdots & 0 \\ \vdots & & & & \vdots \\ 0 & \cdots & 0 & -I & I \end{bmatrix}, \quad F_0^{-1} = \begin{bmatrix} I & 0 & 0 & \cdots & 0 \\ I & I & 0 & \cdots & 0 \\ I & I & I & \cdots & 0 \\ \vdots & & & & \vdots \\ I & \cdots & I & I & I \end{bmatrix}. \tag{8.21}$$

This gives the objective function $J_M(\lambda, u^{(1)}, u^{(2)}, \ldots, u^{(M)}, u_k)$ in the form

$$J(u^e, u_k^{(rep)}) = \sum_{t=0}^{N} \left((e^e(t))^T Q_e(t) e^e(t) + \epsilon^2 (u^e(t) - u_k^{(rep)}(t))^T R_e(t)(u^e(t) - u_k^{(rep)}(t)) \right) \tag{8.22}$$

where

$$Q_e(t) = \begin{bmatrix} Q(t) & 0 & \cdots & 0 \\ 0 & \lambda^2 Q(t) & \cdots & 0 \\ \vdots & & & \vdots \\ 0 & \cdots & 0 & \lambda^{2(M-1)} Q(t) \end{bmatrix} \quad (8.23)$$

and

$$R_e(t) = F_0^T \begin{bmatrix} R(t) & 0 & \cdots & 0 \\ 0 & \lambda^2 R(t) & \cdots & 0 \\ \vdots & & & \vdots \\ 0 & \cdots & 0 & \lambda^{2(M-1)} R(t) \end{bmatrix} F_0. \quad (8.24)$$

The optimization problem represented by (8.22) and (8.18) is the problem to be solved to obtain the optimal solution $u^e_{k+1}(t),\ 0 \le t \le N$ that will form the basis of the construction of the next plant input $u_{k+1}(t),\ 0 \le t \le N$. The methodology for this final choice is discussed later in the chapter.

8.4 Predictive Iterative Control: Properties

The previous section has shown that Predictive Iterative Learning Control can be seen, in part as a Norm-Optimal Iterative Learning problem. This is shown to be true using the Multi-model (8.4)

$$y^e = G_e u^e + d^{(rep)} \quad (8.25)$$

as the constraint on the minimization of a quadratic objective function of NOILC form aimed at improving the tracking of r^e by y^e using u^e. This idea is developed below and ensures that Predictive Iterative Control inherits many of the properties of NOILC.

8.4.1 Predictive Control as NOILC

Suppose that the kth iteration has been completed and the plant data u_k, y_k, e_k have been recorded. To show that $(k+1)$th Predictive Optimal Iterative Control iteration can be formulated as a familiar NOILC iteration, consideration must be given to the form of the model (8.4) and objective function (8.9). The first step is to transform the Multi-model using a matrix operator transformation F_0 of the form of (8.21) and the definition, with $u^{(0)} = u_k$,

$$\begin{bmatrix} u^{(1)} - u^{(0)} \\ \vdots \\ u^{(M)} - u^{(M-1)} \end{bmatrix} = \Delta u^e = F_0 \left(u^e - u_k^{(rep)} \right) , \quad u_k^{(rep)} = \begin{bmatrix} u_k \\ \vdots \\ u_k \end{bmatrix} . \tag{8.26}$$

Substituting into the Multi-model gives

$$y^e = G_e \left(F_0^{-1} \Delta u^e + u_k^{(rep)} \right) + d^{(rep)} = G_e F_0^{-1} \Delta u^e + y_k^{(rep)} , \tag{8.27}$$

which produces the error dynamics

$$\underbrace{e^e = -G_e F_0^{-1} \Delta u^e + e_k^{(rep)}}_{Error-based\ Model} , \quad e_k^{(rep)} = \begin{bmatrix} e_k \\ \vdots \\ e_k \end{bmatrix} . \tag{8.28}$$

With this notation, the objective function takes the NOILC form,

$$\underbrace{J_M(\lambda, u^{(1)}, u^{(2)}, \ldots, u^{(M)}, u_k) = \|e^e\|_e^2 + \epsilon^2 \|\Delta u^e\|_e^2}_{Predictive\ Objective\ Function\ in\ NOILC\ Form} \tag{8.29}$$

which is minimized by

$$\begin{aligned} \Delta u_{k+1}^e &= \epsilon^{-2} (G_e F_0^{-1})^\dagger e_{k+1}^e , \\ i.e., \quad u_{k+1}^e &= \underbrace{u_k^{(rep)} + \epsilon^{-2} F_0^{-1} (G_e F_0^{-1})^\dagger e_{k+1}^e}_{Multi-model\ Optimal\ Inputs} . \end{aligned} \tag{8.30}$$

Remark 8.2 Note the use of the adjoint $\dagger$ notation as $G_e F_0^{-1} : \mathcal{U}_e \to \mathcal{Y}_e$.

The operator $G_e F_0^{-1}$ has the matrix representation

$$G_e F_0^{-1} = \begin{bmatrix} G & 0 & 0 & \cdots & 0 \\ G & G & 0 & \cdots & 0 \\ \vdots & & & & \vdots \\ G & G & G & \cdots & G \end{bmatrix} : \mathcal{U}_e \to \mathcal{Y}_e , \tag{8.31}$$

and its adjoint $(G_e F_0^{-1})^\dagger$ can be deduced from the identity

$$\sum_{j=1}^{M} \lambda^{2(j-1)} \langle y^{(j)}, \sum_{i=1}^{j} G w^{(i)} \rangle_{\mathcal{Y}} = \sum_{i=1}^{M} \lambda^{2(i-1)} \langle \sum_{j=i}^{M} \lambda^{2(j-i)} G^* y^{(j)}, w^{(i)} \rangle_{\mathcal{U}} \tag{8.32}$$

and takes the form

$$(G_e F_0^{-1})^\dagger = \begin{bmatrix} G^* & \lambda^2 G^* & \lambda^4 G^* & \cdots & \lambda^{2(M-1)} G^* \\ 0 & G^* & \lambda^2 G^* & \cdots & \lambda^{2(M-2)} G^* \\ \vdots & & & & \vdots \\ 0 & 0 & 0 & \cdots & G^* \end{bmatrix} : \mathcal{Y}_e \to \mathcal{U}_e , \tag{8.33}$$

where G^* is the adjoint of $G : \mathcal{U} \to \mathcal{Y}$ in the, more familiar, $\mathcal{U}$ and $\mathcal{Y}$ inner product topologies.

It follows that

$$\mathcal{R}[G_e F_0^{-1}(G_e F_0^{-1})^\dagger] = \{ (\gamma_1^T, \ldots, \gamma_M^T)^T \ ; \ \gamma_j \in \mathcal{R}[GG^*], \ 1 \le j \le M \}$$
$$ker[G_e F_0^{-1}(G_e F_0^{-1})^\dagger] = \{ (\gamma_1^T, \ldots, \gamma_M^T)^T \ ; \ \gamma_j \in ker[GG^*], \ 1 \le j \le M \}.$$

Using (8.30) in (8.28) gives the relationship

$$e_{k+1}^e = -\epsilon^{-2} G_e F_0^{-1}(G_e F_0^{-1})^\dagger e_{k+1}^e + e_k^{(rep)} , \tag{8.34}$$

so that, the optimal Multi-model errors $\{e_{k+1}^{(j)}\}_{1 \le j \le M}$ on iteration $k+1$ are described by

$$\underbrace{e_{k+1}^e = (I + \epsilon^{-2} G_e F_0^{-1}(G_e F_0^{-1})^\dagger)^{-1} e_k^{(rep)}}_{Optimal\ Multi-model\ error\ computation} , \quad e_{k+1}^e = \begin{bmatrix} e_{k+1}^{(1)} \\ \vdots \\ e_{k+1}^{(M)} \end{bmatrix} , \tag{8.35}$$

or,

$$e_{k+1}^{(j)} = L_{M,j} e_k , \quad 1 \le j \le M \tag{8.36}$$

where

$$\begin{bmatrix} L_{M,1} \\ \vdots \\ L_{M,M} \end{bmatrix} = (I + \epsilon^{-2} G_e F_0^{-1}(G_e F_0^{-1})^\dagger)^{-1} \begin{bmatrix} I \\ \vdots \\ I \end{bmatrix} \tag{8.37}$$

defines the operators $\{L_{M,j}\}_{1 \le j \le M}$. Note that, as they are algebraic functions of the elements of the self-adjoint operator $G_e F_0^{-1}(G_e F_0^{-1})^\dagger$ and hence of GG^*, each is self-adjoint in the inner product topology of $\mathcal{Y}$.

The above descriptions normally play no role in the computations. Their value lies primarily in making analysis of the algorithm possible.

8.4.2 The Case of $M = 2$

The case of $M = 2$ is likely to be the most practical design choice in many situations. Analysis of this case reveals the essential properties of the algorithm. The approach uses the equations developed in Sect. 8.4.1 with $M = 2$.

The first step is to find the form of $L_{2,j},\ j = 1, 2$ from the description (8.35), i.e.,

$$\begin{bmatrix} L_{2,1} \\ L_{2,2} \end{bmatrix} = (I + \epsilon^{-2} G_e F_0^{-1} (G_e F_0^{-1})^{\dagger})^{-1} \begin{bmatrix} I \\ I \end{bmatrix}, \tag{8.38}$$

where

$$\begin{aligned} G_e F_0^{-1} (G_e F_0^{-1})^{\dagger} &= \begin{bmatrix} G & 0 \\ G & G \end{bmatrix} \begin{bmatrix} G^* & \lambda^2 G^* \\ 0 & G^* \end{bmatrix} \\ &= \begin{bmatrix} GG^* & \lambda^2 GG^* \\ GG^* & (1+\lambda^2) GG^* \end{bmatrix}. \end{aligned} \tag{8.39}$$

As a consequence

$$\begin{aligned} (I + \epsilon^{-2} G_e F_0^{-1} (G_e F_0^{-1})^{\dagger})^{-1} &= \begin{bmatrix} I + \epsilon^{-2} GG^* & \lambda^2 \epsilon^{-2} GG^* \\ \epsilon^{-2} GG^* & I + (1+\lambda^2)\epsilon^{-2} GG^* \end{bmatrix}^{-1} \\ = \begin{bmatrix} H^{-1} & 0 \\ 0 & H^{-1} \end{bmatrix} &\begin{bmatrix} I + (1+\lambda^2)\epsilon^{-2} GG^* & -\lambda^2 \epsilon^{-2} GG^* \\ -\epsilon^{-2} GG^* & I + \epsilon^{-2} GG^* \end{bmatrix} \\ \textit{where} \quad H = (I + \epsilon^{-2} GG^*)^2 &+ \lambda^2 \epsilon^{-2} GG^* . \end{aligned} \tag{8.40}$$

Note that H^{-1} exists and is bounded as $H \geq I$.

It follows that,

$$\begin{aligned} &\underbrace{L_{2,1} = H^{-1} L^{-1},\ L_{2,2} = H^{-1}}_{\textit{The Operators } L_{2,j},\ j=1,2} \quad \textit{where} \quad \underbrace{L = \left(I + \epsilon^{-2} GG^*\right)^{-1}}_{\textit{NOILC Operator}}, \\ &\textit{and} \qquad e_{k+1}^{(1)} = L_{2,1} e_k,\quad e_{k+1}^{(2)} = L_{2,2} e_k . \end{aligned} \tag{8.41}$$

As H and L commute, $GG^* \geq 0$ and $L^{-1} \geq I$, it follows that $0 < L_{2,2} \leq L_{2,1}$ with respect to the inner product in $\mathcal{Y}$. More generally,

$$\underbrace{0 < L_{2,2} \leq L_{2,1} \leq L \leq I}_{\textit{Ordering of } L_{2,j},\ j=1,2, \textit{ and } L}, \tag{8.42}$$

as

$$\begin{aligned} L - H^{-1} L^{-1} = L H^{-1} \left(H - L^{-2}\right) &= \lambda^2 \epsilon^{-2} H^{-1} L G G^* \\ &= \lambda^2 H^{-1} (I - L) \geq 0 . \end{aligned} \tag{8.43}$$

The operator ordering (8.42) implies immediately that

$$\underbrace{\|e_{k+1}^{(2)}\|_{\mathcal{Y}} \le \|e_{k+1}^{(1)}\|_{\mathcal{Y}} \le \|Le_k\|_{\mathcal{Y}}}_{Guaranteed\ Improvements\ on\ NOILC} \tag{8.44}$$

where Le_k is the error obtained using a NOILC iteration using u_k. In addition, as $L_{2,2} = L_{2,1}L$, it follows that

$$\underbrace{L_{2,1} \le L \qquad \& \qquad L_{2,2} \le L^2}_{Better\ than\ one\ and\ two\ NOILC\ Iterations} \tag{8.45}$$

so that the error norm $\|e_{k+1}^{(j)}\|_{\mathcal{Y}}^2$ is smaller than that achieved using j sequential NOILC iterations.

Remark 8.3 A more detailed analysis of error convergence is presented in Sect. 8.4.4, together with important insights into the influence of the parameter $\lambda \ge 1$ on the convergence properties.

8.4.3 Choosing u_{k+1} Using $\{u_{k+1}^{(j)}\}_{1 \le j \le M}$

A knowledge of the M components $\{u_{k+1}^{(j)}\}_{1 \le j \le M}$ in the minimizing u_{k+1}^e does not make the decision on how to choose u_{k+1} (i.e., the actual input to be applied to the system on iteration $k+1$). This decision may require additional computations/considerations. The analysis of the $M = 2$ case suggests the more general properties,

$$\underbrace{0 \le L_{M,M} \le \cdots \le L_{M,2} \le L_{M,1} \le L \le I}_{Ordering\ of\ the\ L_{M,j}} \qquad and$$
$$\underbrace{\|e_{k+1}^{(M)}\|_{\mathcal{Y}} \le \cdots \le \|e_{k+1}^{(2)}\|_{\mathcal{Y}} \le \|e_{k+1}^{(1)}\|_{\mathcal{Y}} \le \|Le_k\|_{\mathcal{Y}}}_{Guaranteed\ Improvements\ on\ NOILC} . \tag{8.46}$$

A detailed analysis confirms this [11] but is not included here. The following observations can be made:

1. Any choice $u_{k+1} = u_{k+1}^{(j)}$ will be an improvement on the use of a single NOILC iteration.
2. Although each error $e_{k+1}^{(j)}$ is an "improvement" on its predecessor, the role of $u_{k+1}^{(j)}$ is also to condition the sequence to optimize the form and effectiveness of its successors $u_{k+1}^{(i)}$, $i > j$. This could imply that the natural choice for u_{k+1} is

$$u_{k+1} = u_{k+1}^{(1)} \quad (The\ Receding\ Horizon\ Choice) . \tag{8.47}$$

If $\lambda > 1$, minimization of J_M places greater numerical weight on the terms $j \geq 1$, so that, intuitively, $u^{(1)}_{k+1}$ should both reduce the error $e^{(1)}_{k+1}$ substantially and use an input that reduces the need for significant updates in inputs.

3. Alternatively the norm inequalities suggest that

$$u_{k+1} = u^{(M)}_{k+1} \quad (Greatest\ Reduction\ in\ Error\ Norm) \tag{8.48}$$

may be a good choice as it produces, on each iteration, the greatest reduction in error norm.

4. More generally, it is possible to choose scalar parameters $\{a_j\}_{1 \leq j \leq M}$, and set

$$u_{k+1} = \sum_{j=1}^{M} a_j u^{(j)}_{k+1} \,. \tag{8.49}$$

If $\sum_{j=1}^{M} a_j = 1$, this generates the error

$$e_{k+1} = \sum_{j=1}^{M} a_j e^{(j)}_{k+1} = \sum_{j=1}^{M} a_j L_{M,j} u^{(j)}_{k+1} \,. \tag{8.50}$$

The choice of $a_1 = 1,\ a_j = 0,\ j > 1$, is the Receding Horizon Choice whilst $a_M = 1,\ a_j = 0\ ,\ j < M$, generates the greatest norm reduction (see [7]). Whatever the choice made, the parameters could remain the same for all iterations. Alternatively, they could then be chosen to minimize the norm of e_{k+1} to achieve the maximum benefit on that iteration. That is, it may be useful to choose the $\{a_j\}_{1 \leq j \leq M}$ using optimization problems of the form, say,

$$\{a_j\}_{1 \leq j \leq M} = argmin\{\|e_{k+1}\|^2_{\mathcal{Y}}\ ;\ \sum_{j=1}^{M} a_j = 1\ ,\ a_j \geq 0,\ j = 1, \ldots, M\}\ , \tag{8.51}$$

where the positivity conditions have been introduced to avoid any ill-conditioning of the problem.

This approach is suited to a feedforward implementation of the approach but it could also be used in a feedback implementation if the errors obtained from empirical considerations or model-based computations are used to choose the parameters.

For example, for $M = 2$, with $a_1 = (1 - a)$ and $a_2 = a$ with a a scalar parameter, this problem is simply the minimization of the quadratic function

$$\|e_{k+1}\|^2_{\mathcal{Y}} = (1-a)^2 \|e^{(1)}_{k+1}\|^2_{\mathcal{Y}} + a^2 \|e^{(2)}_{k+1}\|^2_{\mathcal{Y}} + 2a(1-a)\langle e^{(1)}_{k+1}, e^{(2)}_{k+1}\rangle_{\mathcal{Y}} \tag{8.52}$$

subject to the constraint that $0 \leq a \leq 1$.

5. An alterative to the optimization problem above is the choice of a_j using the following optimization problem,

$$\{a_j\}_{1 \le j \le M} = argmin\{\|e_{k+1}\|^2_{\mathcal{Y}} + \omega(\sum_{j=1}^{M-1} a_j^2 + (a_j^2 - 1)^2)\, ;\, a_j \ge 0,\ j = 1, \ldots, M\}, \tag{8.53}$$

where $\omega > 0$ a small weighting factor to limit the value of a_js to be used.
6. Being a NOILC algorithm, the extrapolation method for NOILC introduced in previous chapter to accelerate its convergence can also be used, i.e.,

$$u_{k+1} = u_k + \lambda_{k+1}(u^{(1)}_{k+1} - u_k) \tag{8.54}$$

where λ_{k+1} is an extrapolation parameter to be chosen (see Sect. 7.2 and [3] for more detail).

8.4.4 Error Convergence Properties

In what follows, the $\{a_j\}$ are assumed to be iteration independent. The conclusions remain true, however, if they vary from iteration to iteration. The details are omitted and left for the reader as an exercise.

General Nature of Algorithm Convergence: Using the notation of Sect. 8.4.3, consider the error evolution in $\mathcal{Y}$,

$$e_{k+1} = L_p e_k\,, \quad (Error\ Evolution)$$

$$\begin{aligned} L_p = L_p^* = \textstyle\sum_{j=1}^M a_j L_{M,j}\,, \quad L_p : \mathcal{Y} \to \mathcal{Y}\,, \\ with\ \textstyle\sum_{j=1}^M a_j = 1\ \&\ a_j \ge 0\,,\ 1 \le j \le M\,,\ \ so\ that \\ 0 \le L_p \le L \le I\,. \end{aligned} \tag{8.55}$$

The equation has the same structure as that seen in NOILC and the same self-adjoint property. The fact that $L_p \le L$ and that L_p commutes with L indicates that $L_p^k \le L^k$ for all $k \ge 1$ and hence that,

$$\begin{aligned} \|e_k\|^2_{\mathcal{Y}} = \langle L_p^k e_0, L_p^k e_0\rangle_{\mathcal{Y}} = \langle e_0, L_p^{2k} e_0\rangle_{\mathcal{Y}} \\ \le \langle e_0, L^{2k} e_0\rangle_{\mathcal{Y}} = \langle L^k e_0, L^k e_0\rangle_{\mathcal{Y}} = \|L^k e_0\|^2_{\mathcal{Y}}\,. \end{aligned} \tag{8.56}$$

It follows that $\{e_k\}_{k \ge 0}$ converges to zero if $\lim_{k\to\infty} L^k e_0 = 0$, i.e., if NOILC converges to zero from the initial error e_0. Using the convergence proofs for NOILC, this always occurs when $e_0 \in \overline{\mathcal{R}[G]}$. If, however, $\overline{\mathcal{R}[G]} \ne \mathcal{Y}$ (that is, $ker[G^*] \ne \{0\}$), the expression (using (8.37))

$$L_p = \left[a_1 I \cdots a_M I \right] (I + \epsilon^{-2} G_e F_0^{-1} (G_e F_0^{-1})^*)^{-1} \begin{bmatrix} I \\ \vdots \\ I \end{bmatrix} , \tag{8.57}$$

and the algebraic expression for matrix inversion, implies that $L_p e_k = e_k$ whenever $e_k \in ker[G^*]$. As $\mathcal{Y} = \overline{\mathcal{R}[G]} \oplus ker[G^*]$ and $ker[G^*]^\perp = \overline{\mathcal{R}[G]}$, it follows that,

$$\underbrace{\lim_{k \to \infty} e_k = P_{ker[G^*]} e_0}_{Error\ Convergence\ Property} \quad , \tag{8.58}$$

where $P_{ker[G^*]}$ is the orthogonal projection operator from $\mathcal{Y}$ onto $ker[G^*]$.

In summary, not only does the Predictive Iterative Learning Control Algorithm improve on the convergence rates of NOILC (in a manner dependent on the choice of $\lambda > 1$), it also inherits the error convergence properties of Norm Optimal Iterative Control with the same limit errors.

Eigenstructure of GG^*: In the basic NOILC Algorithm and in the Error Corrected Reference Algorithms for example, a more detailed picture of the nature of the convergence was obtained by analysis of a function associated with the operator in the error update equation and the spectrum/eigenvalues of GG^*.

The use of a spectral/eigenvalue analysis of GG^* necessitates that, in summary:

1. GG^* has a spectrum that consists only of eigenvalues of $\sigma_0^2 = 0$ (if $ker[G^*] \neq \{0\}$) and positive values $\sigma_j^2 > 0$, ordered in the natural ascending way,

$$0 < \cdots \leq \sigma_{j+1}^2 \leq \sigma_j^2 \leq \cdots \leq \sigma_2^2 \leq \sigma_1^2 \, . \tag{8.59}$$

2. The operator norm has an exact relationship to the largest eigenvalue,

$$\sigma_1^2 = \|GG^*\| = \|G\|^2 = \|G^*\|^2. \ (Operator\ Norm\ and\ Eigenvalues) \tag{8.60}$$

3. The span of the eigenvectors of GG^* corresponding to zero eigenvalues is precisely $ker[GG^*] = ker[G^*]$.
4. Suppose also that GG^* has a complete (finite or infinite) set of eigenvectors $\{w_j\}_{j \geq 1}$ spanning $\overline{\mathcal{R}[GG^*]}$. As GG^* is self-adjoint, they can be assumed to be orthonormal, namely, they satisfy,

$$GG^* w_j = \sigma_j^2 w_j \, , \ j \geq 1 \, , \quad \& \quad \langle w_j, w_p \rangle_{\mathcal{Y}} = \delta_{j,p} \quad (Eigenvector\ Equation\ for\ GG^*) \tag{8.61}$$

where $\delta_{j,p}$ is the Kronecker delta. In addition, $GG^* w_0 = 0$ for all $w_0 \in ker[GG^*]$. All problems with finite dimensional signal space $\mathcal{Y}$ (e.g., those associated with discrete-time state-space models) automatically have this property.
5. The set of eigenvectors form a basis for $\overline{\mathcal{R}[GG^*]}$. In this circumstance, any error, for example, can be expanded in terms of the eigenvectors

$$e_k = a_{0,k}w_0 + \sum_{j\ge 1} a_{j,k}w_j\ , \quad (Eigenvector\ Expansion) \\ w_0 \in ker[G^*]\ , \quad \|w_0\|_{\mathcal{Y}} = 1 \tag{8.62}$$

by suitable choice of the coefficients $a_{j,k}$. The norm of e_k can then be expressed in the form

$$\|e_k\|^2_{\mathcal{Y}} = \sum_{j\ge 0} a^2_{j,k} < +\infty\ . \tag{8.63}$$

Convergence Characteristics and the Effect of λ when $M = 2$: Using the analysis of Sect. 8.4.2 and, in particular, equation (8.41) gives

$$L_p = H^{-1}\left(a_1 L^{-1} + a_2 I\right) = H^{-1}\left((a_1+a_2)I + a_1\epsilon^{-2}GG^*\right) \\ H = (I+\epsilon^{-2}GG^*)^2 + \lambda^2\epsilon^{-2}GG^*\ , \\ a_1 + a_2 = 1\ , \ a_1 \ge 0\ , \ a_2 \ge 0\ . \tag{8.64}$$

Defining the spectral functions for $L_{2,j},\ j = 1, 2,$

$$f_{2,1}(\sigma^2) = \frac{1+\epsilon^{-2}\sigma^2}{(1+\epsilon^{-2}\sigma^2)^2 + \lambda^2\epsilon^{-2}\sigma^2} \quad \& \quad f_{2,2}(\sigma^2) = \frac{1}{(1+\epsilon^{-2}\sigma^2)^2 + \lambda^2\epsilon^{-2}\sigma^2}\ , \tag{8.65}$$

gives the spectral function for $L_p = a_1 L_{2,1} + a_2 L_{2,2}$ as,

$$f_p(\sigma^2) = \frac{a_1 + a_2 + a_1\epsilon^{-2}\sigma^2}{(1+\epsilon^{-2}\sigma^2)^2 + \lambda^2\epsilon^{-2}\sigma^2}\ , \qquad 0 \le \sigma^2 \le \|G\|^2\ . \tag{8.66}$$

As a consequence:

1. The eigenvalues of L_p are obtained by setting $\sigma^2 = 0$ (if $ker[GG^*] \ne \{0\}$) and $\sigma^2 = \sigma_j^2,\ j \ge 1$.
2. The shape of the function f_p provides insight into the convergence of the error sequence in terms of the attenuation of the eigenvectors as described by

$$e_k = L_p^k e_0 = a_{0,0}w_0 + \sum_{j\ge 1} a_{j,0} f_p(\sigma_j^2)^k w_j\ , \\ a_{0,k} = a_{0,0}\ , \quad \& \ a_{k,j} = a_{j,0} f_p(\sigma_j^2)^k\ , \quad j \ge 1\ , \tag{8.67}$$

even if the eigenvalues are unknown.

The functions $f_{2,j}(\sigma^2),\ j = 1, 2,$ and $f_p(\sigma^2)$ have a number of properties, namely:

1. $0 < f_{2,2}(\sigma^2) \le f_{2,1}(\sigma^2) \le 1$ for all $\sigma^2 \ge 0$, implying that the choice of $u_{k+1} = u_{k+1}^{(2)}$ will produce faster convergence than the choice $u_{k+1} = u_{k+1}^{(1)}$ by increased attenuation of *all* eigenvectors of GG^*.
2. $f_p(\sigma_j^2)$ are the eigenvalues of the NOILC algorithm if $\lambda = 0,\ a_1 = 1,\ a_2 = 0$.
3. $f_p(0) = 1$ for all $a_1 + a_2 = 1$ and all λ.
4. $f_p(\sigma^2)$ is bounded for all $\lambda > 0,\ a_1 \ge 0,\ a_2 \ge 0,\ a_1 + a_2 = 1$, as follows,

$$0 < f_{2,2}(\sigma^2) \le f_p(\sigma^2) \le f_{2,1}(\sigma^2) \le \frac{1}{1+\epsilon^{-2}\sigma^2} \le 1\,, \;\; for\, all\; \lambda > 0\,. \tag{8.68}$$

This implies that, for parameters as defined, the convergence speed following on from the choice will lie between that achieved by choosing $u_{k+1} = u_{k+1}^{(1)}$ and $u_{k+1} = u_{k+1}^{(2)}$.

5. For all choices of positive a_1, a_2 and λ, and all $\sigma^2 \ge 0$,

$$\frac{df_p(\sigma^2)}{d\sigma^2} < 0\,, \tag{8.69}$$

so that $f_p(\sigma^2)$ is monotonically decreasing.

6. For all parameter choices,

$$\limsup_{\lambda \to +\infty} \{\, f_p(\sigma^2)\;;\; \|G\|^2 \ge \sigma^2 \ge \delta \,\} = 0\,, \quad for\; all \quad \delta > 0\,. \tag{8.70}$$

This can be interpreted as saying that large values of λ will produce an algorithm that will, ultimately, almost eliminate, in one iteration, all of the eigenvectors corresponding to eigenvalues in any "bandwidth" $\delta \le \sigma^2 \le \|G\|^2$ where $\delta > 0$.

Remark 8.4 Clearly, λ plays a similar role to ϵ^2 in NOILC, fast convergence being obtainable by suitable choice *even if* ϵ^2*is large*.

7. Finally, the choice of parameters will have an effect on the robustness of the algorithm (Sect. 8.5).

Remark 8.5 (Frequency Attenuation Interpretation) Using the observations of Sect. 2.14.5, the identification of approximate eigenvalues with the function $f_p(\sigma^2)$ and frequency domain estimates of σ^2 can be valuable. For SISO systems and interval tracking problems, the relevant representation is $f_p(R^{-1}Q|G(z|^2),\ |z| = 1$. That is, for strictly proper systems small values of the eigenvalues can be interpreted as describing high frequency behaviour whilst larger values can be associated with resonance and/or DC behaviour. In particular, this suggests that the rate of convergence of the high frequency component of the reference time series will be slow whilst that of the low frequency will be much faster.

These observations provide some insight into the form and nature of f_p as a function of parameters, particularly $\lambda > 1$. More detail can be revealed using simple examples, e.g., consider a situation where $\epsilon^2 = \|G\|^2$. Evaluation of the pairs $(f_{2,1}(\sigma^2), f_{2,2}(\sigma^2))$, at several points in the interval $0 \le \sigma^2 \le \|G\|^2$ gives,

σ^2	$\lambda = 1$	$\lambda = 2$	$\lambda = 4$	$(1 + \|G\|^{-2}\sigma^2)^{-1}$
$\|G\|^2$	(0.40, 0.20)	(0.25, 0.125)	(0.10, 0.05)	0.50
$0.5\|G\|^2$	(0.55, 0.36)	(0.35, 0.23)	(0.15, 0.10)	0.67
$0.25\|G\|^2$	(0.69, 0.55)	(0.49, 0.39)	(0.22, 0.18)	0.80
$0.1\|G\|^2$	(0.84, 0.76)	(0.68, 0.62)	(0.4, 0.36)	0.91

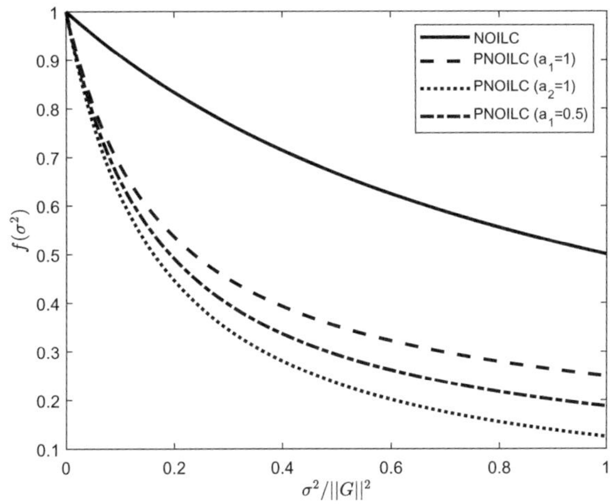

Fig. 8.1 Upper and lower bounds on the spectrum of NOILC and L_p for $M = 2,\ \lambda = 2$

the last column being the equivalent attenuation for the NOILC algorithm. A plot of $f_{2,1}(\sigma^2)$ and $f_{2,2}(\sigma^2)$ for the case of $\lambda = 2$ is shown in the Fig. 8.1 and represents the upper and lower values for $f_p(\sigma^2)$ for choices of $a_1 \geq 0,\ a_2 \geq 0,\ a_1 + a_2 = 1$.

Predictive NOILC converges increasingly rapidly as λ increases and is always better than that achieved by NOILC with the same value of ϵ^2. In particular:

1. As λ increases, the improvement is substantial for eigenvalues in the vicinity of $\sigma^2 = \|G\|^2$, but improvements for smaller eigenvalues gradually decrease.
2. The effect of the choice of ϵ^2 is revealed by noting that that

$$\begin{array}{c} \lim_{\epsilon^2 \to 0+} f_{2,1}(\sigma^2) = \lim_{\epsilon^2 \to 0+} f_{2,2}(\sigma^2) = 0\ , \quad \sigma^2 > 0\ , \\ \& \qquad f_{2,2}(\sigma^2) f_{2,1}^{-1}(\sigma^2) = (1 + \epsilon^{-2}\sigma^2)^{-1}\ . \end{array} \tag{8.71}$$

More precisely, as the value of ϵ^2 decreases, the algorithm converges more rapidly and the relative convergence rates described by $f_{2,1}(\sigma^2)$ and $f_{2,2}(\sigma^2)$ increase.

8.4.5 Implementation

Implementation of the Predictive Algorithm, as in other NOILC-based algorithms, can take either a feedback or feedforward form.

The feedback implementation for state-space systems is a natural consequence of Linear Quadratic Optimal Control Theory and the solution of the minimization of J_M subject to the Multi-model constraint, expressed in its state feedback form. The implementation must parallel the Multi-model structure by replacing one of the M plant models by the plant itself. The experimental phase therefore uses synchronized, real-time simulation of $M - 1$ models in parallel with the plant to create a physical

Multi-model to generate the data required to compute the input signal. The Multi-model state in this situation consists of the states of the $M-1$ models and the estimated state of the plant deduced from input-output measurements obtained from the plant. The optimal state feedback matrix and other components of the control signal can be computed off-line using the Multi-model. If the plant replaces the qth component of the Multi-model, then the state $x^e(t)$ used for feedback consists of the real-time model states $x^{(j)}(t),\ j \neq q$, but with $x^{(q)}(t)$ replaced by the estimated plant state. This data will generate, in real-time, the M feedback inputs $\{u^{(j)}(t)\}_{1\leq j\leq M}$ which are then applied to the Multi-model. As the plant is only one component, this means that the choice of $\{a_j\}_{1\leq j\leq M}$ is constrained to be $a_j = \delta_{jq}$, where δ_{jq} is the Kronecker Delta. That is, $u_{k+1}(t)$ must be set equal to $u^{(q)}(t)$. For example, if $q=1$, the implementation is using the Receding Horizon control choice.

The feedforward form of the algorithm, in general, derives the new input from the measured plant tracking error. It is deduced as follows. Using the input description (8.30), the error model (8.28) and objective function (8.29), the Multi-model "error" e^e_{k+1} is generated by solving the NOILC minimization problem

$$\underbrace{\begin{array}{l} J_M = \|e_k^{(rep)} - z^e\|_e^2 + \epsilon^2\|\Delta u^e\|_e^2 \\ subject\ to \quad z^e = G_e F_0^{-1}\Delta u^e \end{array},}_{Feedforward\ Optimization\ Problem} \tag{8.72}$$

for the solution Δu^e_{k+1} and z^e_{k+1} and setting

$$\underbrace{e^e_{k+1} = e_k^{(rep)} - z^e_{k+1}}_{Optimal\ Multi-model\ Error} . \tag{8.73}$$

For the state-space example of Section 8.3, the feedforward computation is the off-line minimization of the quadratic performance index,

$$J(\Delta u^e, 0) = \sum_{t=0}^{N}\left((e_k^{(rep)}(t) - z^e(t))^T Q_e(t)(e_k^{(rep)}(t) - z^e(t)) + \epsilon^2\Delta u^e(t)^T R_e(t)\Delta u^e(t)\right) , \tag{8.74}$$

subject to the modified Multi-model dynamics,

$$\begin{array}{l} x^e(t+1) = A_e x^e(t) + B_e \underbrace{F_0^{-1}}\Delta u^e(t) , \quad \underbrace{x^e(0) = 0} , \\ z^e(t) = C_e x^e(t) , \quad 0 \leq t \leq N . \end{array} \tag{8.75}$$

Note that the main changes introduced into the model are the inclusion of F_0^{-1} and the use of zero initial conditions.

8.4.6 Computational Examples

In this section, we present two numerical examples to illustrate the improved performance that is obtainable using the PNOILC algorithm. The first example demonstrates the power of using prediction for achieving faster error convergence.

Example 8.1 Consider the following system

$$G(s) = \frac{s+1}{s^2+5s+6}, \tag{8.76}$$

which is sampled using zero-order hold and a sampling time of 0.1 s. The trial length is 20 s and the reference signal is generated by the square-wave input shown in Fig. 8.2. The chosen initial input is $u_0 = 0$. The simulation compares the performance of the Predictive NOILC Algorithm (with the receding horizon choice, i.e., $a_1 = 1, a_j = 0, j > 1$) against the basic NOILC Algorithm over 20 iterations. Simulations were run with $\lambda = 2$, $Q = 1$, $R = 0.1$, $\epsilon^2 = 1$ and $M = 2, 3, 4$, respectively. The results are shown in Fig. 8.3.

From Fig. 8.3, it can be seen that the predictive NOILC algorithm with $M = 2, 3, 4$ outperforms the NOILC algorithm and the improvement is very significant. It can be found that the convergence rate of the predictive NOILC algorithm (error norm reduction) in the first few iterations is much faster than NOILC, which is of great value in practice. The norms of the tracking error in the first eight iterations are shown in Table 8.1 to provide a more detailed picture. Moreover, the initial benefits of prediction are maintained with the final error at 20th iteration with the choice of $M = 4$ being less than 0.4% of that achieved by NOILC.

The second example illustrates the effect of parameter choices on the convergence performance.

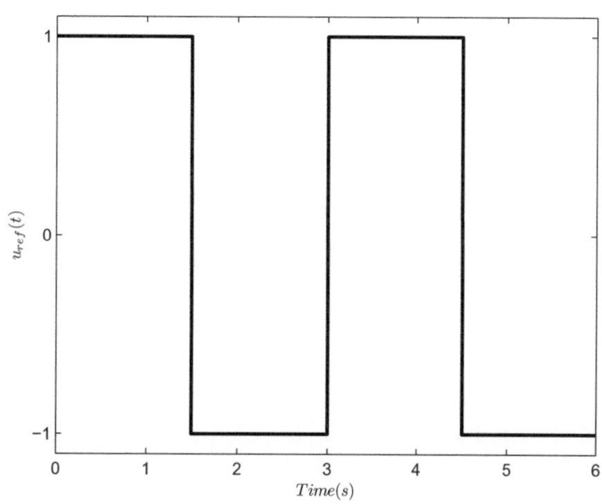

Fig. 8.2 The input signal $u^*(t)$ for Example 8.1

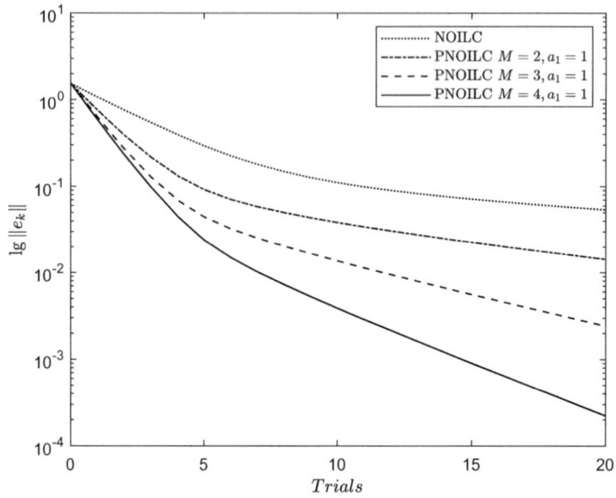

Fig. 8.3 Comparison of convergence for Example 8.1

Table 8.1 Tracking error norms for Example 8.1

Iteration (k)	NOILC	PNOILC $M = 2, a_1 = 1$	PNOILC $M = 3, a_1 = 1$	PNOILC $M = 4, a_1 = 1$
0	1.5530	1.5530	1.5530	1.5530
1	1.0816	0.7676	0.6399	0.5972
2	0.7604	0.3933	0.2738	0.2351
3	0.5419	0.2158	0.1274	0.0974
4	0.3937	0.1321	0.0690	0.0448
5	0.2933	0.0919	0.0445	0.0241
6	0.2256	0.0709	0.0325	0.0150
7	0.1797	0.0585	0.0253	0.0102
8	0.1483	0.0499	0.0204	0.0073

Example 8.2 Consider the same setting as in the previous example and compare the convergence performance of the algorithm for different prediction horizons M and with different choices of selecting the input. In particular, consider the choice of $a_1 = 1$ (i.e., the receding horizon choice) and $a_M = 1$ (with the greatest reduction in error norm), as well as the optimal non-negative linear combination a_M^* (i.e., the solution of (8.53)). The result is shown in Fig. 8.4 and Table 8.2.

From Fig. 8.4, it can be seen again that the Predictive NOILC Algorithm outperforms the NOILC algorithm. Furthermore, a different choice of input for the next trial can produce significantly different convergence performance. From the figure, it can be seen that the convergence rate of the optimal non-negative linear combination a_M^* clearly outperforms all other choices, and converges much faster than others. What is more, it can be seen that the choice of the optimal non-negative linear combination

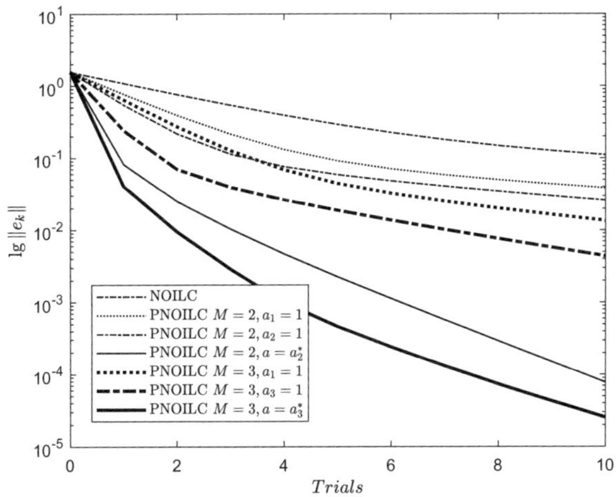

Fig. 8.4 Comparison of convergence for Example 8.2 with $M = 2, 3$

Table 8.2 Tracking error norms for Example 8.2

Iteration (k)	NOILC	PNOILC $M = 2, a_1 = 1$	PNOILC $M = 2, a_2 = 1$	PNOILC $M = 2, a = a_2^*$	PNOILC $M = 3, a_1 = 1$	PNOILC $M = 3, a_3 = 1$	PNOILC $M = 3, a = a_3^*$
0	1.5530	1.5530	1.5530	1.5530	1.5530	1.5530	1.5530
1	1.0816	0.7676	0.5444	0.0818	0.6399	0.2411	0.0404
2	0.7604	0.3933	0.2184	0.0251	0.2738	0.0702	0.0095
3	0.5419	0.2158	0.1137	0.0104	0.1274	0.0393	0.0029
4	0.3937	0.1321	0.0765	0.0047	0.0690	0.0265	0.0010
5	0.2933	0.0919	0.0590	0.0023	0.0445	0.0188	0.0005
6	0.2256	0.0709	0.0483	0.0011	0.0325	0.0137	0.0002
7	0.1797	0.0585	0.0406	0.0006	0.0253	0.0102	0.0001
8	0.1483	0.0499	0.0346	0.0003	0.0204	0.0076	0.0001

with prediction horizon $M = 2$ even outperforms Predictive NOILC with prediction horizon $M = 3$ and choices of $a_1 = 1$ and $a_3 = 1$, suggesting a substantial reduction in the computational load and great potential for practical applications.

8.5 Robustness Characterization

Robustness of the Predictive Algorithm can be approached using the same general methodology as that used for NOILC and Error Corrected NOILC.

8.5.1 Problem Formulation

A feedforward implementation is assumed and the plant input u_k, the measured error e_k and the parameters $\{a_j\}_{1 \le j \le M}$, $\sum_{j=1}^{M} a_j = 1, \ a_j \ge 0, \ 1 \le j \le M$, are assumed to be available. Using (8.30), (8.36), and (8.37),

$$u^e_{k+1} = u_k^{(rep)} + \epsilon^{-2} F_0^{-1} (G_e F_0^{-1})^\dagger e^e_{k+1}$$
$$\textit{so that}$$

$$\begin{aligned} u_{k+1} &= u_k + \epsilon^{-2} \begin{bmatrix} a_1 I \ a_2 I \ \cdots \ a_M I \end{bmatrix} F_0^{-1} (G_e F_0^{-1})^\dagger \begin{bmatrix} L_{M,1} \\ \vdots \\ L_{M,M} \end{bmatrix} e_k \\ &= u_k + \epsilon^{-2} G^* \begin{bmatrix} a_1 I \ a_2 I \ \cdots \ a_M I \end{bmatrix} F_0^{-1} (F_0^{-1})^\dagger \begin{bmatrix} L_{M,1} \\ \vdots \\ L_{M,M} \end{bmatrix} e_k \\ &= u_k + \epsilon^{-2} G^* \Omega_{\{a_j\}} e_k \end{aligned} \tag{8.77}$$

where $\Omega_{\{a_j\}} : \mathcal{Y} \to \mathcal{Y}$ is defined by

$$\Omega_{\{a_j\}} = \begin{bmatrix} a_1 I \ a_2 I \ \cdots \ a_M I \end{bmatrix} F_0^{-1} (F_0^{-1})^\dagger \begin{bmatrix} L_{M,1} \\ \vdots \\ L_{M,M} \end{bmatrix} . \tag{8.78}$$

This operator has two important properties, namely:

1. It is both self-adjoint and positive definite in the inner product topology in $\mathcal{Y}$ (as, the $\{L_{M,j}\}$ have this property), $a_j \ge 0, \ 1 \le j \le M$ and

$$F_0^{-1} (F_0^{-1})^\dagger = \begin{bmatrix} I & 0 & 0 & \cdots & 0 \\ I & I & 0 & \cdots & 0 \\ \vdots & & & \vdots & 0 \\ I & I & I & \cdots & I \end{bmatrix} \begin{bmatrix} I & \lambda^2 I & \lambda^4 I & \cdots & \lambda^{2(M-1)} \\ 0 & I & \lambda^2 I & \cdots & \lambda^{2(M-2)} I \\ \vdots & & & \vdots & \lambda^2 I \\ 0 & 0 & 0 & \cdots & I \end{bmatrix} , \tag{8.79}$$

   which is a matrix of blocks that are strictly positive multiples of the identity I.
2. It can be expressed as a rational polynomial $f_\Omega(\sigma^2)$ of the familiar operator GG^*. That is,

$$\Omega_{\{a_j\}} = f_\Omega(GG^*) . \tag{8.80}$$

Example 8.3 In the case of $M = 2$, this takes the form

$$\Omega_{\{a_j\}} = f_\Omega(GG^*)$$
$$= \begin{bmatrix} a_1 I & a_2 I \end{bmatrix} \begin{bmatrix} I & 0 \\ I & I \end{bmatrix} \begin{bmatrix} I & \lambda^2 I \\ 0 & I \end{bmatrix} \begin{bmatrix} L_{2,1} \\ L_{2,2} \end{bmatrix} \tag{8.81}$$
$$= L_{2,1} + (a_2 + \lambda^2) L_{2,2}$$

so that:

1. $f_\Omega(\sigma^2)$ takes the form

$$f_\Omega(\sigma^2) = f_{2,1}(\sigma^2) + (a_2 + \lambda^2) f_{2,2}(\sigma^2) = \frac{1+\epsilon^{-2}\sigma^2+(a_2+\lambda^2)}{(1+\epsilon^{-2}\sigma^2)^2+\epsilon^{-2}\lambda^2\sigma^2} \tag{8.82}$$

   where $f_{2,j}$, $j = 1, 2$ are calculated in Sect. 8.4.2.
2. f_Ω depends on $a_2 \geq 0$ only and hence satisfies the inequality

$$f_\Omega(\sigma^2) \geq f_{2,1}(\sigma^2) + \lambda^2 f_{2,2}(\sigma^2) \,, \quad \sigma^2 \geq 0 \,, \tag{8.83}$$

   i.e., it is uniformly minimized if the Receding Horizon Choice of parameters, $a_1 = 1$ and $a_2 = 0$, is used.

8.5.2 Robustness Conditions

Suppose that the plant is described precisely by replacing G by GU where the bounded linear operator $U : \mathcal{U} \to \mathcal{U}$ is a right multiplicative perturbation. The observed error evolution in this case is described by

$$\underbrace{e_{k+1} = \left(I - \epsilon^{-2} G U G^* \Omega_{\{a_j\}}\right) e_k \,, \quad k \geq 0}_{\textit{Plant Error Evolution}} \,. \tag{8.84}$$

It is used to define a new inner product $\langle \cdot, \cdot \rangle_0$ and associated norm in $\mathcal{Y}$, i.e., for all y, z in $\mathcal{Y}$

$$\langle y, z \rangle_0 = \langle y, \Omega_{\{a_j\}} z \rangle_{\mathcal{Y}} \,, \qquad \|y\|_0 = \sqrt{\langle y, y \rangle_0} \,. \tag{8.85}$$

Robustness is defined by requiring that the sequence of observed errors $\{e_k\}_{k \geq 0}$ has the monotonicity property

$$\|e_{k+1}\|_0^2 < \|e_k\|_0^2 \,, \quad k \geq 0 \quad (\textit{Monotonic Norm Reduction}). \tag{8.86}$$

Using an identical argument to that used in the robustness analysis of both NOILC (Sect. 5.5) and Error Corrected NOILC (Sect. 7.4.5), this is true if

$$\underbrace{U + U^* > \epsilon^{-2} U^* G^* \Omega_{\{a_j\}} G U \quad \textit{on } \mathcal{R}[G^*] \textit{ in } \mathcal{U}}_{\textit{Robust Monotonicity Condition in General Form}} \,, \tag{8.87}$$

or, if U has a bounded inverse $\hat{U}$,

$$\underbrace{\hat{U} + \hat{U}^* > \epsilon^{-2} G^* \Omega_{\{a_j\}} G \quad on \ \mathcal{R}[UG^*] \ in \ \mathcal{U}}_{Robust\ Monotonicity\ Condition\ in\ Inverse\ Form} . \tag{8.88}$$

These relation can be simplified, firstly, by noting that the operator $G^*\Omega_{\{a_j\}}G$ can be replaced by

$$\underbrace{G^* \Omega_{\{a_j\}} G = G^* f_\Omega(GG^*) G = \underbrace{f_\Omega(G^*G) G^* G}}_{Rational\ Polynomial\ Form} . \tag{8.89}$$

Two other changes make the conditions increasingly conservative, namely:

1. The range $\mathcal{R}[G^*]$ or $\mathcal{R}[UG^*]$ can be replaced by the whole space $\mathcal{U}$ removing the need to characterize the nature of $\mathcal{R}[G^*]$ or $\mathcal{R}[UG^*]$.
2. In addition, the dependence on the details of the complex operator $\Omega_{\{a_j\}}$ can be removed by using the inequality $\epsilon^{-2} G^* \Omega_{\{a_j\}} G \leq \zeta I$ to replace the robust monotonicity conditions by the familiar inequalities

$$\begin{aligned} &\underbrace{U + U^* > \zeta U^* U \qquad or \qquad \hat{U} + \hat{U}^* > \zeta I}_{Simplified\ Robustness\ Conditions} \\ &\zeta = \sup_{0 \leq \sigma^2 \leq \|G\|^2} \epsilon^{-2} \sigma^2 f_\Omega(\sigma^2) . \end{aligned} \tag{8.90}$$

A sufficient condition for this to be true is that

$$\underbrace{\left\| \frac{1}{\zeta} I - U \right\| < \frac{1}{\zeta}}_{Robustness\ Condition\ in\ Norm\ Form} , \tag{8.91}$$

again identical in structure to previously described conditions but with a different value for the parameter ζ.

The simplifications show that:

1. The structure of the robustness theory is identical to that of other optimization approaches described in previous chapters but each problem requires a different definition for the norm $\| \cdot \|_0$ and different values for the parameter ζ.
2. Robustness is predicted to improve as ζ decreases.
3. The case of $M = 2$, discussed in the last section, suggests that the use of the Receding Horizon Choice for a_1 and a_2 is the most robust choice, as ζ is then minimized. However, it will not lead to the greatest acceleration of convergence. In this case, $a_1 = 1$ and $a_2 = 0$ and:
   a. After a little calculus, $\frac{d(\sigma^2 f_\Omega(\sigma^2))}{d\sigma^2} > 0$, $\sigma^2 > 0$, and hence,

$$
\begin{aligned}
0 < \zeta \quad &= \sup_{0 \le \sigma^2 \le \|G\|^2} \left(\frac{\epsilon^{-2}\sigma^2(1+\epsilon^{-2}\sigma^2+\lambda^2)}{(1+\epsilon^{-2}\sigma^2)^2+\epsilon^{-2}\sigma^2\lambda^2} \right) , \\
&= \sup_{0 \le \sigma^2 \le \|G\|^2} \left(1 - \underbrace{\frac{(1+\epsilon^{-2}\sigma^2)}{(1+\epsilon^{-2}\sigma^2)^2+\epsilon^{-2}\sigma^2\lambda^2}}_{Positive\ and\ decreasing} \right) \\
&= \underbrace{\epsilon^{-2}\|G\|^2 f_\Omega(\|G\|^2)}_{Value\ of\ \zeta} < 1 ,
\end{aligned} \tag{8.92}
$$

i.e., the maximum value is less than unity and is achieved when $\sigma^2 = \|G\|^2$. It can be estimated using the function f_Ω and an estimate of $\|G\|$. In particular $\frac{\epsilon^{-2}\|G\|^2}{1+\epsilon^{-2}\|G\|^2} < \zeta$ so that Predictive NOILC is predicted to be less robust than the basic NOILC algorithm.

b. Also, $\frac{d\zeta}{d\lambda} > 0$ so that robustness decreases as λ increases. Some robustness is retained even if λ is very large as $\zeta \to 1-$ as $\lambda \to +\infty$ and the set of U satisfying the inequality $\|I - U\| < 1$ contains a wide range of permitted modelling errors U.

8.6 Discussion and Further Reading

This chapter introduces the idea of using optimization-based prediction concepts in iterative learning control design. The predictive algorithm developed is based on a multi-model description of the prediction process and the minimization of an objective function that has close relationships to that used in NOILC in Chap. 5. This close relationship is by design as it means that the computational infrastructure and implementation options are almost identical to those proposed in NOILC. Readers may wish to look at the more detailed material in the companion text [11] and references [1, 3, 4], if they so wish.

Two additional parameters are introduced into the design framework, namely M, the Prediction Horizon, and a weight parameter $\lambda > 0$ (and normally chosen to be ≥ 1) that describes the weighting of future virtual iterations in the predictive process. The NOILC process then offers M options for control signals to be used on the next iteration. This results in a number of choices of way forward. All of the resulting Predictive NOILC Algorithms described have significantly improved convergence performance when compared with the basic NOILC Algorithm (Chap. 5). A rigorous analysis provides detail of that convergence in terms of operator inequalities and an eigenvalue/eigenvector analysis. There are several possible design choices for the engineer to consider although, if computational complexity needs to be limited, the choice of $M = 2$ provides most of the benefits. For this reason, the presentation places emphasis on this case and reveals the properties of importance.

One of these properties is that of robustness. A general analysis follows the approach used for NOILC in Sect. 5.5 using an error characterization based on right, multiplicative, linear modelling errors U. Errors can take many other forms so more research is needed in this area to obtain a fuller picture of the robustness property. The general conclusion is that using prediction retains robustness but, as compared with NOILC, robustness margins are reduced. Despite this statement, it is emphasized that Predictive NOILC retains the same "core" robustness to the same defined class of modelling errors already seen in NOILC, i.e., those U satisfying $\|I - U\| < 1$. This class is not insignificant. In this sense, a substantial degree of robustness is, indeed, retained for this class but is reduced for errors outside this class.

Remark 8.6 It is worth mentioning that, being a NOILC-based design concept, all the methods introduced in Chap. 7 on accelerating convergence speed could be considered, in principle, to further improve the performance of PNOILC algorithms. This includes the use of pre-compensation, extrapolation, including off-line iterations as well as the use of error corrected references. Further research may clarify any benefits from such options and any details are not discussed further here.

Remark 8.7 It is also worth pointing out that this chapter considers the use of prediction for the NOILC algorithm. The idea can also be applied, in principle, to other ILC design methods, e.g., gradient-based ILC, inverse-based ILC etc., and the convergence and robustness analysis should follow in a similar manner to that described in this chapter. Results on predictive gradient ILC design can be found in [5].

All the key algorithms to be included in this text have now been described. In the next few chapters, the application of some of the algorithms is described in three representative case studies on robotic systems, non-minimum systems as well as healthcare. Additionally, the application of ILC extends more widely —interested readers are referred to [2, 6, 8–10, 12–16] for more examples.

References

1. Amann N, Owens DH, Rogers E (1998) Predictive optimal iterative learning control. Int J Control 69(2):203–226
2. Blanken L, Koekebakker S, Oomen T (2020) Data-driven feedforward tuning using non-causal rational basis functions: with application to an industrial flatbed printer. Mechatronics 71(102):424
3. Chu B, Owens DH (2011) Accelerated predictive norm-optimal iterative learning control. Proc Inst Mech Eng Part I: J Syst Control Eng 225(6):744–759
4. Chu B, Owens DH, Freeman CT (2015) Iterative learning control with predictive trial information: convergence, robustness, and experimental verification. IEEE Trans Control Syst Technol 24(3):1101–1108
5. Chu B, Owens DH, Freeman CT (2015) Predictive gradient iterative learning control. In: 2015 54th IEEE Conference on Decision and Control (CDC). IEEE, pp 2377–2382
6. Gunnarsson S, Norrlof M (2001) On the design of ILC algorithms using optimization. Automatica 37(12):2011–2016

7. Hatonen J, Owens DH (2004) Convex modifications to an iterative learning control law. Automatica 40(7):1213–1220
8. Jensen TN, Kallesøe CS, Bendtsen JD, Wisniewski R (2018) Iterative learning pressure control in water distribution networks. In: 2018 IEEE Conference on Control Technology and Applications (CCTA). IEEE, pp 583–588
9. Johansen SV, Jensen MR, Chu B, Bendtsen JD, Mogensen J, Rogers E (2019) Broiler FCR optimization using norm optimal terminal iterative learning control. IEEE Trans Control Syst Technol 29(2):580–592
10. Nowicka WN, Chu B, Tutty OR, Rogers E (2022) Reducing aerodynamic load fluctuation in wind turbines using iterative learning control laws designed using reduced order models of the flow. In: 2022 American Control Conference (ACC). IEEE, pp 4242–4247
11. Owens DH (2015) Iterative learning control—an optimization paradigm. Springer
12. Rogers E, Owens D, Werner H, Freeman C, Lewin P, Kirchhoff S, Schmidt C, Lichtenberg G (2010) Norm-optimal iterative learning control with application to problems in acceleration-based free electron lasers and rehabilitation robotics. Eur J Control 5:496–521
13. Rogers E, Chu B, Freeman C, Lewin P (2023) Iterative learning control algorithms and experimental benchmarking. Wiley
14. Wang Z, Pannier CP, Barton K, Hoelzle DJ (2018) Application of robust monotonically convergent spatial iterative learning control to microscale additive manufacturing. Mechatronics 56:157–165
15. Wu RB, Chu B, Owens DH, Rabitz H (2018) Data-driven gradient algorithm for high-precision quantum control. Phys Rev A 97(4):042122
16. Xu Z, Chu B, Geng H, Nian X, Zhang C (2024) Model-guided learning for wind farm power optimization. IEEE Trans Control Syst Technol 32(2):428–439

Chapter 9
Application to Robotic Systems

Application of Iterative Learning control (ILC) to practical systems is an important part of the development and evaluation of the subject area. A number of results are available, including [1, 7, 8, 10–13, 15–18]. It is also clearly necessary to test the practical value and robustness of these results in physical experiments where, inevitably, the transfer function used in analysis is only an approximation to the actual plant dynamics.

In this chapter, we present a case study on two robotic testing platforms, namely, a gantry robot system (described in Sect. 9.1) and an anthropomorphic robotic arm (introduced in Sect. 9.4). Experimental results for some of the key algorithms introduced in this book, namely norm optimal iterative learning control (NOILC) and its accelerated version, constrained NOILC algorithms for both hard and soft constraints, as well as NOILC for point-to-point tracking tasks, are presented to demonstrate their performance in practice, in Sects. 9.2, 9.3, and 9.5, respectively.

9.1 Gantry Robot: System Description and Modelling

Figure 9.1 shows a three-axis gantry robot experimental facility where the robot head performs a "pick-and-place" task and is similar to systems which can be found in many industrial applications. These include food canning, bottle filling, or automotive assembly, all of which require accurate tracking control each time the operation is performed, with a minimum level of error in order to maximize production rates. This is an obvious general area for the application of ILC.

The gantry robot has three axes, which are perpendicular to each other. The X-axis and the Y-axis move in the horizontal plane and are driven by linear brush-less DC motors. The Z-axis with vertical moving direction is placed on the top of the other two axes and comprises a linear ball-screw stage driven by a rotary brushless DC motor. For each axis, a BNC channel sends the input voltage signal to the motor amplifier,

B. Chu and D. H. Owens, *Optimal Iterative Learning Control*, Advances in Industrial Control, https://doi.org/10.1007/978-3-031-80236-2_9

and the displacement of each axis is measured as the output trajectory signal by an optical incremental encoder installed on the same axis. The hybrid motion of the three axes gives rise to a path of the gantry robot end effector in 3D space. This test platform consists of a dSPACE DS1103 microcontroller, Aerotech model BA10 linear amplifiers, and Renishaw RGH22 linear encoders with a resolution of 1 μm for X-axis and Y-axis and Aerotech rotary encoder for Z-axis with a resolution of 0.5 μm.

The gantry robot is designed to repeatedly complete a pick-and-place motion on the conveyor. A reference trajectory for the gantry movement has been pre-defined with the purpose of synchronizing its motion with that of the conveyor, which is running at a constant speed. Each axis is controlled individually and has its own reference trajectory and they are combined to form the 3D reference trajectory given in Fig. 9.2 which clearly shows the "pick-and-place" action. The signal length is 2 s in the default setting.

Each axis of the gantry robot has been modelled based on frequency response tests where, since the axes are orthogonal, it is assumed that there is minimal interaction between them. Here we first consider the X-axis (the one parallel to the conveyor in Fig. 9.1) and frequency response tests (via Bode approximate gain plots in Fig. 9.3) result in a seventh-order continuous-time transfer function as an adequate model of the dynamics on which to base control systems design:

Fig. 9.1 The multi-axis gantry robot

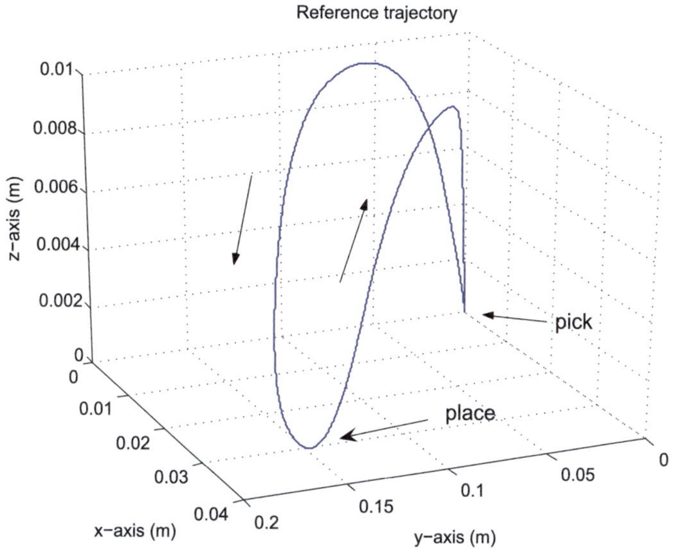

Fig. 9.2 3D combined reference trajectory

$$G_X(s) = \frac{13077183.4436(s+113.4)}{s(s^2+61.57\,s+1.125\times10^4)} \times \cdots$$
$$\frac{(s^2+30.28\,s+2.13\times10^4)}{(s^2+227.9\,s+5.647\times10^4)(s^2+466.1\,s+6.142\times10^5)}. \qquad (9.1)$$

This has then been sampled at $T_s = 0.01$ seconds to yield the following discrete state-space model which can be used to compute the model matrix G:

$$A_x = \begin{bmatrix} 0.0353 & 1.0000 & 0 & 0 & 0 & 0 & 0 \\ -0.0082 & 0.0353 & -0.0714 & 0.1556 & 0 & 0 & 0 \\ 0 & 0 & -0.1575 & 1.0000 & 0 & 0 & 0 \\ 0 & 0 & -0.0776 & -0.1575 & 0.5692 & 0.6978 & 1.3194 \\ 0 & 0 & 0 & 0 & 0.3879 & 1.0000 & 0 \\ 0 & 0 & 0 & 0 & -0.3898 & 0.3879 & 1.0799 \\ 0 & 0 & 0 & 0 & 0 & 0 & 1.0000 \end{bmatrix}$$
$$B_x = \begin{bmatrix} 0 & 0 & 0 & 0.0164 & 0 & 0.0134 & 0.0197 \end{bmatrix}^T$$
$$C_x = \begin{bmatrix} -0.0003 & -0.0054 & -0.0145 & 0.0316 & 0 & 0 & 0 \end{bmatrix}.$$

A zero-phase low-pass filter has been applied on the X-axis to prevent the high frequency components in the signal from building up which will make the learning process unstable.

The same procedure has been applied to the Y-axis and Z-axis. For more details, please refer to [2, 15].

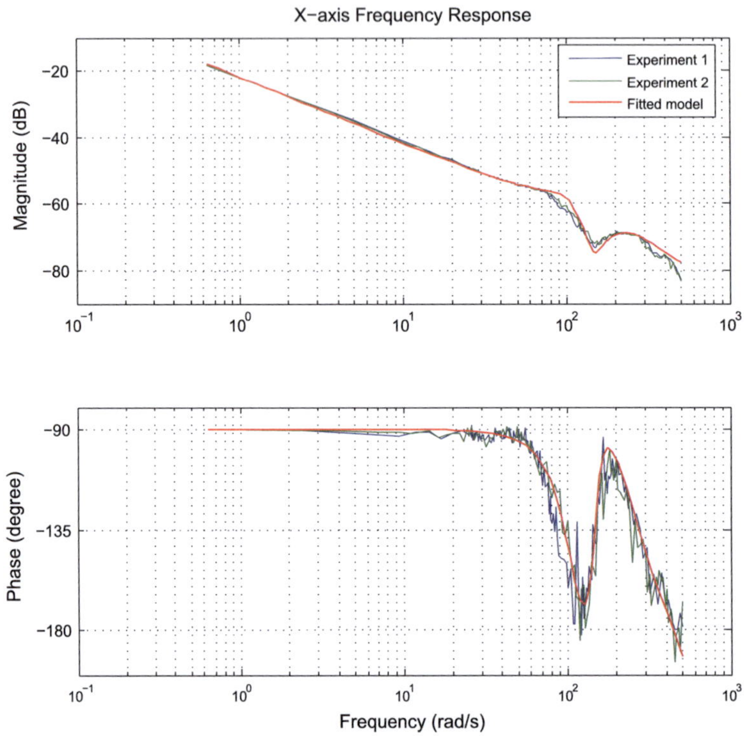

Fig. 9.3 Frequency response test results and fitted model for X-axis

9.2 NOILC and Accelerated Algorithm

The performance of NOILC algorithm and the accelerated algorithm (ANOILC) using extrapolation with acceleration step $\lambda_k = \lambda_{k,best}$ (introduced in Sect. 7.2) has been compared [4]. First, the weighting matrices are chosen as $Q = 10 \times I$, $R = 0.1 \times I$, and $\epsilon^2 = 1$ for all three axes, where I denotes the identity matrix with appropriate dimensions and the results are given in Fig. 9.4.

Inspection of these results confirms both algorithms achieve improved tracking over the trials. Furthermore, accelerated algorithm outperforms NOILC and the improvements are substantial. In particular, the improvement in the early trials is significant where on trial 10, for example, the accelerated algorithm outperforms NOILC by a factor of approximately 10^2 for the X- and Y-axes and a factor of 10 for the Z-axis. Moreover, this benefit continues to accrue over further trials, resulting in an improvement factor of 10^2, 10^2, 10^3 for X-, Y-, and Z-axes, respectively.

The acceleration steps are shown in Fig. 9.5 and contain oscillations that also appear in numerical simulations [3]. These have no known detrimental effects since the input signal will be still updated gradually but further research is required to explain their presence.

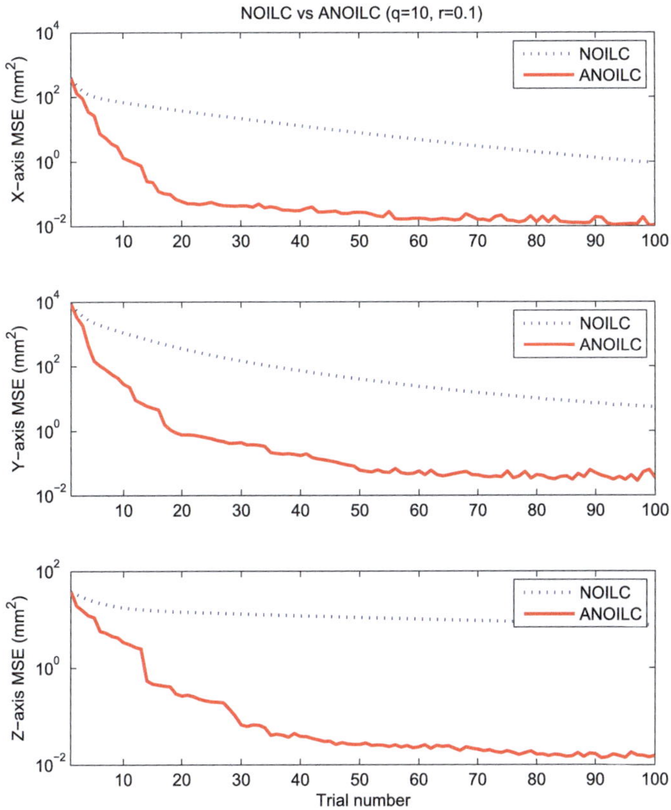

Fig. 9.4 Comparison of Convergence for NOILC and Accelerated NOILC using Extrapolation with $Q = 10I$, $R = 0.1I$, $\epsilon^2 = 1$

To investigate the effects of the selection of the weighting matrices, Figs. 9.6, 9.7, and 9.8 show the effects of varying these. Note that this is equivalent to keeping Q, R fixed but varying the value of ϵ^2.

From these figures, it can be seen the accelerated algorithm outperforms NOILC algorithm, which again verifies the theoretical predictions. It can also be seen that the improvements in the performance become smaller as the ratio Q/R, where now Q denotes the scalar multiplying the identity matrix, increases (or equivalently, Q, R being fixed, ϵ^2 decreases). The explanation for this is as this ratio increases, NOILC converges faster and the scope for improving the convergence becomes smaller. However, even in this case, the improvements at initial stage are substantial, which is of great interest in practical applications.

Previous work [15] investigated varying the ratio Q/R to obtain better performance in the NOILC case. However, optimizing the performance of NOILC requires careful selection of Q and R and a large number of experimental tests are needed to achieve optimal values. Figure 9.9 shows the comparison of accelerated NOILC

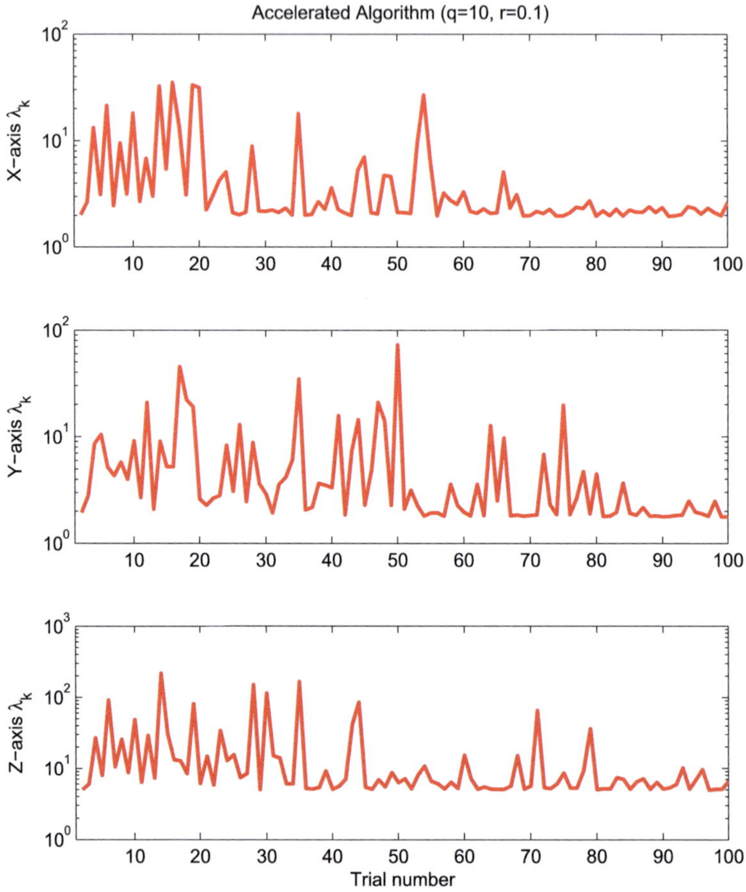

Fig. 9.5 Accelerated steps in accelerated algorithm with $Q = 10I$, $R = 0.1I$, $\epsilon^2 = 1$

and NOILC over only 20 trials by using $Q = 1000I$ and $R = 0.01I$ which were the suggested optimal values for X-axis. Figure 9.10 shows the corresponding values of $\lambda_{k,\text{best}}$ and for the X-axis the computed accelerations are very close to 1. This means the performance of accelerated NOILC is very close to that of NOILC and there is hardly any improvement as the parameters of NOILC have already approached the optimal ones.

The above results confirm that the accelerated NOILC algorithm using extrapolation can outperform the original NOILC algorithm and that the improvements can be substantial, especially in the early trials. This is of great practical interest and suggests the accelerated NOILC algorithm is a powerful design method for practical applications.

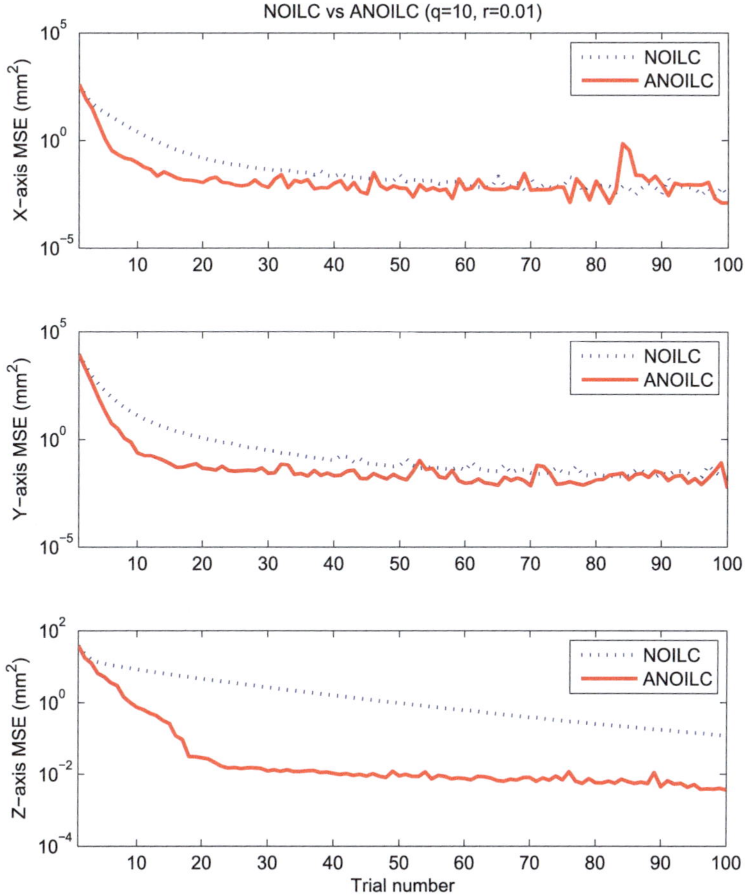

Fig. 9.6 Comparison of convergence for NOILC and accelerated NOILC using extrapolation with $Q = 10I$, $R = 0.01I$, $\epsilon^2 = 1$

9.3 NOILC with Hard Input Constraints

This section gives the results of experimental performance of the two hard constrained ILC algorithms given in Chap. 6. Input constraints were imposed such that for each axis, the amplitude of the input voltage was limited to be 90% that of the optimal input u^* (which produces perfect tracking). Note that under these constraints perfect tracking of the reference is not possible, which is more practically relevant than the trivial case where perfect tracking is achievable.

The cost function weighting matrices are chosen as diagonal matrices with common diagonal entries of $R = 100I$, $Q = 0.01I$, and $\epsilon^2 = 1$ for all three axes and the experimental results are given in Fig. 9.11.

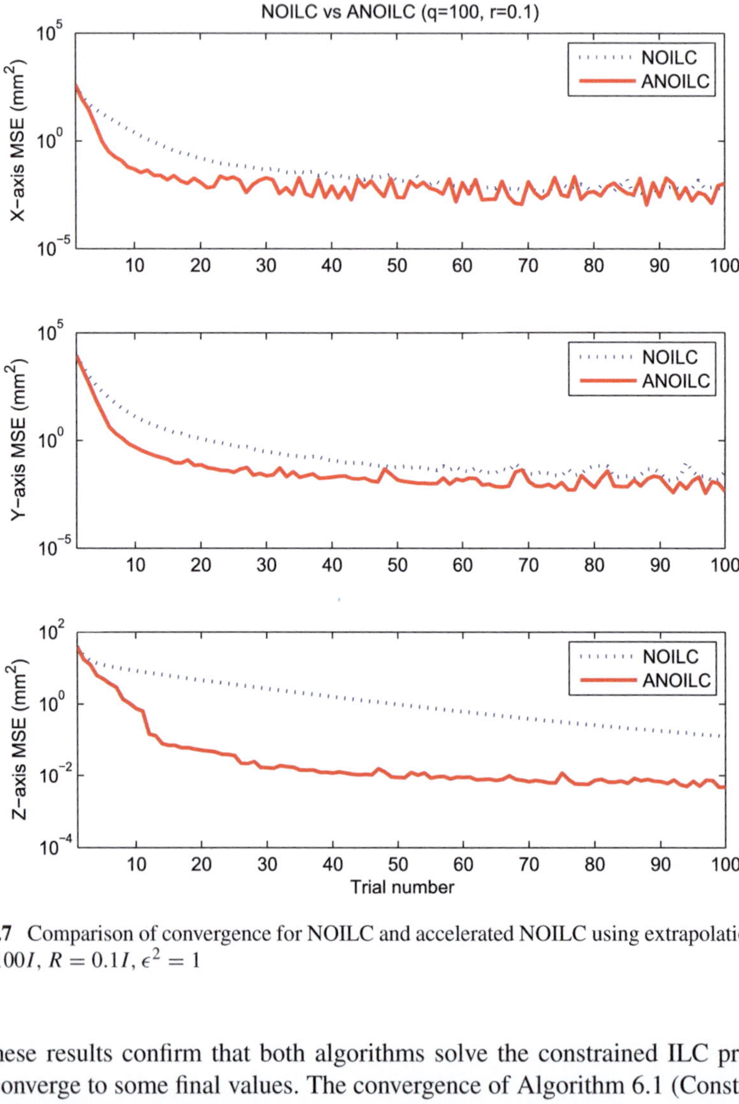

Fig. 9.7 Comparison of convergence for NOILC and accelerated NOILC using extrapolation with $Q = 100I$, $R = 0.1I$, $\epsilon^2 = 1$

These results confirm that both algorithms solve the constrained ILC problem and converge to some final values. The convergence of Algorithm 6.1 (Constrained ILC Algorithm 1) is monotonic in the tracking error norm, whereas Algorithm 6.2 (Constrained ILC Algorithm 2) is not. Moreover, the final tracking error norm of Algorithm 6.1 (Constrained ILC Algorithm 1) is smaller than that of Algorithm 6.2 (Constrained ILC Algorithm 2), which is consistent with the theoretical predictions.

To expand the discussion of these results, the input, output, and tracking errors on 50th trial for X-axis are given in Figs. 9.12, 9.13, and 9.14.

Using Algorithm 6.1 (Constrained ILC Algorithm 1), a smaller tracking error (on average) is obtained, compared to Algorithm 6.2 (Constrained ILC Algorithm 2). Note also that the input computed by Algorithm 6.2 (Constrained ILC Algorithm 2)

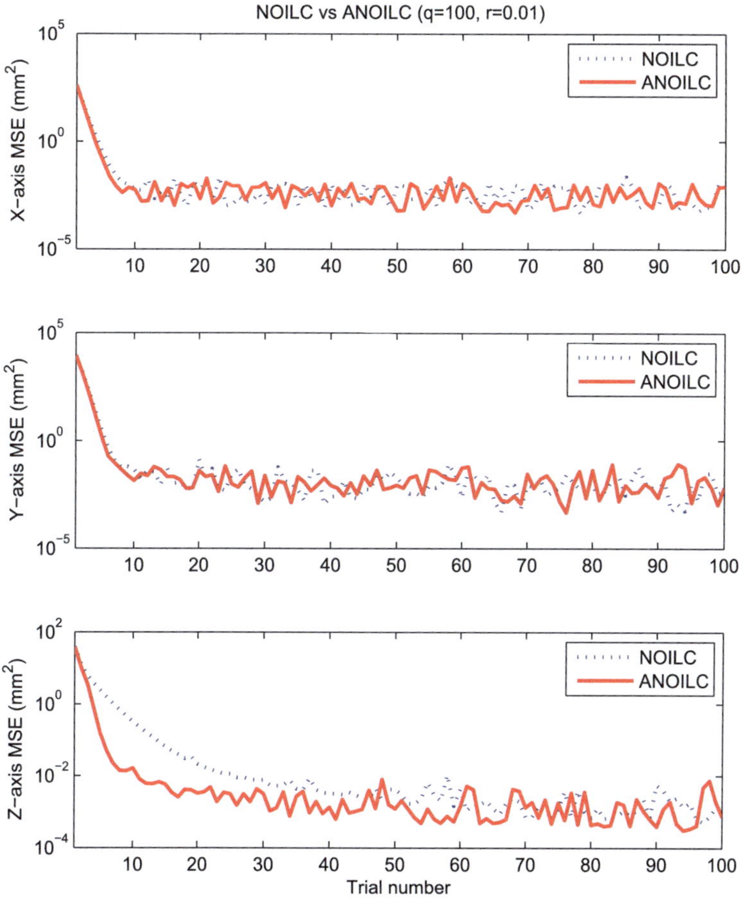

Fig. 9.8 Comparison of convergence for NOILC and accelerated NOILC using extrapolation with $Q = 100I$, $R = 0.01I$, $\epsilon^2 = 1$

does not just enforce saturation on the original input but adds some compensation (Fig. 9.12).

Figures 9.15 shows the effects of varying the selection of the weighting matrices keeping the diagonal structure but changing the control input weighting to 0.1 (or equivalently keeping Q, R unchanged but setting $\epsilon^2 = 10$). Compared to the results in Fig. 9.11, this new choice puts larger weighting on the input change, leading to slower convergence. Varying this weighting value has no effect on the final tracking error for Algorithm 6.1 (Constrained ILC Algorithm 1) but in Algorithm 6.2 (Constrained ILC Algorithm 2) it results in a better (almost optimal) tracking accuracy, which verifies the theoretical results.

Using these results, it can be concluded that Algorithm 6.1 (Constrained ILC Algorithm 1) performs better than Algorithm 6.2 (Constrained ILC Algorithm 2).

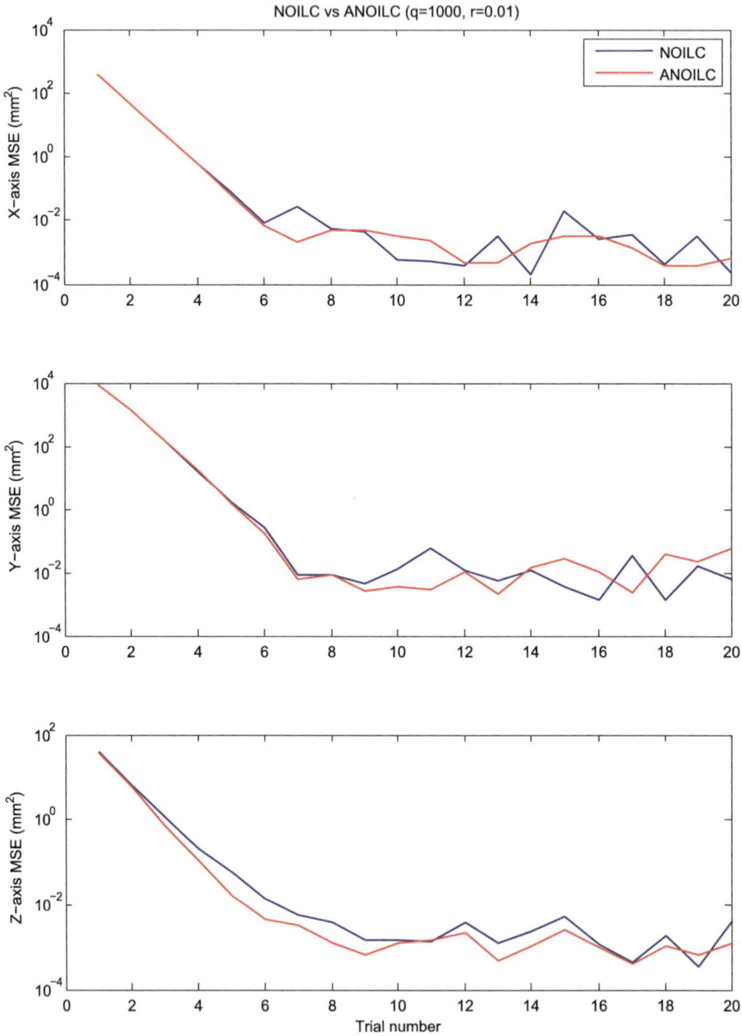

Fig. 9.9 Comparison of convergence for NOILC and accelerated NOILC using extrapolation with $Q = 1000I$, $R = 0.01I$, $\epsilon^2 = 1$

However, this is achieved at the expense of higher computational load, which may be not acceptable in some applications but Algorithm 6.2 (Constrained ILC Algorithm 2) achieves nearly optimal performance using simple computations, which is equally (if not more) important in many cases.

The results confirm that both algorithms can solve the constrained ILC problem efficiently, whilst the computation of one algorithm is much less than the other at the cost of slightly sacrificed convergence performance. This requires a compromise between the performance/accuracy and the computational cost when selecting the

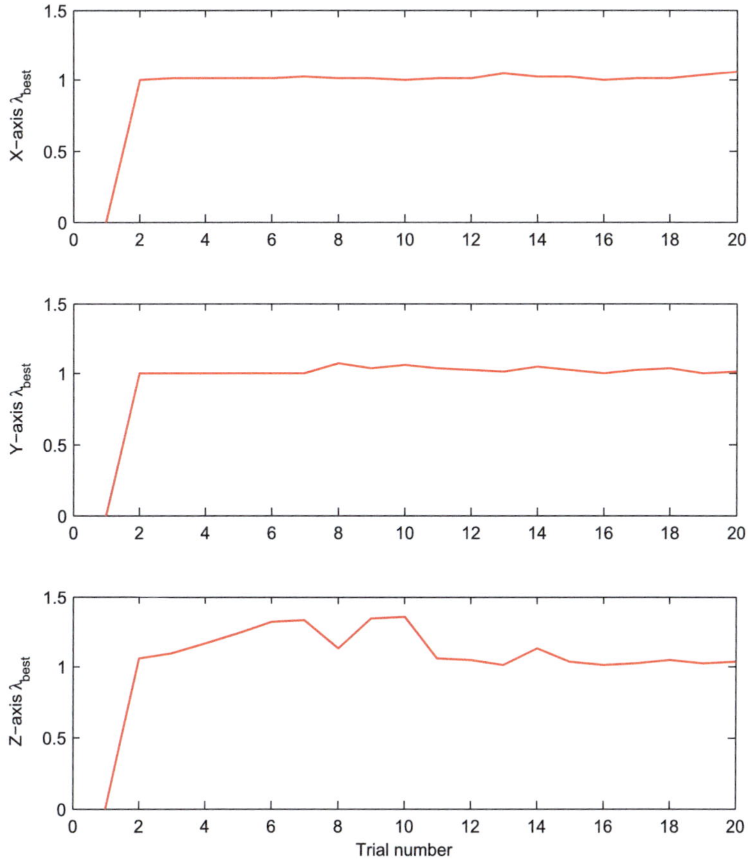

Fig. 9.10 Computed λ_{best} with $Q = 100I$, $R = 0.01I$, $\epsilon^2 = 1$

algorithm to use in practice. The two constrained algorithms have also been tested over longer (100) trials to investigate their long-term performance—see [5] for more details.

9.4 Anthropomorphic Robotic Arm: System Description and Modelling

We now move to the second robotic platform. This section introduces a six degree of freedom anthropomorphic robotic arm whose five rotary joints are composed of PowerCubes (Schunk GmbH & Co.) incorporating brushless servomotors with integrated power electronics and transmission (shown in Fig. 9.16). Each servomotor includes cascaded current and velocity control loops with a control input supplied

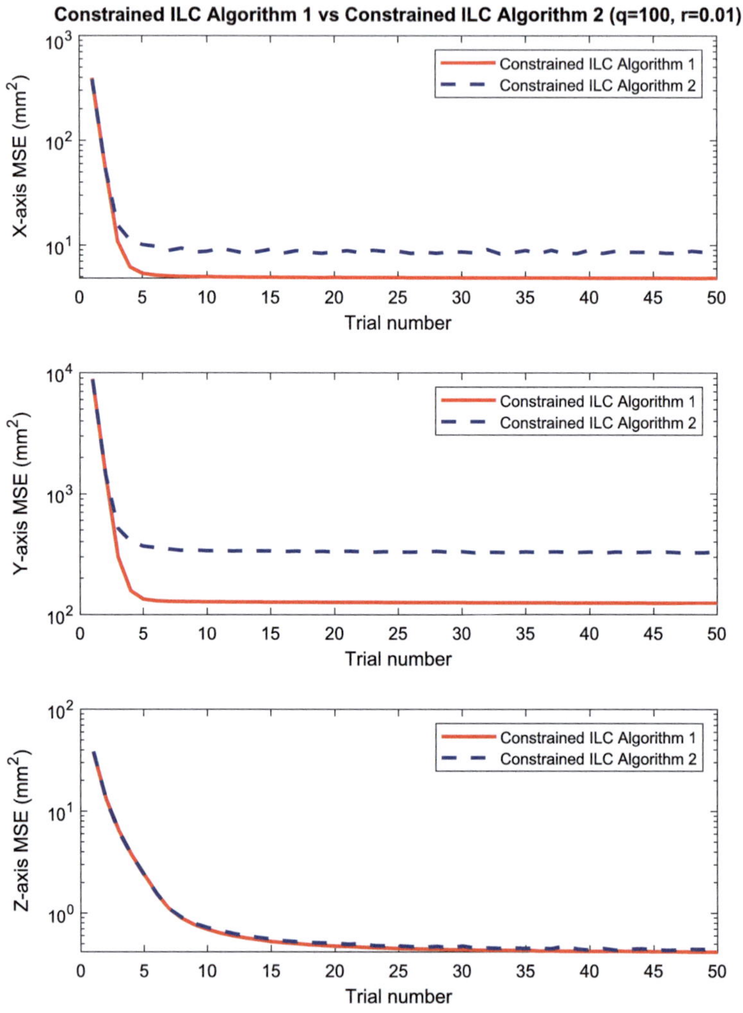

Fig. 9.11 Comparison of Convergence with $Q = 100I$, $R = 0.01I$, $\epsilon^2 = 1$

by a dSPACE DS1103 real-time board with communication via a CAN bus at a rate of 500 kbit/s.

In the following two sections, NOILC designs for point-to-point tracking tasks and with hard and soft constraints are tested on this platform. Results are presented for the first joint which is aligned in the horizontal plane as illustrated in Fig. 9.16.

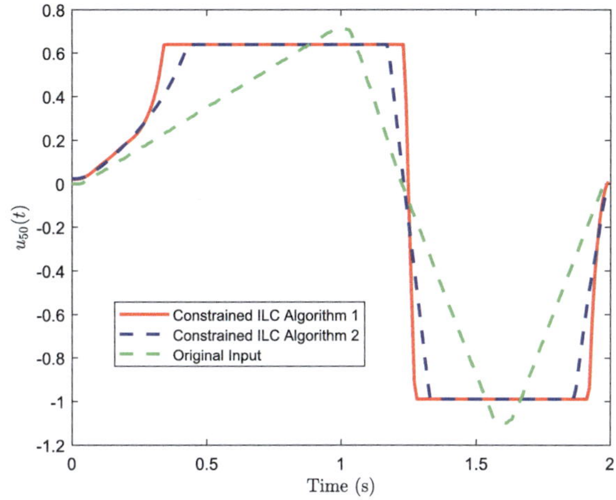

Fig. 9.12 Input on the 50th trial for the X-axis

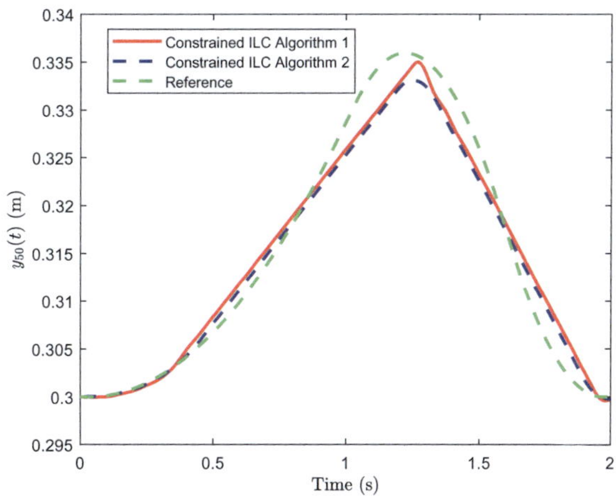

Fig. 9.13 Output on the 50th trial for the X-axis

Frequency response tests have established that the linear model (9.2) adequately represents the system dynamics, with angular input and output specified in degrees (see [9, 14] for more details):

$$G(s) = \frac{400788.1582(s + 12.14)(s + 24.01)^2}{(s + 31.52)(s + 22.97)(s + 2.178)(s^2 + 36.59\,s + 363.7)(s^2 + 124.5\,s + 4076)}. \tag{9.2}$$

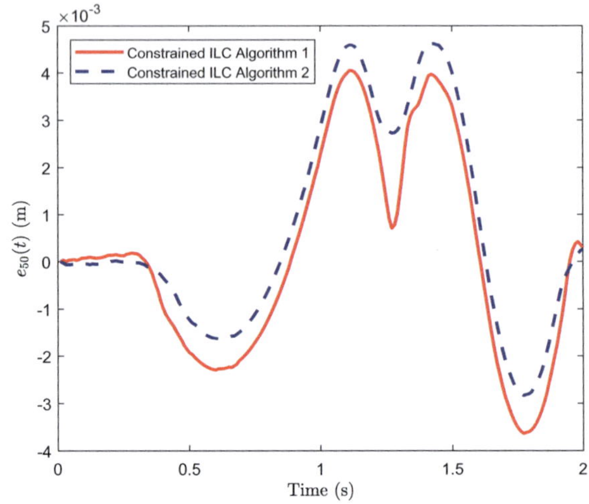

Fig. 9.14 Tracking error on the 50th trial for the X-axis

9.5 NOILC for Point-to-Point Tracking Tasks

The point-to-point task selected replicates an industrial multiple "pick-and-place" process in which payloads are manipulated during an assembly operation. The reference is given by the vector

$$r^e = \begin{bmatrix} 25 & -30 & 15 & 10 & 20 \end{bmatrix}^T \tag{9.3}$$

defining the desired outputs at the $M = 5$ time points

$$t_1 = 1s, \quad t_2 = 3s, \quad t_3 = 4.5s, \quad t_4 = 5s, \quad t_5 = 6s \tag{9.4}$$

on the time interval $[0, 6]$. NOILC algorithm and Algorithm 6.2 (Constrained ILC Algorithm 2) as "representative algorithms" for the constraint free and constrained scenarios are tested as the NOILC algorithm has the desirable physical property of converging to minimum norm (energy) solution and Algorithm 6.2 (Constrained ILC Algorithm 2) has much lower computational complexity [6]. In all tests a sampling frequency of 100 Hz is employed.

The NOILC algorithm is first applied for the unconstrained case. For comparison the predicted limiting solution is given using the nominal model. Results for the scalar weighting choices $Q = 1$, $R = 0.2, 0.1$, and 0.02, $\epsilon^2 = 1$ (or equivalently, $Q = R = 1$ and $\epsilon^2 = 0.2, 0.1, 0.02$) are shown in Fig. 9.17 which illustrate the effect on convergence speed of reducing the weighting on the input norm term in the cost. In all cases the algorithm enforces rapid convergence to similar low levels of input and error norm. The minimum energy property of the approach results in an input norm which does not match that of the nominal solution, due to the presence of modelling uncertainty and noise.

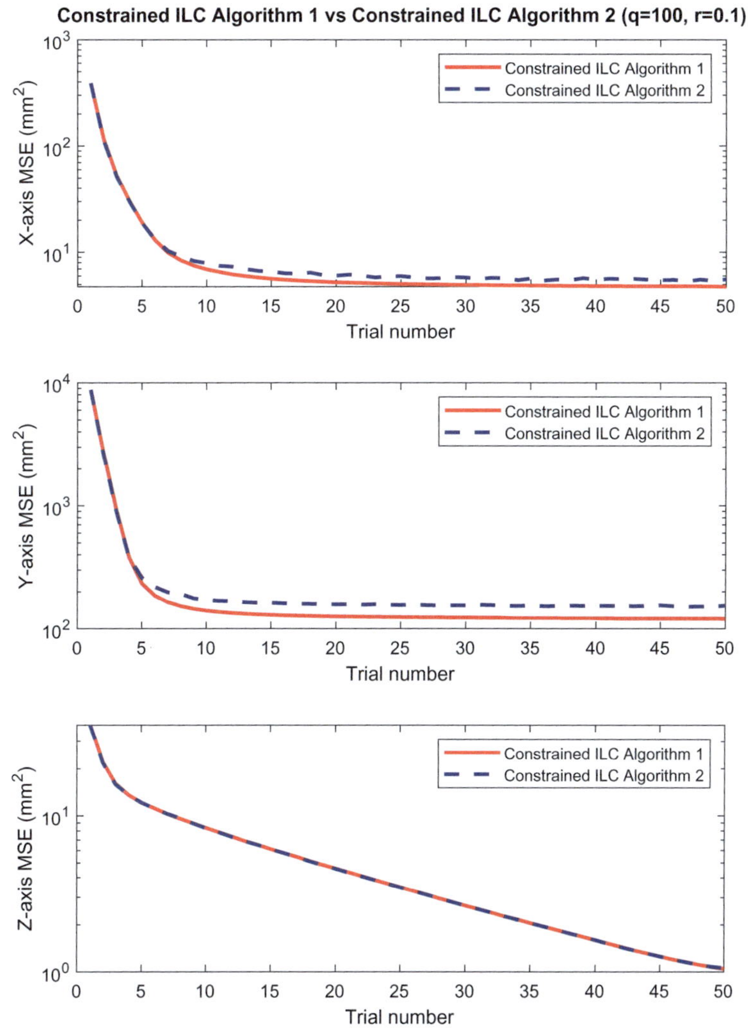

Fig. 9.15 Comparison of Convergence with control weighting increased to 0.1

Algorithm 6.2 (Constrained ILC Algorithm 2) is next applied to illustrate how input constraints can be successfully tackled in the projection framework. In the present application these commonly arise due to actuator and transmission limitations such as lead screw length, unidirectional drives, or maximum power/force/supply characteristics. The unconstrained input solutions had a maximum magnitude of over 40 for each R, and in the next experiment a constraint of $|u(t)| \leq 30$ is imposed. Smaller R values than for the NOILC algorithm are appropriate due to the input "clipping" effect which reduces the update increment on each trial. Results are shown in Fig. 9.18 for $R = 0.1, 0.02$, and 0.01 and confirm that the tracking of r^e is still

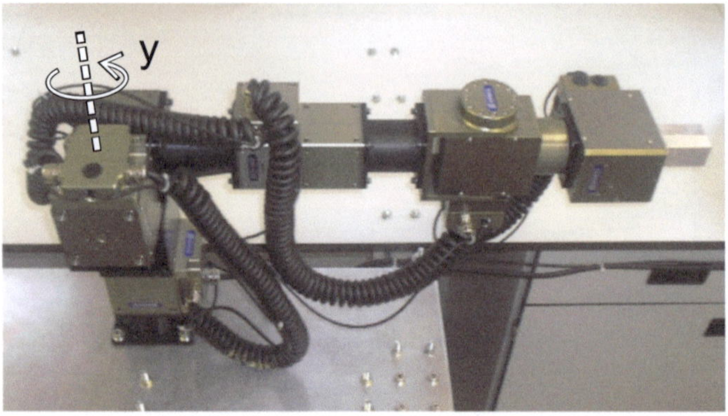

Fig. 9.16 Robotic manipulator system showing actuated output, y

possible, resulting in similar error norm levels as the unconstrained case. However, the converged input and output are clearly different, and correspond to an increased input norm, as less flexibilities are allowed than the unconstrained case (which converges to the minimum norm (energy) solution).

Next the input magnitude constraint is reduced to $|u(t)| \leq 20$, with results shown in Fig. 9.19. In this case perfect tracking of r^e is not feasible, and each choice of $R = 0.1, 0.02$, and 0.01 converges to a limiting solution which does not pass through the point-to-point locations. Smaller values of R have also been tested and also converge to similar limiting tracking error norms, which represent the "best" solution that is possible given the applied constraint.

9.6 NOILC with Soft Constraints

In this section, we consider the ILC design problem with soft constraints, i.e., to optimize some auxiliary performance. The two switching algorithms, i.e., Algorithm 6.3 (Switching Algorithm 1) and Algorithm 6.4 (Switching Algorithm 2) introduced in Chap. 6, are applied [14].

As in the previous section, the intermediate point task replicates an industrial multiple "pick-and-place" process in which payloads are manipulated during an assembly operation. The selected reference is given by the vector of output values

$$r^e = \begin{bmatrix} 20, & -30, & 10, & 20 \end{bmatrix}^T \tag{9.5}$$

at the $M = 4$ time points

$$t_1 = 1s, \quad t_2 = 3s, \quad t_3 = 5s, \quad t_4 = 6s \tag{9.6}$$

on the interval [0, 6].

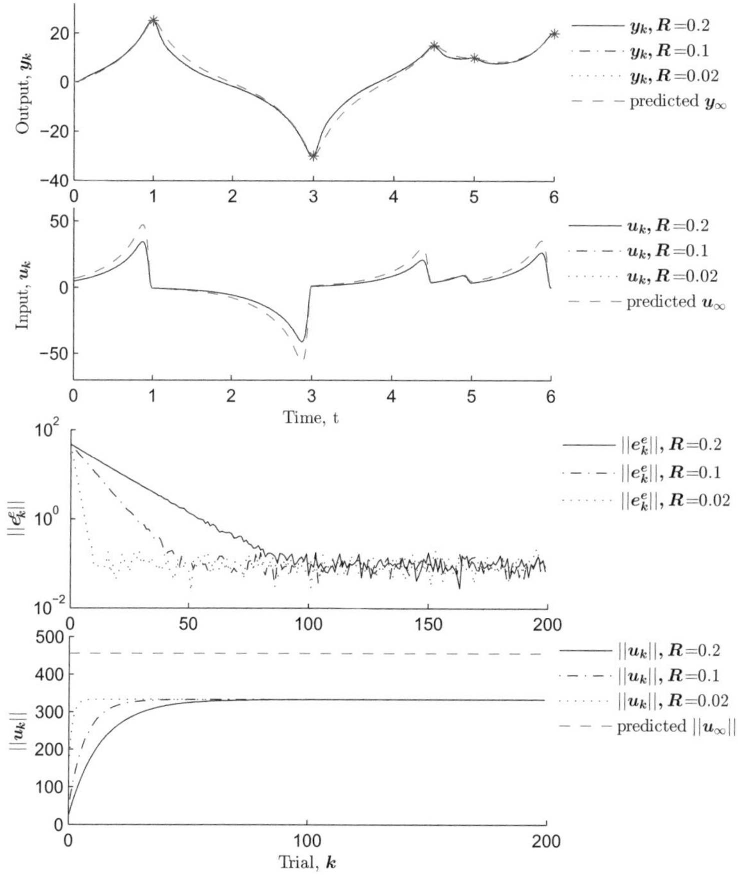

Fig. 9.17 Experimental final trial output and input, together with error norm and input norm results using NOILC. Reprinted with permission from [6], ©2014 IEEE

To produce a smooth output response at all points $t \in [0, 6]$, and hence reduce vibration and potential payload damage/spillage between the specified intermediate points, some degree of minimization of the mean square joint acceleration is proposed. The auxiliary variable is hence $z(t) = \ddot{y}(t)$, and, since system (9.2) has relative degree > 2, the auxiliary variable system operator G_1 can be written in continuous-time state-space form with $S(A_z, B_z, C_z, D_z) = S(A, B, CA^2, 0)$. In the results which follow, all algorithms have been calculated in continuous time and then discretized using a sampling time of 500 Hz.

Algorithm 6.3 (Switching Algorithm 1) is first implemented. 200 trials have been performed and results are shown in Fig. 9.20. Norm results comparing performance during the initial convergence stage are shown in Fig. 9.21 and it can be seen that negligible degradation of performance occurs as the number of trials increases.

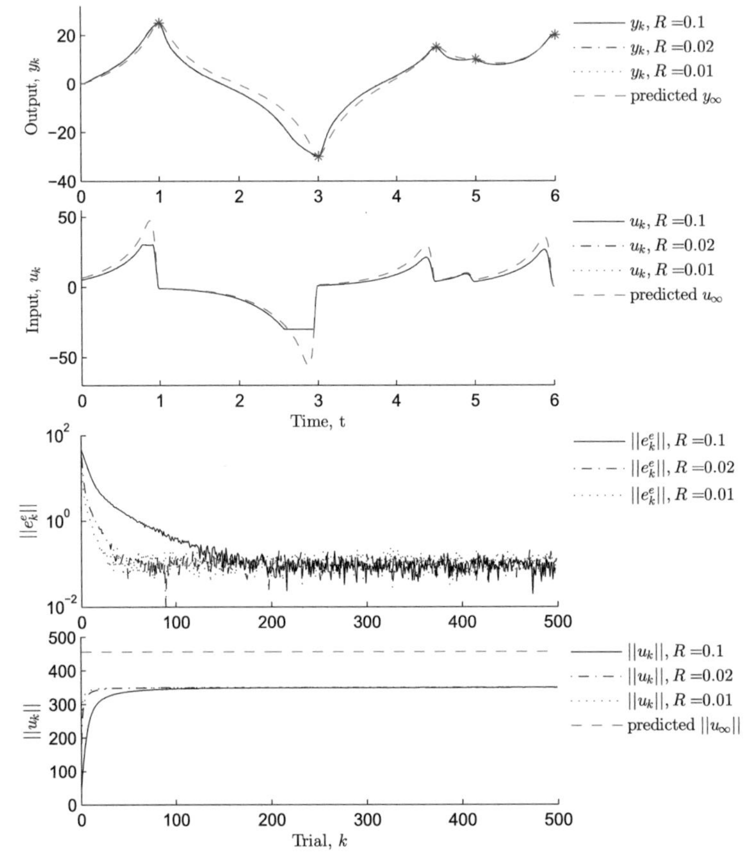

Fig. 9.18 Experimental final trial output and input, together with error norm and input norm results using Algorithm 6.2 (Constrained ILC Algorithm 2) with input bound $|u(t)| \leq 30$. Reprinted with permission from [6], ©2014 IEEE

To quantify convergence rates, the iteration index, k^*, is recorded that corresponds to when the intermediate point tracking criterion

$$\|r^e - y_k^e\| < \varepsilon \|r^e\| \tag{9.7}$$

is first met. Here y_k^e is the output of the constraint trial in Step 4. A value of $\varepsilon = 0.01$ has been taken in all the results that follow. Experimental norm values recorded on Step 4 of trial k^* are given in Table 9.1 for a variety of Q values and compared with the theoretical limit values obtained from the model. The excellent agreement is self-evident.

As Q increases the theoretical optimal value of $J(u_\infty)$ increases. At the same time the theoretical optimal value of $||z_\infty||/\sqrt{Q}$ decreases as more weight is given to it in the optimization problem (6.3). The experimental results match the predicted

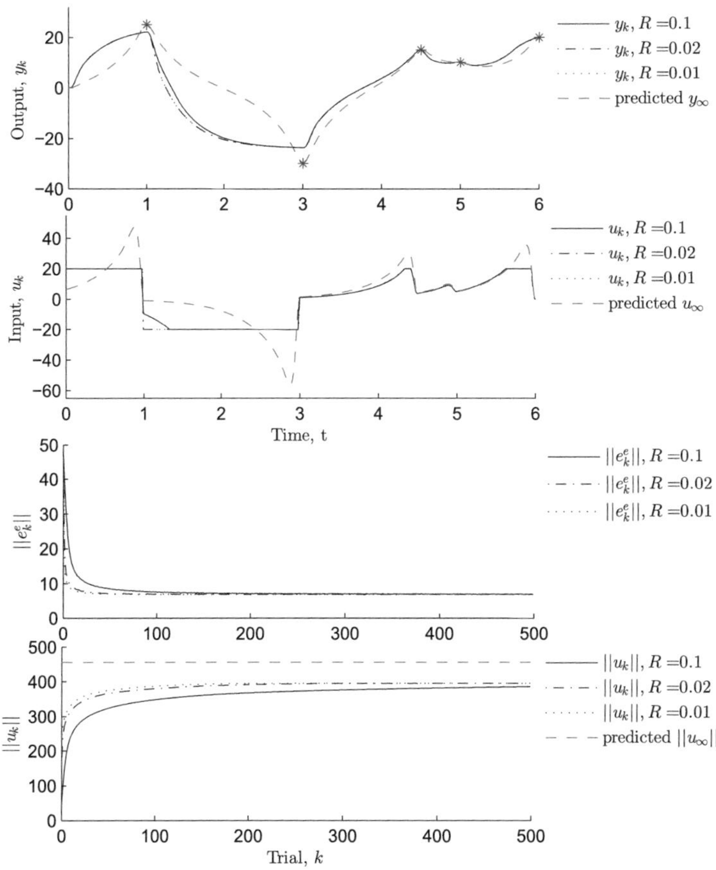

Fig. 9.19 Experimental final trial output and input, together with error norm and input norm results using Algorithm 6.2 (Constrained ILC Algorithm 2) with input bound $|u(t)| \leq 20$. Reprinted with permission from [6], ©2014 IEEE

results reasonably well, although it has been found that for high values of Q ($\gg$1) there is a reduction in performance, with fluctuation in norms and eventual lack of convergence. This reflects Theorem 6.8 in Sect. 6.3.3 which bounds the admissible plant uncertainty as $\|\Delta G_0\| < (1 - r(W_\infty))(1 + \underline{\delta}^2)$ (the value of which decreases as Q increases implying reduced robustness).

Since robustness reduces as Q increases, excessive values should be avoided, especially as such Q values produce negligible improvement in cost, as reflected in the fact that the theoretical value of $||z_\infty||/\sqrt{Q}$ corresponding to $Q = 0.2$ (95.925) is close to the limiting value as $Q \to \infty$ (94.266). As Q is increased, the convergence of the plant outputs shown in Fig. 9.20 is close to the limiting case as $Q \to \infty$. To mitigate the effect of noise and model uncertainty, it is common practice in ILC to apply a zero-phase low-pass filter to the control input u_{k+1} (and/or $\tilde{u}_{k+1}$) prior to its

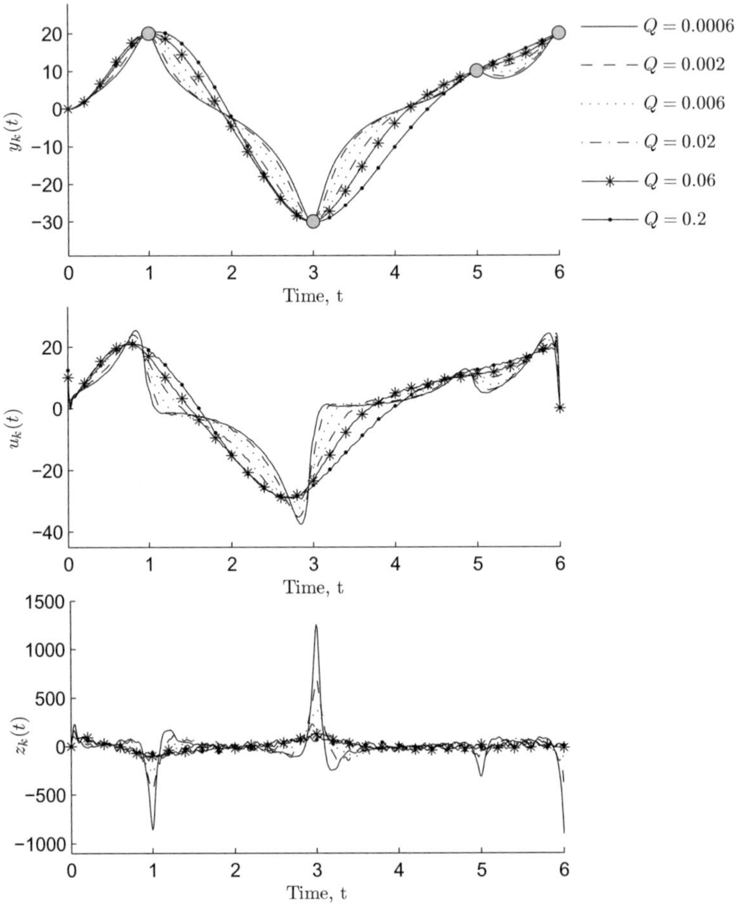

Fig. 9.20 Algorithm 6.3 (Switching Algorithm 1)—experimental results recorded during Step 4. Final trial input, output, and auxiliary signals for a range of Q. Reprinted from [14] by permission of the publisher (Taylor and Francis Ltd, http://www.tandfonline.com)

application in the subsequent trial; however, in the present case, this has not been done in order to assess the effects of noise and uncertainty.

Algorithm 6.4 (Switching Algorithm 2) has also been implemented and similar results have been observed. The results are not included here due to space reasons. The readers are referred to [14] for more details.

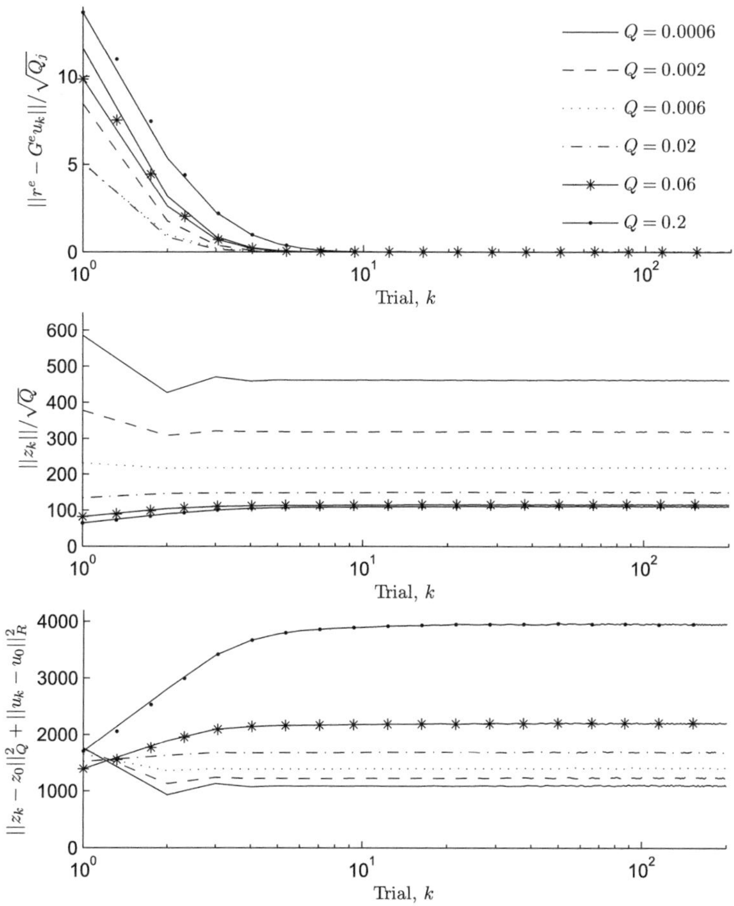

Fig. 9.21 Algorithm 6.3 (Switching Algorithm 1)—experimental results recorded during Step 4. Intermediate point tracking, auxiliary signal, and cost function norms for a range of Q. Reprinted from [14] by permission of the publisher (Taylor and Francis Ltd, http://www.tandfonline.com)

9.7 Summary and Discussion

In this chapter, some of the key algorithms introduced in this book are tested experimentally on two robotic platforms, namely, a multi-axis gantry robot and a anthropomorphic robotic arm. The results demonstrate that these algorithms can solve the iterative learning control design problem effectively and that the availability of different algorithms offers a freedom to choose between the computational complexity and the possible sacrifice in the performance.

The experimental results in this chapter are obtained using linear models of the robots, where nonlinearities and noise/disturbances are neglected and can be treated as model uncertainties. The experimental results clearly demonstrate a certain degree

Table 9.1 Algorithm 6.3 (Switching Algorithm 1)—experimental results recorded during Step 4. Convergence data using $\varepsilon = 0.01$, and corresponding auxiliary and cost function norms. Predicted values for the limit $k \to \infty$ using the nominal plant model are shown on the right-hand side. Reprinted from [14] by permission of the publisher (Taylor and Francis Ltd, http://www.tandfonline.com)

Q	k^*	$\|\|u_{k^*}\|\|$	$\|\|z_{k^*}\|\|/\sqrt{Q}$	$J(z_{k^*}, u_{k^*})$	$\|\|u_\infty\|\|$	$\|\|z_\infty\|\|/\sqrt{Q}$	$J(z_\infty, u_\infty)$
0.0006	4	30.774	457.994	1072.946	34.650	321.143	1262.510
0.002	3	32.256	304.160	1225.506	35.451	228.599	1361.341
0.006	3	33.462	207.854	1378.941	36.652	164.190	1505.176
0.02	3	35.369	141.851	1653.406	38.413	120.152	1764.333
0.06	4	37.215	112.847	2149.056	40.071	102.355	2234.326
0.2	6	38.681	108.103	3833.482	41.653	95.925	3575.325

of robustness of the algorithms used, as predicted by the robustness analysis results presented in previous chapters. Despite this, further theoretical robustness analysis still needs to be done, e.g., robustness of the ILC design with hard constraints, robustness against nonlinear model uncertainties, as well as detailed analysis on the effect of noises/disturbances.

References

1. Blanken L, Koekebakker S, Oomen T (2020) Data-driven feedforward tuning using non-causal rational basis functions: with application to an industrial flatbed printer. Mechatronics 71(102):424
2. Cai ZL, Freeman CT, Lewin PL, Rogers E (2008) Iterative learning control for a non-minimum phase plant based on a reference shift algorithm. Control Eng Pract 16(6):633–643
3. Chu B, Owens DH (2009) Accelerated norm-optimal iterative learning control algorithms using successive projection. Int J Control 82(8):1469–1484
4. Chu B, Cai Z, Owens DH, Rogers E, Freeman CT, Lewin PL (2010) Experimental verification of accelerated norm-optimal iterative learning control. In: UKACC International Conference on Control. IET, pp 1–6
5. Chu B, Cai Z, Owens DH, Rogers E, Freeman CT, Lewin PL (2012) Experimental verification of constrained iterative learning control using successive projection. In: Proceedings of 2012 UKACC International Conference on Control. IEEE, pp 351–356
6. Chu B, Freeman CT, Owens DH (2015) A novel design framework for point-to-point ILC using successive projection. IEEE Trans Control Syst Technol 23(3):1156–1163
7. Chu B, Rauh A, Aschemann H, Rogers E, Owens DH (2021) Constrained iterative learning control for linear time-varying systems with experimental validation on a high-speed rack feeder. IEEE Trans Control Syst Technol 30(5):1834–1846
8. Daley S, Owens DH, Hatonen J (2007) Application of optimal iterative learning control to the dynamic testing of mechanical structures. Proc Inst Mech Eng Part I-J Syst Control Eng 221(2):211–222
9. Freeman CT, Tan Y (2013) Iterative learning control with mixed constraints for point-to-point tracking. IEEE Trans Control Syst Technol 21(3):604–616
10. Freeman CT, Rogers E, Hughes AM, Burridge JH, Meadmore KL (2012) Iterative learning control in health care: electrical stimulation and robotic-assisted upper-limb stroke rehabilitation. IEEE Control Syst Mag 32(1):18–53

11. Gunnarsson S, Norrlof M (2001) On the design of ILC algorithms using optimization. Automatica 37(12):2011–2016
12. Johansen SV, Jensen MR, Chu B, Bendtsen JD, Mogensen J, Rogers E (2019) Broiler FCR optimization using norm optimal terminal iterative learning control. IEEE Trans Control Syst Technol 29(2):580–592
13. Kirchhoff S, Schmidt C, Lichtenberg G, Werner H (2008) Iterative learning algorithm for control of an accelerator based free electron laser. In: Proceedings of 47th IEEE Conference on Decision and Control, vols 1–6, pp 3032–3037
14. Owens DH, Freeman CT, Chu B (2013) Multivariable norm optimal iterative learning control with auxiliary optimisation. Int J Control 86(6):1026–1045
15. Ratcliffe JD, Lewin PL, Rogers E, Hatonen JJ, Owens DH (2006) Norm-optimal iterative learning control applied to gantry robots for automation applications. IEEE Trans Robot 22(6):1303–1307
16. Rogers E, Owens D, Werner H, Freeman C, Lewin P, Kirchhoff S, Schmidt C, Lichtenberg G (2010) Norm-optimal iterative learning control with application to problems in acceleration-based free electron lasers and rehabilitation robotics. Eur J Control 5:496–521
17. Wang Z, Pannier CP, Barton K, Hoelzle DJ (2018) Application of robust monotonically convergent spatial iterative learning control to microscale additive manufacturing. Mechatronics 56:157–165
18. Wu RB, Chu B, Owens DH, Rabitz H (2018) Data-driven gradient algorithm for high-precision quantum control. Phys Rev A 97(4):042122

Chapter 10
Application to a Rotary Electro-mechanical Non-minimum Phase System

Non-minimum phase (NMP) systems are known to cause difficulties for iterative learning control (ILC) design. In this chapter, the theoretical results on the analysis of NMP systems and unconstrained Norm Optimal Iterative Learning control algorithms introduced in this book, namely, Norm Optimal Iterative Learning Control (NOILC), Predictive Norm Optimal Iterative Learning Control (PNOILC), and their acceleration mechanisms, are applied to a non-minimum phase rotary electro-mechanical system to demonstrate their effectiveness and to reveal the potential difficulties met when dealing with non-minimum phase systems. Robustness properties of the algorithms are also illustrated.

The chapter is organized as follows. First, the rotary electro-mechanical NMP system is described in Sect. 10.1. Two different choices of reference signals are used to illustrate the effect of non-minimum phase zeros experimentally in Sect. 10.2. Then the unconstrained algorithms are verified numerically in Sect. 10.3 using exact system model. Their robustness properties with respect to model uncertainty and noise are demonstrated in Sects. 10.4 and 10.5, respectively. Finally, concluding remarks are given in Sect. 10.6.

10.1 System Description

In this chapter, the rotary electro-mechanical, non-minimum phase test facility shown in Fig. 10.1 (and illustrated schematically in Fig. 10.2) is used to assess the validity of the theory and design methods of this text.

The experimental test-bed consists of a rotary mechanical system of inertias, dampers, torsional springs, a timing belt, pulleys, and gears. The non-minimum phase characteristic is achieved by using the arrangement shown in Fig. 10.2 where θ_i and θ_o are the input and output positions, J_r and J_g are inertias, B_r is a damper, K_r is a spring, and G_r represents the gearing. A further spring–mass–damper system

B. Chu and D. H. Owens, *Optimal Iterative Learning Control*, Advances in Industrial Control, https://doi.org/10.1007/978-3-031-80236-2_10

Fig. 10.1 Experimental setup of a rotary electro-mechanical non-minimum phase test facility

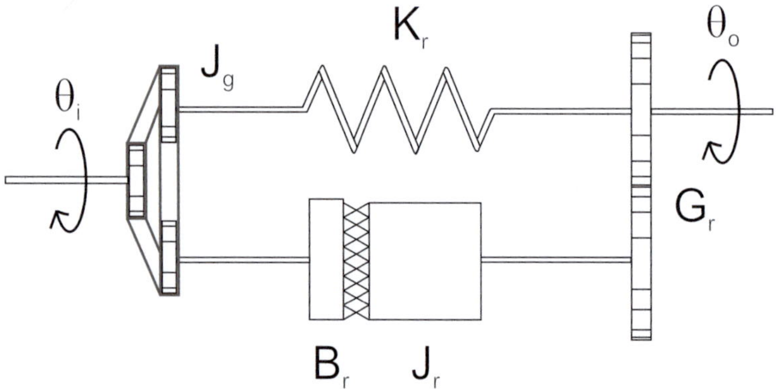

Fig. 10.2 Non-minimum phase characteristic

is connected to the input in order to increase the relative degree and complexity of the system. Frequency response test measurements (see [2] for details) have been used to obtain a continuous-time plant transfer function (including a proportional plus integral plus derivative stabilizer) as follows:

$$G(s) = \frac{165.95(4 - s)}{s^4 + 21.5\,s^3 + 170.28\,s^2 + 368.52\,s + 663.82}. \tag{10.1}$$

From this, it can be clearly seen that this is a NMP system with a NMP zero (at $z = 4$).

10.2 The Slow Convergence Behaviour: Experimental Verification

First, experimental results are presented to illustrate and verify the slow convergence behaviour for NMP systems.

This system is non-minimum phase with one zero at $s = 4$ (giving $z = 4$) in the right-half of the complex plane. The one-dimensional subspace corresponding to the small eigenvalue of the system (due to the NMP zero) is hence spanned by the single vector $v_1 = e^{-4t}$. Choosing a time interval $t \in [0, 6]$, the factor $\alpha z = 4 \times 6 = 24$ so $e^{-\alpha z} = e^{-24}$ is extremely small suggesting that an analogy (for continuous-time systems) of the theoretical results introduced in Chap. 5 of the book (also see [3, 5] for more details) should apply with good accuracy.

To show the close agreement existing between predicted and measured results, 40 trials of NOILC (with zero initial input), using cost function weights $Q = 5$, $R = 1$, $\epsilon^2 = 1$, and a sampling frequency of 250Hz, have been performed, using the reference, $r(t)$, shown in Fig. 10.3 plot (d). The reference comprises a sinusoidal positional movement (in radians), which is both preceded and followed by periods where it is set at zero. Following the discussion under (1) and (2) at the end of Sect. 5.6.1 on the estimated plateau tracking error, there exists a time interval for this case $[0, 1.5] \subset [0, 6]$ where the reference r is zero and the plant has zero initial conditions. Hence the error e_0 will be zero in this interval leading to a low predicted final error, e_∞ (as $\langle v_1, e_0\rangle_Q$ is small), and associated norm $\|e_\infty\|$. These predicted values are shown in Fig. 10.3 plots (a) and (b), respectively, and accurately match the experimental results obtained.

In order to examine a case where a significant non-zero final error is predicted, the reference is now replaced with that shown in Fig. 10.4 plot (d). This corresponds to the same movement as the previous reference; however, the starting point has now been changed to half way along its length. The initial period of $r(t)$ is now not zero so, recalling again the discussion under (1) and (2) at the end of Sect. 5.6.1, the e_0 will *not* be small in this interval, leading to predicted error values significantly different from zero, as shown in Fig. 10.4 plots (a) and (b). Again, 40 trials of NOILC, using cost function weights $Q = 5$, $R = 1$, and $\epsilon^2 = 1$ have been performed, and the experimental error results e_k and $\|e_k\|$ obtained can be seen to closely match their predicted counterparts e_∞ and $\|e_\infty\|$. In particular, the error norm $||e_k||$ enters the plateau at $||e_k|| \approx 2$ (plot b) in Fig. 10.4 and the predicted error is modelled by $e_\infty \approx 6e^{-4t}$ (plot a) in Fig. 10.4.

Having illustrated the slow convergence behaviour, the following sections apply the key unconstrained algorithms introduced in this book, namely NOILC, NOILC with accelerated convergence by extrapolation (with the optimal extrapolation factor (7.7)), PNOILC with various parameter choices as introduced in Sect. 8.4.3, in particular, the receding horizon choice (RHC) of (8.47), the optimal combination (8.53), and the accelerated PNOILC using extrapolation (8.54) with optimal extrapolation factor, to demonstrate their effectiveness and robustness numerically. In all

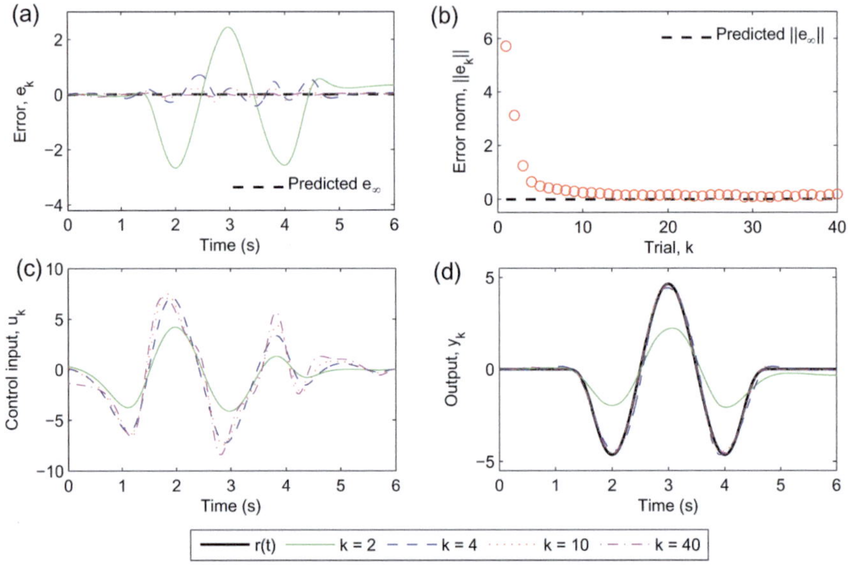

Fig. 10.3 Experimental results with initial zero period: b error, e_k, **b** error norm, $\|e_k\|$, **c** control input, u_k, and **d** output, y_k. Reprinted with permission from [5], ©2013 IEEE

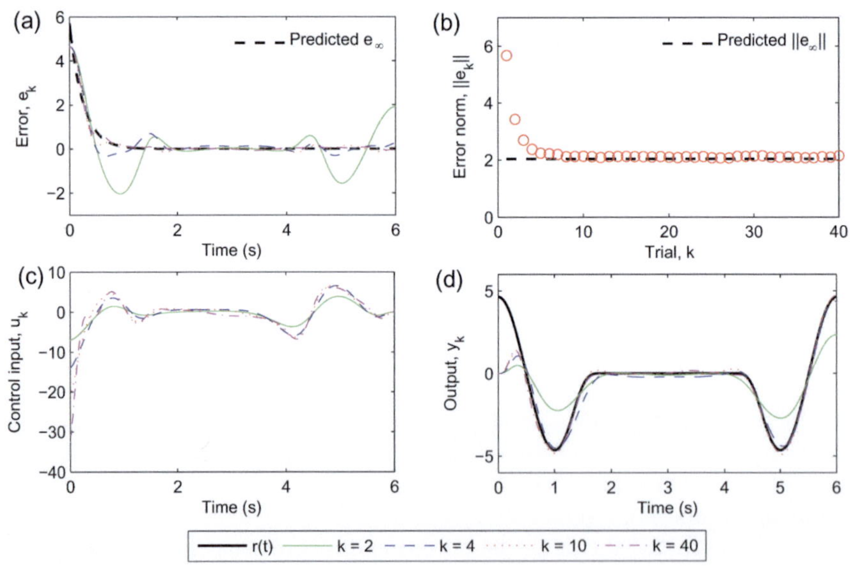

Fig. 10.4 Experimental results without initial zero period: **a** error, e_k, **b** error norm, $\|e_k\|$, **c** control input, u_k, and **d** output, y_k. Reprinted with permission from [5], ©2013 IEEE

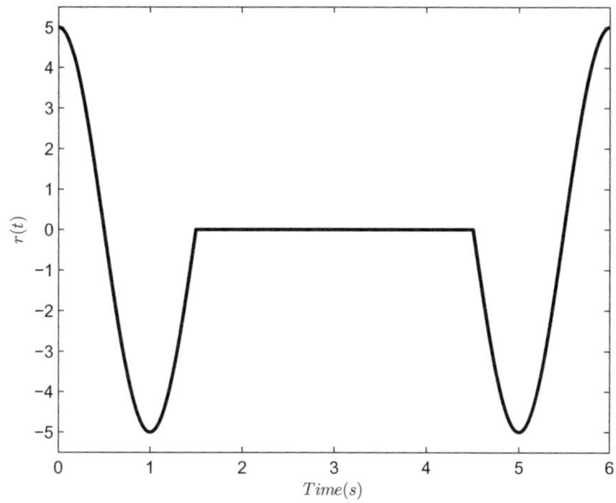

Fig. 10.5 The reference signal $r_1(t)$

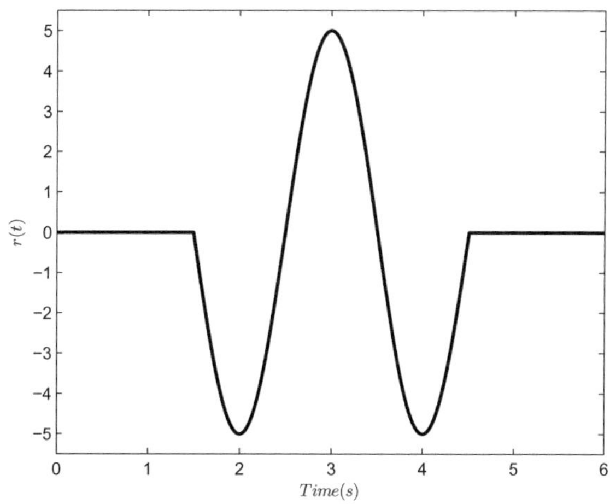

Fig. 10.6 The reference signal $r_2(t)$

the following simulations, the system is sampled using a zero-order hold and a sampling time of 0.1 s. Zero initial input is assumed, and the weightings are chosen as $Q = R = I, \epsilon^2 = 1$ (unless specified otherwise). In PNOILC, the parameter choice of $\lambda = 2$ and prediction horizon $M = 2, 3$ are considered. The two references used in this section, i.e., those shown in Fig. 10.5 (i.e., plot d) in Fig. 10.4 and Fig. 10.6 (i.e., plot d) in Fig. 10.3, are chosen.

10.3 Convergence Performance Using Exact System Model

This section demonstrates the convergence performance of the aforementioned ILC algorithms using exact system model to demonstrate their convergence performance.

First, for this non-minimum phase system the results in Chap. 5 are used to predict the possible performance of the algorithms. The reference signal $r_1(t)$ comprises a sinusoidal positional movement (in radians) at the initial and final stages, between which the signal is set at zero. Following the discussion under (1) and (2) at the end of Sect. 5.6.1, the initial tracking error e_0 will *not* be small in time interval $[0, 1.5] \subset [0, 6]$, leading to predicted final error value $\|e_\infty\| = 7.757$ significantly different from zero. The aforementioned algorithms are applied to the system for this reference and the simulation results are shown in Figs. 10.7, 10.8and Tables 10.1, 10.2, which verify

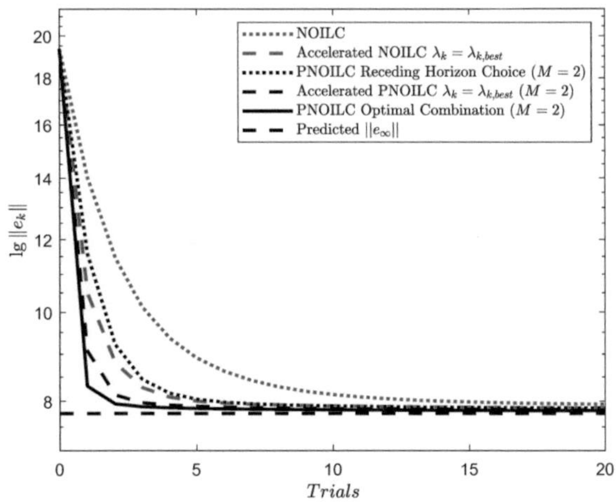

Fig. 10.7 Convergence performance for $r_1(t)$ using NOILC and PNOILC with prediction horizon $M = 2$

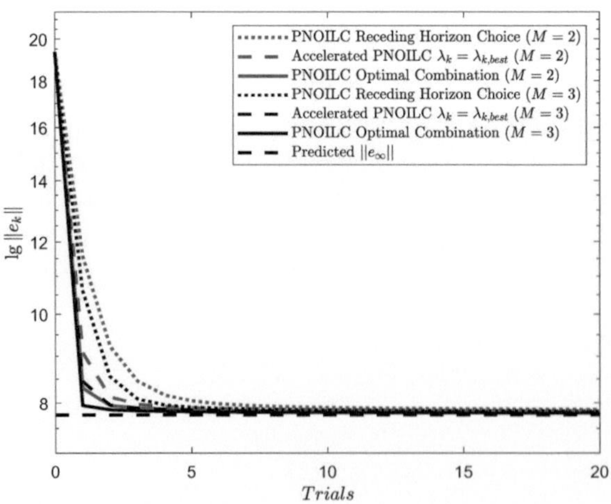

Fig. 10.8 Convergence performance for $r_1(t)$ using PNOILC with prediction horizon $M = 2, 3$

Table 10.1 Tracking error norms for $r_1(t)$ using NOILC and PNOILC with prediction horizon $M = 2$

Iteration (k)	NOILC	Accelerated NOILC $\lambda_{k,best}$	PNOILC $M = 2$ receding horizon choice	Accelerated PNOILC $M = 2$ $\lambda_{k,best}$	PNOILC $M = 2$ optimal combination
0	19.3649	19.3649	19.3649	19.3649	19.3649
1	14.0107	10.4872	11.5840	9.0770	8.3081
2	11.4678	8.8038	9.2256	8.1267	7.9467
3	10.1304	8.2745	8.4519	7.9679	7.8896
4	9.3696	8.0797	8.1622	7.9224	7.8690
5	8.9096	8.0003	8.0385	7.8999	7.8566
6	8.6171	7.9625	7.9790	7.8855	7.8482
7	8.4234	7.9402	7.9468	7.8754	7.8418
8	8.2909	7.9248	7.9274	7.8679	7.8369

Table 10.2 Tracking error norms for $r_1(t)$ using PNOILC with prediction horizon $M = 2, 3$

Iteration (k)	PNOILC $M = 2$ receding horizon choice	Accelerated PNOILC $M = 2$ $\lambda_{k,best}$	PNOILC $M = 2$ optimal combination	PNOILC $M = 3$ receding horizon choice	Accelerated PNOILC $M = 3$ $\lambda_{k,best}$	PNOILC $M = 3$ optimal combination
0	19.3649	19.3649	19.3649	19.3649	19.3649	19.3649
1	11.5840	9.0770	8.3081	10.6296	8.4528	7.9474
2	9.2256	8.1267	7.9467	8.5543	7.9528	7.8612
3	8.4519	7.9679	7.8896	8.0797	7.8938	7.8393
4	8.1622	7.9224	7.8690	7.9531	7.8711	7.8284
5	8.0385	7.8999	7.8566	7.9096	7.8584	7.8214
6	7.9790	7.8855	7.8482	7.8898	7.8498	7.8164
7	7.9468	7.8754	7.8418	7.8782	7.8435	7.8125
8	7.9274	7.8679	7.8369	7.8703	7.8385	7.8094

our expectations in the sense that the final tracking error norm is predicted accurately using the theory.

Now the reference is replaced by $r_2(t)$ shown in Fig. 10.6. This corresponds to the same movement as the previous reference; however, the starting point has now been changed to half way along its length. The initial period of $r_2(t)$ is now zero so, recalling again the discussion under (1) and (2) at the end of Sect. 5.6.1, there exists a time interval for this case $[0, 1.5] \subset [0, 6]$ where the reference $r_2(t)$ is zero and the plant has zero initial conditions. Hence the error e_0 will be zero in this interval leading to a low predicted final error, e_∞, and associated norm $\|e_\infty\| = 0.0125$. To save simulation time, the input weighting matrix was set to be $R = 0.001 \times I$

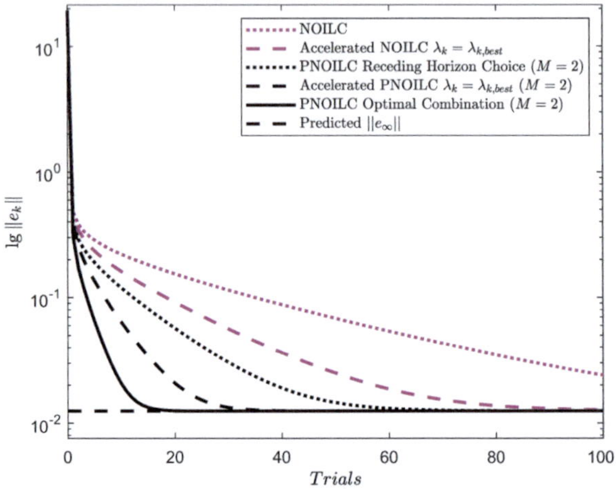

Fig. 10.9 Convergence performance for $r_2(t)$ using NOILC and PNOILC with prediction horizon $M = 2$

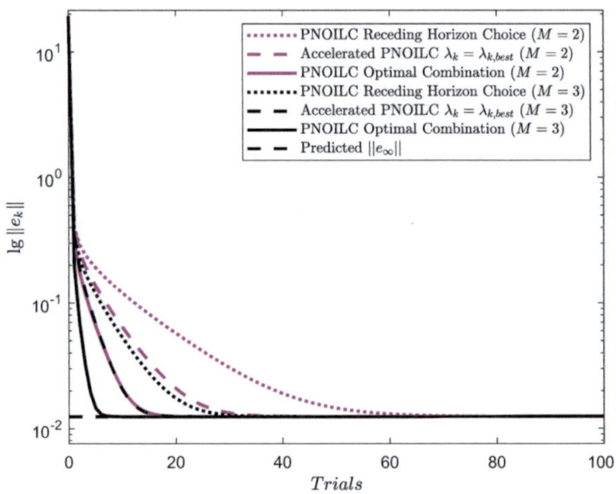

Fig. 10.10 Convergence performance for $r_2(t)$ using PNOILC with prediction horizon $M = 2, 3$

to get faster convergence in this example (choosing $R = I$ needs more than 500 iterations for the tracking error norms to stagnate). The simulation results are shown in Figs. 10.9, 10.10 and Tables 10.3, 10.4, which are, again, consistent with theoretical expectations.

From the simulation results, it can be found that the introduced ILC algorithms can be applied to the complex non-minimum phase system successfully. Note that the PNOILC algorithm outperforms the NOILC algorithm (even with the simplest receding horizon parameter choice). The accelerated algorithms further improve the convergence performance of NOILC and PNOILC algorithms. It can also be seen that in PNOILC with proper parameter choice, the convergence performance can be improved significantly, particularly in the first few iterations. Taking the optimal combination choice for example, as seen in Fig. 10.9 and Table 10.3, to achieve a reduction by factor of 100 in the tracking error norm, the optimal combination

Table 10.3 Tracking error norms for $r_2(t)$ using NOILC and PNOILC with prediction horizon $M = 2$

Iteration (k)	NOILC	Accelerated NOILC $\lambda_{k,best}$	PNOILC $M = 2$ receding horizon choice	Accelerated PNOILC $M = 2$ $\lambda_{k,best}$	PNOILC $M = 2$ optimal combination
0	19.3649	19.3649	19.3649	19.3649	19.3649
1	0.4802	0.4780	0.3965	0.3939	0.3041
2	0.3788	0.3421	0.2889	0.2534	0.1722
3	0.3355	0.2909	0.2434	0.1973	0.1185
4	0.3072	0.2584	0.2131	0.1614	0.0863
5	0.2860	0.2342	0.1900	0.1351	0.0646
6	0.2688	0.2148	0.1713	0.1146	0.0494
7	0.2544	0.1985	0.1555	0.0982	0.0384
8	0.2418	0.1845	0.1420	0.0846	0.0304

Table 10.4 Tracking error norms for $r_2(t)$ using PNOILC with prediction horizon $M = 2, 3$

Iteration (k)	PNOILC $M = 2$ receding horizon choice	Accelerated PNOILC $M = 2$ $\lambda_{k,best}$	PNOILC $M = 2$ optimal combina-tion	PNOILC $M = 3$ Receding horizon choice	Accelerated PNOILC $M = 3$ $\lambda_{k,best}$	PNOILC $M = 3$ optimal combina-tion
0	19.3649	19.3649	19.3649	19.3649	19.3649	19.3649
1	0.3965	0.3939	0.3041	0.3514	0.3485	0.1971
2	0.2889	0.2534	0.1722	0.2266	0.1878	0.0694
3	0.2434	0.1973	0.1185	0.1750	0.1270	0.0329
4	0.2131	0.1614	0.0863	0.1419	0.0911	0.0193
5	0.1900	0.1351	0.0646	0.1177	0.0673	0.0145
6	0.1713	0.1146	0.0494	0.0990	0.0508	0.0130
7	0.1555	0.0982	0.0384	0.0839	0.0390	0.0126
8	0.1420	0.0846	0.0304	0.0717	0.0306	0.0125

choice for prediction horizon $M = 2$ needs 2 iterations, whereas the receding horizon choice needs 6 iterations and NOILC needs 14 iterations. If an improvement of factor 1000 is required, the optimal combination choice needs 11 iterations whilst the receding horizon choice needs 42 iterations and the number for NOILC is 116! This improvement implies a significant saving of time and experimental cost, which is of great interest in practice, provided that the system model is accurate enough.

It can also be seen, even for the accelerated algorithms with faster convergence, that the slow convergence phenomenon hasn't been improved much, except for the convergence rate in the initial stage, in the sense that the "final" tracking error almost remains the same. This indicates the difficulties of ILC for non-minimum phase systems, as in all other control paradigms.

10.4 Robustness with Respect to Model Uncertainty

In this section, robustness issue against model uncertainty is illustrated. It can be expected that the algorithms will still converge if the model uncertainty is "sufficiently small", which will be shown in the following simulations.

As mentioned earlier, the system model was identified using frequency response data [2] and its frequency response is shown in Fig. 10.11with the following transfer function:

$$G(s) = \frac{165.95(4-s)}{s^4 + 21.5\,s^3 + 170.28\,s^2 + 368.52\,s + 663.82}. \tag{10.2}$$

Simple computations show that the system has four poles $-1.0711 \pm 2.0569i$, $-9.6789 \pm 5.4546i$, and $-1.0711 \pm 2.0569i$ are dominant. In what follows, two uncertain models with different nondominant poles will be used to test the robustness of the algorithms.

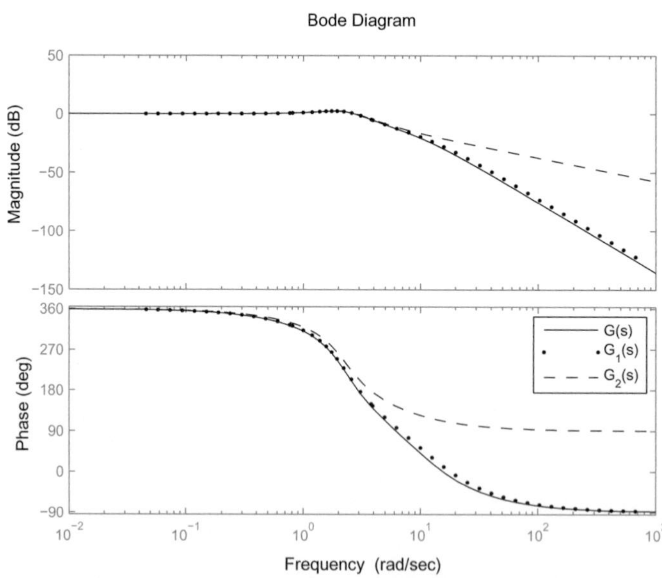

Fig. 10.11 The frequency response of the exact system model $G(s)$ and uncertain models $G_1(s)$, $G_2(s)$

10.4.1 Robustness Against "Small" Model Uncertainty

First, suppose that the nondominant poles $-9.6789 \pm 5.4546i$ were not identified precisely and only an approximation of $-12 \pm 6i$ was obtained. The resulting system is denoted as G_1, whose transfer function is

$$G_1(s) = \frac{241.9(4 - s)}{s^4 + 26.14\,s^3 + 236.8\,s^2 + 514.7\,s + 968}$$

and its frequency response is shown in Fig. 10.11.

From the figure, it can be seen G_1 is a good approximation of the exact model G and the gap between them is very small. The simulation then uses this uncertain model to test the robustness of the algorithms. The simulation results are shown in Figs. 10.12, 10.13 for $r_1(t)$ and Figs. 10.14, 10.15 for $r_2(t)$, respectively.

From the figures, it can be seen that for the chosen references the algorithms tested still converge and demonstrate some degree of robustness, which can be understood by noticing that in this example the mismatch between the uncertain model and the system is very small. The next simulation increases this mismatch to show what happens for "large" model uncertainty.

10.4.2 Robustness Against "Large" Model Uncertainty

Now, suppose that during the identification process, the nondominant poles $-9.6789 \pm 5.4546i$ were not identified at all and only the dominant poles $-1.0711 \pm 2.0569i$ were included. The resulting model is denoted as G_2 with transfer function

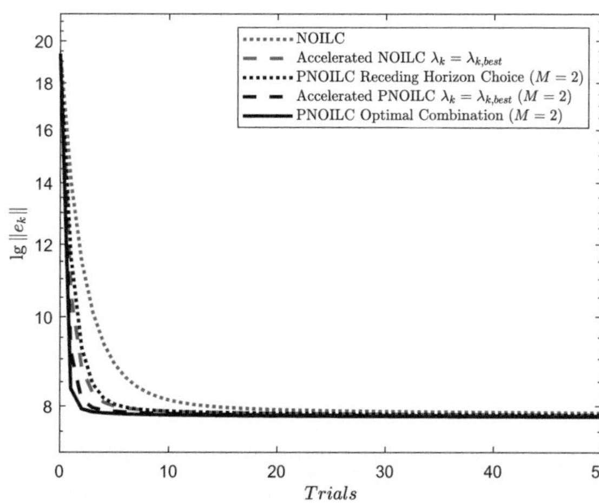

Fig. 10.12 Convergence performance for $r_1(t)$ using uncertain model G_1 for NOILC and PNOILC with prediction horizon $M = 2$

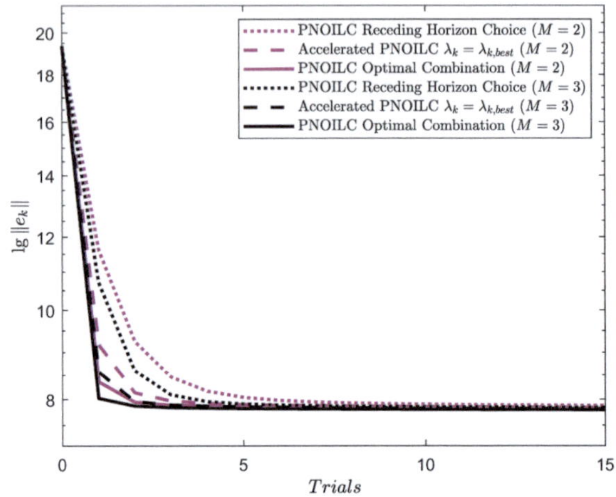

Fig. 10.13 Convergence performance for $r_1(t)$ using uncertain model G_1 for PNOILC with prediction horizon $M = 2, 3$

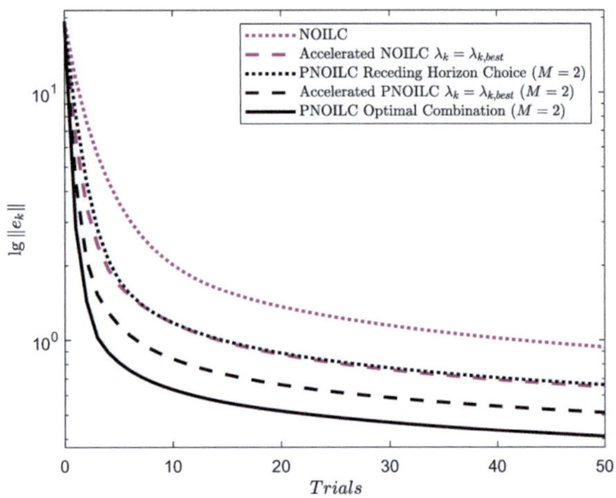

Fig. 10.14 Convergence performance for $r_2(t)$ using uncertain model G_1 for NOILC and PNOILC with prediction horizon $M = 2$

$$G_1(s) = \frac{1.344(4 - s)}{s^2 + 2.142\,s + 5.378}$$

and its frequency response is shown in Fig. 10.11, from which it can be found that the gap between G_2 and the true system model is substantial. The simulation was run using the uncertain model G_2 and the results are shown in Figs. 10.16, 10.17 for $r_1(t)$ and Figs. 10.18, 10.19 for $r_2(t)$.

The above results verify our expectation: the algorithms will eventually diverge if the model uncertainty is significant although the ILC algorithms will be terminated when the divergence occurs in practice. It is noticed that the algorithms still maintain some degree of convergence during the initial stage. If satisfactory performance is achieved before divergence occurs, the ILC algorithms are still successful in the

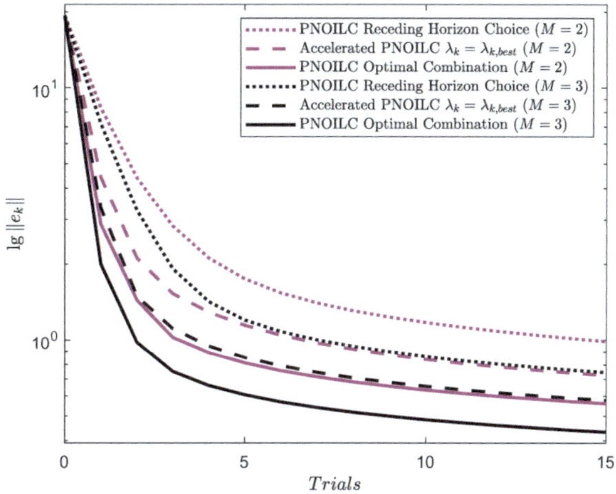

Fig. 10.15 Convergence performance for $r_2(t)$ using uncertain model G_1 for PNOILC with prediction horizon $M = 2, 3$

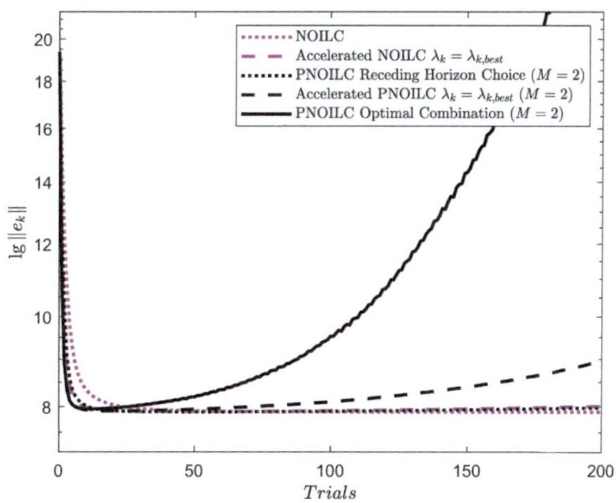

Fig. 10.16 Convergence performance for $r_1(t)$ using uncertain model G_2 for NOILC and PNOILC with prediction horizon $M = 2$

sense that the ILC algorithms can be stopped when the required accuracy is met and the divergence will never be seen.

From the results, it can be also seen that the faster the algorithm converges, the quicker the divergence happens, which implies the algorithm possessing the better convergence performance has the worse robustness capability and this is consistent with the well-known performance/robustness tradeoff in other control situations, as well as that predicted in previous chapters.

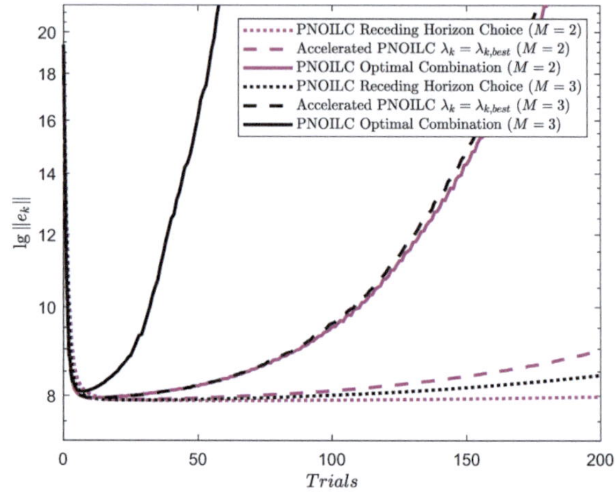

Fig. 10.17 Convergence performance for $r_1(t)$ using uncertain model G_2 for PNOILC with prediction horizon $M = 2, 3$

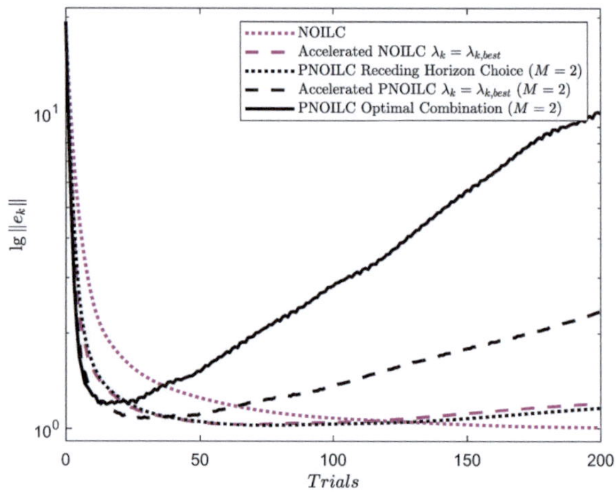

Fig. 10.18 Convergence performance for $r_2(t)$ using uncertain model G_2 for NOILC and PNOILC with prediction horizon $M = 2$

10.5 Robustness with Respect to Noise/Disturbances

In this section, robustness with respect to noise/disturbances is discussed. As noticed in [4]: "In the optimization framework, robustness with respect to bounded disturbances is amenable to geometric interpretation. Clearly convergence to zero error is not, in general, possible but a degree of learning can still take place. The basis for believing this is based on the observation that optimization methods generate descent directions for the norm of the error signal. Hence, provided the disturbance does not 'reverse' the descent direction, learning is still possible. Intuition says that the condition for learning is related to the error/disturbance signal/noise ratio (SNR). As a consequence, learning is achieved until the error and disturbance have similar

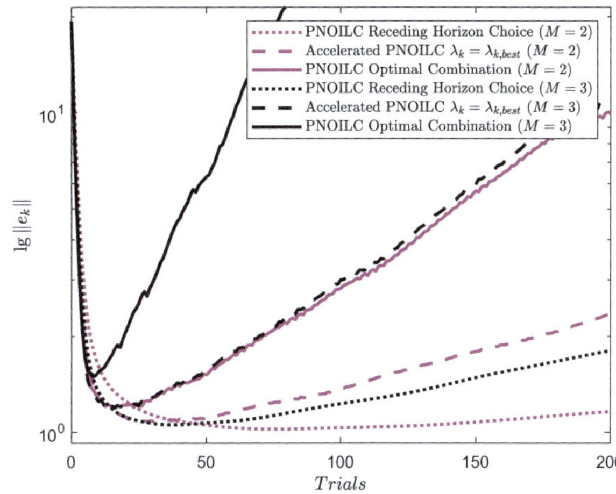

Fig. 10.19 Convergence performance for $r_2(t)$ using uncertain model G_2 for PNOILC with prediction horizon $M = 2, 3$

magnitude". Hence, it can be expected that as the noise/disturbances become larger, the effect of noise emerges and finally the divergence will happen.

The following example illustrates the robustness of the chosen algorithms against measurement noise written in the form

$$y = G(s)u + H(s)n, \tag{10.3}$$

where n is white noise and $H(s)$ is a filter defining the spectral characteristics of the noise seen at the output. In practice, $H(s)$ is usually a high-pass filter (characterizing the high frequency nature of the noise). Intuitively, if the bandwidth of $H(s)$ is high enough, since $G(s)$ is generally low-pass, their frequency response will be clearly separated (as seen in Fig. 10.20). The noise will be filtered by the system and will not pass through the next iteration. Hence, the effect of noise term $H(s)n$ on the performance of ILC algorithm will be sufficiently small. As the bandwidth of $H(s)$ decreases, there will be overlapping between the frequency response of $G(s)$ and $H(s)$. The effect of noise/disturbances begins to arise and divergence will finally happen. To verify this, take $H(s)$ to be the specific form of

$$H(s) = \frac{0.1(1 + Ts)}{1 + 0.1Ts}. \tag{10.4}$$

Increasing T decreases the bandwidth of $H(s)$ and consequently increases the effect of the noise/disturbances.

Simulation was run using different T to demonstrate the effect of noise/disturbances. The bandwidth of the system is 3.40 rad/s and hence T is chosen from 0.1 to 1 in the simulation to compare the different convergence performance. During the simulation, the signal/noise ratio (SNR), which is defined here as the power ratio of the reference signal to the noise signal, is set to be 10 dB. The noise n is zero mean Gaussian white noise generated using MATLAB® toolbox. The simplest case of

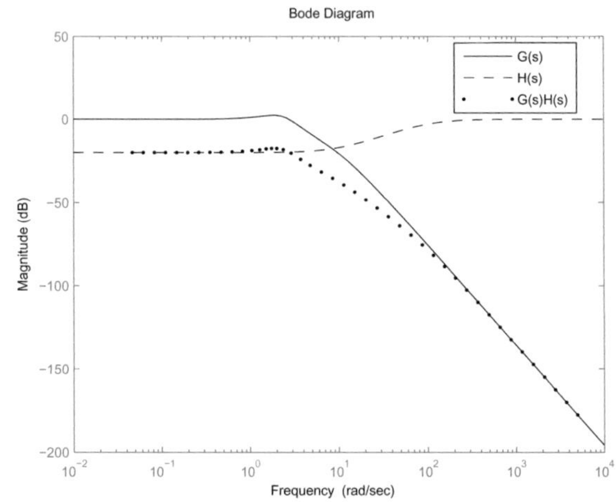

Fig. 10.20 Frequency response of $G(s)$, $H(s)$, and $G(s)H(s)$ with $T = 0.1$

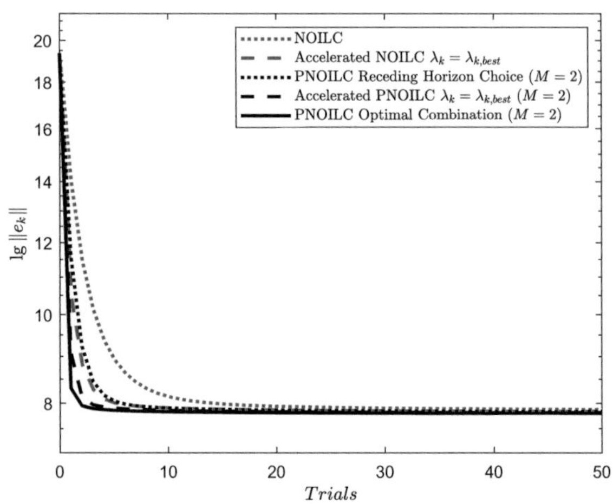

Fig. 10.21 Convergence performance for $r_1(t)$ with $T = 0.1$ using NOILC and PNOILC with prediction horizon $M = 2$

prediction horizon $M = 2$ in predictive NOILC-related algorithms is considered to show the noise/disturbance effect and the simulation results are shown in Figs. 10.20, 10.21, 10.22, 10.23, 10.24, 10.25, 10.26, and 10.27 for $r_1(t)$.

From the figures, it can be seen that when T is small, the frequency response of $H(s)$ is clearly separated from that of $G(s)$, and the spectrum of the coloured noise $H(s)n$ is well beyond the bandwidth of the plant $G(s)$. Hence, the noise/disturbances will not be able to enter the system and their effect will be small, as seen from Fig. 10.21. As T increases, the noise/disturbances begin to enter the system and their effect emerges, and finally, divergence occurs, which can be seen from Fig. 10.27. This verifies our expectations. Similar patterns can be found for reference $r_2(t)$ and the results are omitted here for brevity.

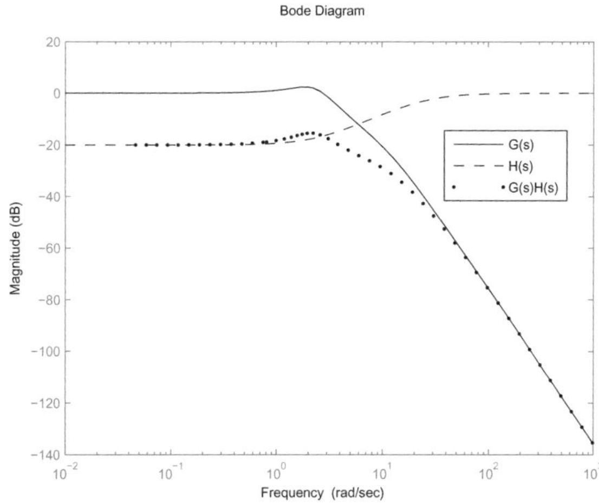

Fig. 10.22 Frequency response of $G(s)$, $H(s)$, and $G(s)H(s)$ with $T = 0.4$

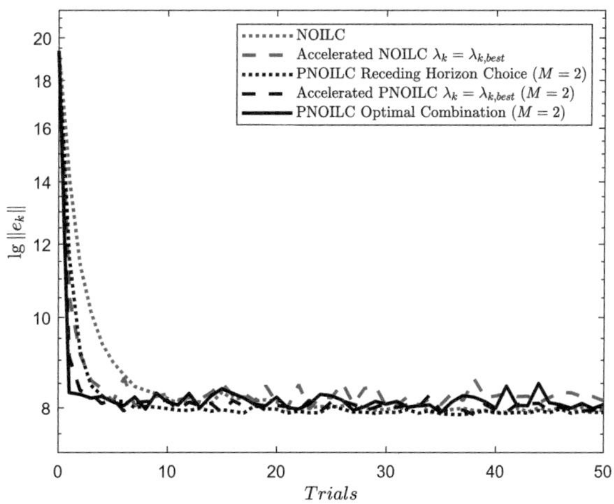

Fig. 10.23 Convergence performance for $r_1(t)$ with $T = 0.4$ using NOILC and PNOILC with prediction horizon $M = 2$

10.6 Summary and Discussion

It is well known that NMP zeros can cause difficulties in control system design. In this chapter, the theoretical analysis of the NMP systems and unconstrained ILC algorithms introduced in this book has been applied to an electro-mechanical, non-minimum phase test-bed model to demonstrate their effectiveness and algorithm's robustness. The results show that the theory/chosen algorithms can be applied to this complex non-minimum phase system successfully with good prediction/convergence performance. The simulation results also demonstrate that the algorithms have a certain degree of robustness with respect to the model uncertainty and noise/disturbances.

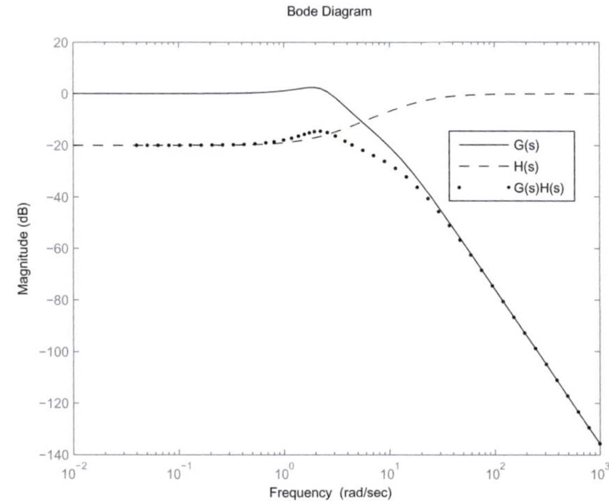

Fig. 10.24 Frequency response of $G(s)$, $H(s)$, and $G(s)H(s)$ with $T = 0.5$

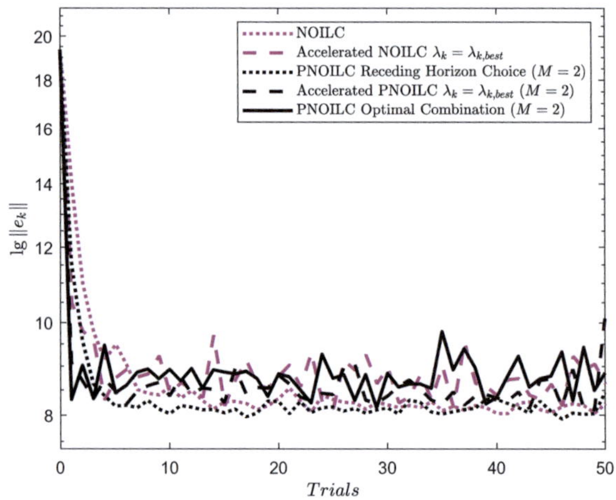

Fig. 10.25 Convergence performance for $r_1(t)$ with $T = 0.5$ using NOILC and PNOILC with prediction horizon $M = 2$

Whilst experimental results are presented to demonstrate the slow convergence behaviour for NOILC algorithm, the other results in the chapter are based on numerical simulations. Additional practical experiments are needed to further verify these results for non-minimum phase systems (some results on experimental verification of the predictive NOILC algorithm can be found in [1]).

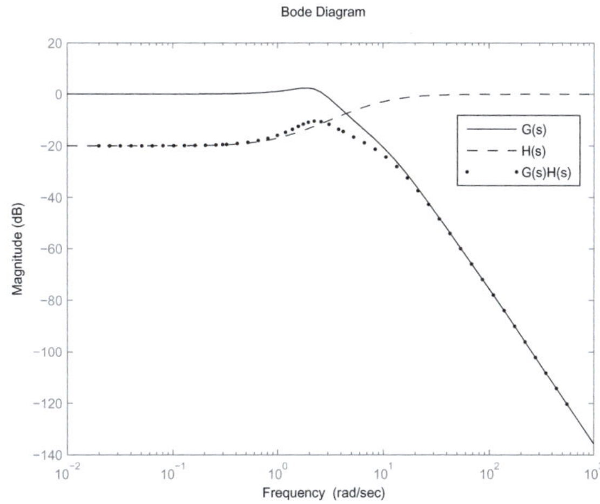

Fig. 10.26 Frequency response of $G(s)$, $H(s)$, and $G(s)H(s)$ with $T = 1$

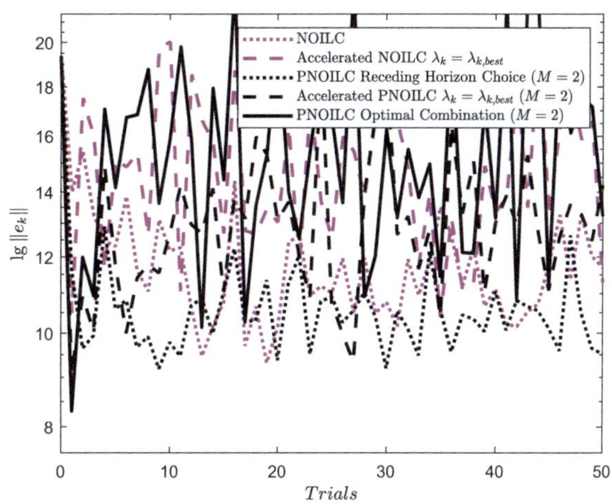

Fig. 10.27 Convergence performance for $r_1(t)$ with $T = 1$ using NOILC and PNOILC with prediction horizon $M = 2$

References

1. Chu B, Owens DH, Freeman CT (2015) Iterative learning control with predictive trial information: convergence, robustness, and experimental verification. IEEE Trans Control Syst Technol 24(3):1101–1108
2. Freeman CT, Lewin PL, Rogers E (2005) Experimental evaluation of iterative learning control algorithms for non-minimum phase plants. Int J Control 78(11):826–846
3. Owens DH, Chu B (2010) Modelling of non-minimum phase effects in discrete-time norm optimal iterative learning control. Int J Control 83(10):2012–2027
4. Owens DH, Hatonen J (2005) Iterative learning control—an optimization paradigm. Ann Rev Control 29(1):57–70
5. Owens DH, Chu B, Rogers E, Freeman CT, Lewin PL (2014) Influence of nonminimum phase zeros on the performance of optimal continuous-time iterative learning control. IEEE Trans Control Syst Technol 22(3):1151–1158

Chapter 11
Application to Robot-Assisted Upper Extremity Stroke Rehabilitation

This chapter develops a combined feedback and Iterative Learning Control (ILC)-based scheme for regulating the assistive stimulation applied in robot-assisted upper limb stroke rehabilitation. The feedback controller is first designed and the resulting controlled system is used in an iterative control setting where the objective is that the patient gradually recovers lost functionality by repeated attempts at a finite duration task assisted by applied electrical stimulation. Once each attempt is complete the limb is returned to the starting location and the tracking error is used in an iterative learning control law to compute the stimulation applied on the next attempt, where the objective is that if the patient is improving with successive attempts the level of stimulation used should decrease and voluntary effort increase.

The chapter contains results from a simulation-based assessment of the design for both Norm Optimal Iterative Learning Control (NOILC) algorithm and its acceleration using error corrected reference, as well as the use of different feedback controllers. These simulation studies are a necessary requirement before patient-based trials can be considered.

11.1 Robot-Assisted Upper Extremity Stroke Rehabilitation

Every year about 15 million people world-wide suffer a stroke and up to a third of these are left with permanent impairment. One common cause of a stroke is blockage of a blood vessel in the brain that starves regions downstream of blood. This, in turn, means that the connecting nerve cells die, usually resulting in partial paralysis on one side of the body, termed hemiplegia. These dead cells cannot regrow but new connections can be made using the brain's spare capacity. In fact, the brain is continually and rapidly changing as new skills are learned, new connections are formed, and redundant ones disappear. A person who re-learns skills after a stroke

B. Chu and D. H. Owens, *Optimal Iterative Learning Control*, Advances in Industrial Control, https://doi.org/10.1007/978-3-031-80236-2_11

goes through the same process as someone learning an everyday task, requiring sensory feedback during repeated practice of a task. The problem is that they have very poor movement skills so they cannot practice therefore do not receive feedback on their performance.

People who survive a stroke commonly have a complex pattern of upper limb motor impairments following stroke, resulting in a loss of functional abilities such as reaching. The relationship between reaching and independence is reflected in measures of functional independence such as the Barthel Index where the ability to reach is required for over half of the activity of daily living tasks. The current prognosis for upper limb recovery following stroke is poor, with a review reporting that complete recovery occurs in less than 15% of patients with initial paralysis. Stroke is also an age-related disease, placing an increasing burden on long-term health and related resources unless improvements are made in achieving independence. Consequently there is a pressing need to improve the effectiveness of treatments. Enabling rehabilitation outside the hospital, supported by mobile technology, may lead to reduced cost, increased intensity of therapy, and a shift in the emphasis of responsibility for good health from health professionals to patients. References to the supporting literature are given in [10].

Conventional therapy and motor learning theory provide evidence that functional recovery can be achieved through the facilitation of motor control and skill acquisition, and restoration of muscle power can be regained through repetitive resistance exercises [16], in addition to the variety of tasks and feedback [3]. This knowledge has motivated the development of novel treatments, such as robot-aided therapy, which not only provides a transformation of rehabilitation clinics from labour-intensive work to technology-assisted operations, but also provides an opportunity for repetitive movement practice. Reviews of the robotic therapy literature for upper limb suggest that robot-assisted treatment improves motor control of the proximal upper limb and may improve functional outcomes [10].

Rehabilitation robots are power-driven or mechanically supported devices that assist a patient with constrained/limited physical capability in performing repetitive exercises. The resulting sensory feedback is known to be associated with cortical changes that facilitate the recovery of functional movement. Functional electrical stimulation (FES) has been found to be applicable as another promising method in driving cortical changes to enable recovery, which is motivated by a growing body of clinical evidence [5] and theoretical support from neurophysiology [2], and motor learning research [20]. FES drives the muscle by causing electrical impulses to travel along the nerves in much the same way as electrical impulses from the brain and, if carefully controlled stimulation is applied, useful movement can be obtained.

A range of FES control methods have been employed for movement control for both lower and upper limbs, including optimal control [13], H_∞ control [14], fuzzy control of standing [4], sliding mode control of shank movement [15], data-driven control of knee joint [19], and multi-channel PID control of the wrist [22]. Few of these model-based schemes have been applied to upper limb movement, or transferred into clinical practice [24], especially in the case of upper limb of stroke rehabilitation. This is mainly due to restrictive conditions such as minimal setup time during the clinical setting, reduced control over environmental constraints, and little possibility of repeating any one test in the programme of treatment undertaken.

Given the repetitive nature of the rehabilitation process, iterative learning control presents itself as a suitable approach for controller design. In the research described in [8, 12] the task was to reach out in the 2D plane to an object, such as a cup, and the reference trajectory, determined in a pre-assessment by a healthcare specialist, is a light beam beamed down from the above onto a surface. The patient's affected arm is connected to a robot and the task is to track the light beam with stimulation applied to the relevant muscle, in this case the triceps. Once each attempt is complete, the arm is returned to the starting location, and in the time to complete this operation, plus rest time to lessen the effects of muscle fatigue, the previous trial error is used in an ILC law to compute the stimulation to be applied on the next trial. Hence ILC directly regulates the level of stimulation applied and in a clinical trial [12] the essential property was confirmed that if the patient is improving with each trial his/her voluntary effort increases with each successive trial and the level of applied stimulation decreases.

In this chapter, we consider the 3D movement such as reaching out and lifting the arm. This involves the stimulation of more than one muscle and also much more involved dynamics. Again, previous research [11] has led to clinical trials [18] that confirm the potential of this approach. There remains, however, much to be done in terms of control law design for this application to enable its full potential to be assessed and this includes the structure and design of the control system. In the following sections, we will introduce the system setup, modelling of the system, and both feedback and iterative learning controller design.

11.2 System Modelling

In this section, the system setup and dynamic modelling are introduced.

11.2.1 The System Setup

The application of ILC to stroke rehabilitation starts from an assessment of the patient by a health professional to determine their current capabilities. From this a reference trajectory for a prescribed task, e.g., reaching out over a table top to a cup in the 2D work, is formulated, which is beyond the current unaided ability of the patient, and then presented as an illuminated path. Guided by a robot, the subject attempts to follow the prescribed path with FES assisting the relevant muscle and the error between the reference trajectory and that followed by the subject is measured. Once the attempt is complete, the arm is returned to the starting location and an ILC algorithm used to compute the stimulation is applied on the next attempt. The underlying idea is that if the patient improves with each successive attempt his/her voluntary effort increases and the level of applied stimulation decreases. Exactly this effect was seen in the clinical trials reported in the rehabilitation research literature [12].

In the previous research, the patient's forearm is constrained to lie in a horizontal plane. The resulting simplification to the underlying biomechanical model enables the development of a simple linearizing controller and the subsequent implementation of linear ILC algorithms which generally have well-defined convergence properties. To maximize the treatment's potential for rehabilitation, it is necessary to use a wider range of more functional movements, which more closely resemble the tasks necessary for daily living and are aligned with the activity-based measures. Relative to the previous work, the first step is to remove the planar forearm constraint, permitting unconstrained movement.

When providing assistive FES during unconstrained 3D upper limb reaching movements, FES must be applied using a controlled environment to reduce fatigue and ensure safety and comfort across a broad spectrum of patient abilities. There exist many exoskeletal robotic systems capable of providing such support, although very few have been combined with FES, and fewer still with model-based FES control schemes. Here a commercially available device is selected, but the ILC algorithms may be applied to a range of such supports. The mechanical exoskeleton employed is shown in Fig. 11.1 [10, 11] and has two springs incorporated in the mechanism to provide support to overcome gravity. This passive unweighing device allows patients to focus practice on the impaired muscles rather than those acting against gravity. Whilst it is supplied with its own broad range of virtual reality tasks, these are not suitable for controller evaluation, and a custom task display system has instead been developed. The patient is seated with their impaired arm strapped into the mechanical unweighing device, whose segmental lengths and degree of anti-gravity support are adjusted for each user. Joint positional data provided by resolvers mounted on the support mechanism are transferred to the interface module, which connects external devices to the real-time hardware. To increase potential for rehabilitation, FES is applied to both the anterior deltoid and triceps muscles in accordance with the clinical objective of providing more assistance, a greater degree of feedback, enabling more accurate movements, and ensuring more muscles are usefully activated.

11.2.2 Mechanical Support

The dynamic model in [9–11] has been developed for a system involving both the stimulated muscles and a combination of the human arm and a specialized rehabilitation robot. With reference to Fig. 11.2, the dynamics of the exoskeleton robot with rigid links is

$$\boldsymbol{B}_a(\Theta)\ddot{\Theta} + \boldsymbol{C}_a(\Theta, \dot{\Theta})\dot{\Theta} + \boldsymbol{F}_a(\Theta, \dot{\Theta}) + \boldsymbol{G}_a(\Theta) + \boldsymbol{K}_a(\Theta) = \boldsymbol{0} \tag{11.1}$$

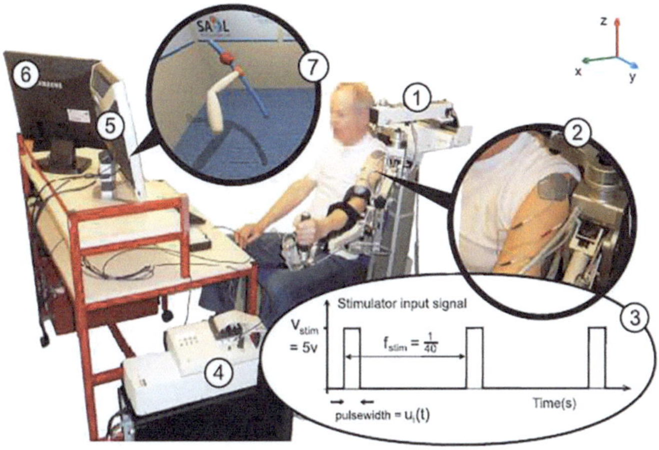

Fig. 11.1 3D ILC system components: (1) unweighing robotic device, (2) surface electrodes on triceps and anterior deltoid, (3) FES module, (4) real-time processor and interface module, (5) monitor displaying task, (6) therapist monitor, and (7) real-time 3D graphics. Reprinted with permission from [6], ©2014 IEEE

where $\Theta = [\theta_1, \theta_2, \theta_3, \theta_4, \theta_5]^T$ are the joint variables of the mechanical support, and $\boldsymbol{B}_a(\cdot)$ and $\boldsymbol{C}_a(\cdot)$ are 5-by-5 inertial and Coriolis matrices, and $\boldsymbol{F}_a(\cdot)$ and $\boldsymbol{G}_a(\cdot)$ frictional and gravitational vectors. The vector $\boldsymbol{K}_a(\cdot)$ denotes the moments arising from gravity compensation provided by the two springs, each of which is the function of θ_3 and θ_5, respectively, thus making $\boldsymbol{K}_a(\cdot)$ take the form of $[0, 0, k_3(\theta_3), 0, k_5(\theta_5)]$.

11.2.3 Muscle and Arm

Spasticity in stroke patients typically produces a resistance to arm extension associated with the overactivity of muscles, like the biceps, wrist, and finger flexors, and with loss of activity of muscles such as the triceps, anterior deltoid, wrist, and finger extensors [17]. In order to provide effective treatment, it is the latter group of muscles that must be activated during the functional reaching tasks. It is first assumed that FES stimulation to the triceps (u_5) produces a moment about an axis orthogonal to both the forearm and upper arm, and stimulation to the anterior deltoid (u_2) generates a moment about an axis that is fixed corresponding to the shoulder. The joint variables with respect to the two kinds of muscles are ϕ_5 and ϕ_2, respectively, with additional joint axes encompassing the remaining degrees of freedom. The dynamics of the human arm with FES applied to the two muscles, as shown in Fig. 11.2.(b), is represented by

$$\boldsymbol{B}_h(\boldsymbol{\Phi})\ddot{\boldsymbol{\Phi}} + \boldsymbol{C}_h(\boldsymbol{\Phi}, \dot{\boldsymbol{\Phi}})\dot{\boldsymbol{\Phi}} + \boldsymbol{F}_h(\boldsymbol{\Phi}, \dot{\boldsymbol{\Phi}}) + \boldsymbol{G}_h(\boldsymbol{\Phi}) = \boldsymbol{\tau}(\boldsymbol{u}, \boldsymbol{\Phi}, \dot{\boldsymbol{\Phi}}) \tag{11.2}$$

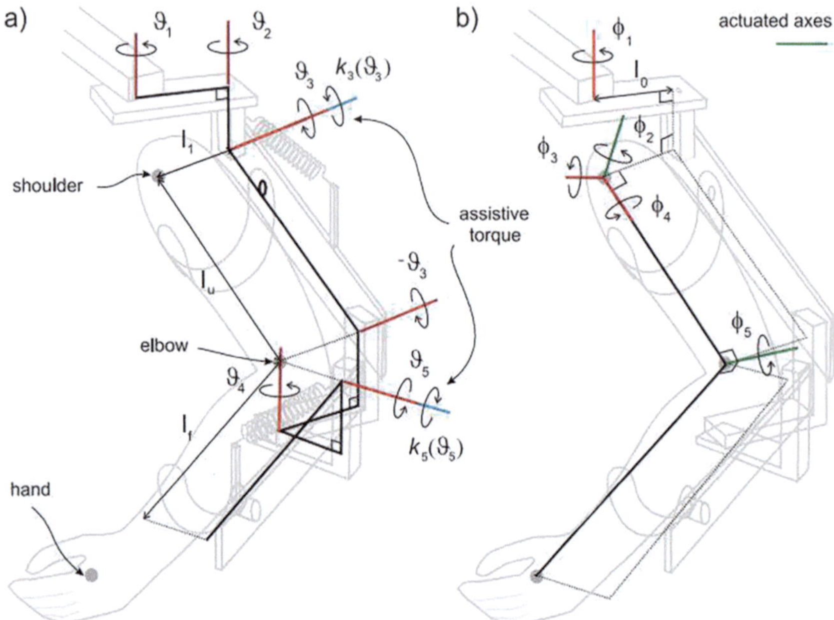

Fig. 11.2 Kinematic system relations: **a** mechanical support and **b** human arm. Reprinted with permission from [6], ©2014 IEEE

where $\mathbf{\Phi} = [\phi_1, \phi_2, \phi_3, \phi_4, \phi_5]^T$ denote the anthropomorphic joints, comprising those actuated by stimulated muscles, and $\tau(\cdot)$ are the torques produced by the stimulated muscles, therefore taking the form

$$\boldsymbol{\tau}(\boldsymbol{u}, \mathbf{\Phi}, \dot{\mathbf{\Phi}}) = [0, \tau_2(u_2, \phi_2, \dot{\phi}_2), 0, 0, \tau_5(u_5, \phi_5, \dot{\phi}_5)]^T. \tag{11.3}$$

This problem setup replicates reaching out and up with the shoulder and elbow.

The muscle models reported in the literature to represent the response to applied simulation and hence the development of model-based controller design for both upper and lower limb research have considerable structural variations. Of these, the Hill-type model, referenced in [10], is widely used where the output torque of the stimulated muscle is modelled as the product of three independent experimentally identified elements: the force–length property, the force–velocity property, and the nonlinear muscle activation dynamics under isometric conditions.

The activation dynamics represents the dominant part of the muscle dynamics and usually modelled by a Hammerstein structure [1], comprising of a static nonlinearity, $h_{IRC}(u)$, and a linear activation dynamics, $h_{LAD}(t)$. These are linked to the steady-state torque T_m that is produced by the static nonlinearity and takes the form

$$T_m = c_1 \left| \frac{e^{c_1 u} - 1}{e^{c_2 u} + c_3} \right| \tag{11.4}$$

where c_1, c_2, c_3 are muscle-dependent parameters to be identified, and u denotes the externally applied stimulation impulses. The linear activation part is modelled as a critically damped second-order system

$$T = \frac{1}{\frac{1}{w_n^2}s^2 + \frac{2}{w_n}s + 1} T_m \tag{11.5}$$

with s the Laplace transform parameter, T is the resultant torque generated by the muscle, and natural frequency w_n is also a muscle-dependent term to be identified as well. As a valid first approximation, a simplified Hill-type model with only the Hammerstein structure is considered here.

In order to decouple the model with respect to the input and joint trajectories, the combined system must be written in terms of Φ. This is possible only if, within the necessary joint ranges, there exists a bijective transformation between these coordinate sets, which is given by $\boldsymbol{\Phi} = \boldsymbol{k}(\boldsymbol{\Theta})$ and hence the Lagrangian equation in one variable can be expressed in terms of the other via the chain rule. Hence the combined model of the mechanical support and the stimulated human arm is written as

$$\begin{aligned} &\boldsymbol{B}(\boldsymbol{\Phi})\ddot{\boldsymbol{\Phi}} + \boldsymbol{C}(\boldsymbol{\Phi}, \dot{\boldsymbol{\Phi}})\dot{\boldsymbol{\Phi}} + \boldsymbol{F}(\boldsymbol{\Phi}, \dot{\boldsymbol{\Phi}}) + \boldsymbol{G}(\boldsymbol{\Phi}) + \boldsymbol{K}(\boldsymbol{\Phi}) \\ &= \boldsymbol{\tau}(\boldsymbol{u}, \boldsymbol{\Phi}, \dot{\boldsymbol{\Phi}}) - \boldsymbol{J}_h^T(\boldsymbol{\Phi})\boldsymbol{h} \end{aligned} \tag{11.6}$$

where $\boldsymbol{J}_h(\boldsymbol{\Phi})$ is the system Jacobian and $\boldsymbol{h}$ represents a vector of external forces and torques applied—for a more detailed description including experimental procedures to identify the parameters please refer to [6, 9].

11.3 Feedback Control Design

The design objective is to control the pulse width inputs $u_2(t)$ and $u_5(t)$ to assist tracking performance of the supported human arm system such that the outputs $\Phi_2(t)$ and $\Phi_5(t)$ of the vector $\boldsymbol{\Phi}(t)$ in (11.6) track given reference trajectories $\hat{\Phi}_2(t)$ and $\hat{\Phi}_5(t)$ as shown in Fig. 11.3 (corresponding to lifting and extending the upper arm and forearm), whilst the remaining joint angles remain bound-input bounded-output stable.

This section details the control design, starting with input output (I/O) linearization method of the muscle model, as well as the supported arm.

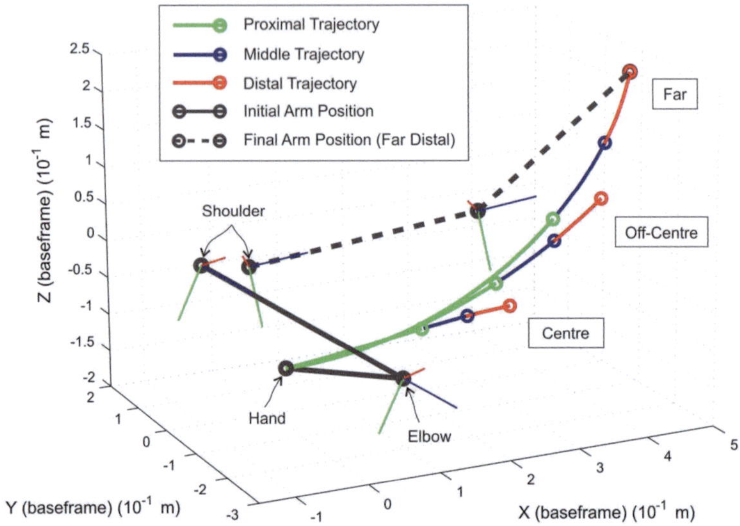

Fig. 11.3 Reference trajectories corresponding to arm extension. Reprinted with permission from [6], ©2014 IEEE

11.3.1 A Feedback Linearizing Controller

It is assumed that the nonlinear part of the muscle, the isometric recruitment curve, is ideal and thus its influence can be cancelled by the direct multiplication with its inverse, $h_{IRC}^{-1}(u)$, which constitutes the first component of the linearizing controller.

If the dynamics are written in the form

$$\dot{x} = f(x) + g(x)u \tag{11.7}$$

$$y = h(x) \tag{11.8}$$

with x the state variable, u the input, and y the output, the I/O linearizing controller is

$$u = \beta(x)^{-1}(-\alpha(x) + v) \tag{11.9}$$

where v is the new control input to achieve stability and trajectory tracking, and α, β are defined by

$$\alpha(x) = (L_f^{\rho} h_i(x))_i \tag{11.10}$$

$$\beta(x) = (L_{g_j} L_f^{\rho-1} h_i(x))_{ij} \tag{11.11}$$

with $i, j = 1, ..., m$, m the number of inputs of the system, and ρ the relative degree of the system. $L_f h_i(x)$, $L_{g_j} h_i(x)$ are, respectively, the Lie Derivative of $h_i(x)$ along

$f(x)$ and of $h_i(x)$ along $g_j(x)$, which are defined as

$$L_f h_i(x) = \frac{\partial h_i}{\partial x} f(x) \tag{11.12}$$

$$L_{g_j} h_i(x) = \frac{\partial h_i}{\partial x} g_j(x). \tag{11.13}$$

Substituting the linearizing controller (11.9) into (11.7) and (11.8) reduces the input–output map to

$$y^{(\rho)} = v \tag{11.14}$$

which is a chain of ρ integrators.

It can be shown that both the nonlinear muscle model and the supported arm model can be transformed into the form of (11.7) and (11.8), with the muscle model taking the state-space equation

$$\dot{\boldsymbol{x}}_i = \begin{bmatrix} 0 & 1 \\ -w_n^2 & -2w_n \end{bmatrix} \boldsymbol{x}_i + \begin{bmatrix} 0 \\ 1 \end{bmatrix} h_{IRU,i}(u_i) \tag{11.15}$$

$$\boldsymbol{y}_i = \begin{bmatrix} w_n^2 & 0 \end{bmatrix} \boldsymbol{x}_i \qquad i \in \{2, 5\} \tag{11.16}$$

and (11.6) being rewritten in the form of

$$\dot{\boldsymbol{x}} = \boldsymbol{f}(\boldsymbol{x}) + \boldsymbol{g}(\boldsymbol{x})\boldsymbol{h}_{LAD}(\boldsymbol{u}) \tag{11.17}$$

$$\boldsymbol{y} = \boldsymbol{h}(\boldsymbol{x}) \tag{11.18}$$

where $\boldsymbol{x} = [\boldsymbol{\Phi}^T \; \dot{\boldsymbol{\Phi}}^T \; \boldsymbol{x}_2^T \; \boldsymbol{x}_5^T]^T$ and

$$\boldsymbol{f}(\boldsymbol{x}) = \begin{bmatrix} \dot{\boldsymbol{\Phi}} \\ p_1(\boldsymbol{\Phi}, \dot{\boldsymbol{\Phi}}) \\ p_2(\boldsymbol{\Phi}, \dot{\boldsymbol{\Phi}}) + (\boldsymbol{B}(\boldsymbol{\Phi})^{-1})_{2,2} \, w_n^2 x_{2,1} \\ p_3(\boldsymbol{\Phi}, \dot{\boldsymbol{\Phi}}) \\ p_4(\boldsymbol{\Phi}, \dot{\boldsymbol{\Phi}}) \\ p_5(\boldsymbol{\Phi}, \dot{\boldsymbol{\Phi}}) + (\boldsymbol{B}(\boldsymbol{\Phi})^{-1})_{5,5} \, w_n^2 x_{5,1} \\ -w_n^2 x_{2,1} - 2w_n x_{2,2} \\ x_{2,1} \\ -w_n^2 x_{5,1} - 2w_n x_{5,2} \\ x_{5,1} \end{bmatrix} \tag{11.19}$$

with

$$p_i(\boldsymbol{\Phi}, \dot{\boldsymbol{\Phi}}) = -\left(\boldsymbol{B}(\boldsymbol{\Phi})^{-1}\left(\boldsymbol{C}(\boldsymbol{\Phi}, \dot{\boldsymbol{\Phi}})\dot{\boldsymbol{\Phi}} + \boldsymbol{F}(\boldsymbol{\Phi}, \dot{\boldsymbol{\Phi}}) + \boldsymbol{G}(\boldsymbol{\Phi}) + \boldsymbol{K}(\boldsymbol{\Phi})\right)\right)_i \tag{11.20}$$

and

$$g(x) = \begin{bmatrix} g_1(x)^T \\ g_2(x)^T \end{bmatrix}^T, \quad h(x) = \begin{bmatrix} \phi_2 \\ \phi_5 \end{bmatrix}. \tag{11.21}$$

According to the development of the input–output linearization, the linearizing controllers about the combined muscle and supported arm system are given by

$$u_2^{io} = h_{IRC,2}^{-1}\left(\frac{\frac{\delta f_7(x)}{\delta x} f(x) - u_2}{(B(\Phi)^{-1})_{2,2}\, w_n^2 x_{2,1}}\right) \tag{11.22}$$

$$u_5^{io} = h_{IRC,5}^{-1}\left(\frac{\frac{\delta f_{10}(x)}{\delta x} f(x) - u_5}{(B(\Phi)^{-1})_{5,5}\, w_n^2 x_{5,1}}\right) \tag{11.23}$$

yielding the linearized muscle and supported arm model

$$y^{(4)} = u \tag{11.24}$$

with $y = [\phi_2\ \phi_5]^T$ denoting the output joint angles and $u = [u_2\ u_5]^T$ the input stimulation levels to be designed.

11.3.2 Optimal Feedback Controller

The results of applying the input–output linearization give a linear system model, for which we choose the following control structure:

$$\phi_2^{(4)} = u_2 = u_2^{ff} + u_2^{fb} + v_2 \tag{11.25}$$

$$\phi_5^{(4)} = u_5 = u_5^{ff} + u_5^{fb} + v_5 \tag{11.26}$$

where ϕ_2, ϕ_5 denote the actual trajectories of shoulder and elbow, u_2^{ff}, u_5^{ff} are the feedforward compensation terms, u_2^{fb}, u_5^{fb} denote the state feedback signals, and v_2, v_5 are the additional iterative learning control input to be designed in the next section.

Feedforward compensation is applied before trajectory tracking, where

$$u_2^{ff} = \hat{\phi}_2^{(4)} \tag{11.27}$$

$$u_5^{ff} = \hat{\phi}_5^{(4)} \tag{11.28}$$

gives error equations

$$e_2^{(4)} = u_2^{fb} + v_2 \tag{11.29}$$

$$e_5^{(4)} = u_5^{fb} + v_5 \tag{11.30}$$

with the trajectory tracking errors defined as

$$e_2 = \phi_2 - \hat{\phi}_2 \tag{11.31}$$

$$e_5 = \phi_5 - \hat{\phi}_5. \tag{11.32}$$

Various methods are available for stabilization of the (11.29)–(11.30). However, given the fact that the patients are encouraged to achieve satisfied trajectory tracking with minimum support from FES, thus the optimal state feedback control algorithm is utilized, with the form of

$$u_i^{fb} = -a_{i3}e_i^{(3)} - a_{i2}e_i^{(2)} - a_{i1}e_i^{(1)} - a_{i0}e_i \tag{11.33}$$

with $a_{i3}, a_{i2}, a_{i1}, a_{i0}$, $i \in \{2, 5\}$ denoting the feedback scalar gains to be designed using optimal control.

In more detail, the selection of the feedback gains $a_{i3}, a_{i2}, a_{i1}, a_{i0}$, $i \in \{2, 5\}$ can be obtained using the optimal state feedback $\boldsymbol{u}^{fb} = -\boldsymbol{K}\boldsymbol{\xi}$ obtained by minimizing the quadratic cost function (with symmetric positive definite cost function matrices $\boldsymbol{Q}$ and $\boldsymbol{R}$)

$$J(\boldsymbol{u}^{fb}) = \int_0^\infty (\boldsymbol{\xi}^T \boldsymbol{Q}\boldsymbol{\xi} + \boldsymbol{u}^{fb^T} \boldsymbol{R}\boldsymbol{u}^{fb})dt \tag{11.34}$$

subject to the error dynamics

$$\dot{\boldsymbol{\xi}} = \begin{bmatrix} 0 & \boldsymbol{I} & 0 & 0 \\ 0 & 0 & \boldsymbol{I} & 0 \\ 0 & 0 & 0 & \boldsymbol{I} \\ 0 & 0 & 0 & 0 \end{bmatrix} + \begin{bmatrix} 0 \\ 0 \\ 0 \\ \boldsymbol{I} \end{bmatrix} \boldsymbol{u}^{fb} \tag{11.35}$$

where $\boldsymbol{u}^{fb} = [u_2^{fb}, u_5^{fb}]^T$, $\boldsymbol{\xi} = [\boldsymbol{e}, \boldsymbol{e}^{(1)}, \boldsymbol{e}^{(2)}, \boldsymbol{e}^{(3)}]^T$, $\boldsymbol{e} = [e_2, e_5]^T$. This controller stabilizes the actuated joints and stability of the unactuated joint dynamics is guaranteed by the Centre Manifold Theorem and relative stability analysis, the details of which including the use of a Kalman filter to estimate the system states can be found in [6] and omitted here for space reasons.

Applying the control laws (11.27)–(11.28) and (11.33) to the linearized system (11.25)–(11.26) gives the error dynamics

$$e_i^{(4)} = -a_{i3}e_i^{(3)} - a_{i2}e_i^{(2)} - a_{i1}e_i^{(1)} - a_{i0}e_i + v_i \tag{11.36}$$

with $i \in \{2, 5\}$ corresponding to the closed-loop tracking error equation of shoulder and elbow, respectively, and hence with the state-space model of (11.36)

$$\underbrace{\begin{bmatrix} e^{(1)} \\ e^{(2)} \\ e^{(3)} \\ e^{(4)} \end{bmatrix}}_{\dot{\xi}} = \underbrace{\begin{bmatrix} 0 & I & 0 & 0 \\ 0 & 0 & I & 0 \\ 0 & 0 & 0 & I \\ -A_0 & -A_1 & -A_2 & -A_3 \end{bmatrix}}_{A} \underbrace{\begin{bmatrix} e^{(0)} \\ e^{(1)} \\ e^{(2)} \\ e^{(3)} \end{bmatrix}}_{\xi} + \underbrace{\begin{bmatrix} 0 \\ 0 \\ 0 \\ I \end{bmatrix}}_{B} v \tag{11.37}$$

$$y = \underbrace{\begin{bmatrix} I \; 0 \; 0 \; 0 \end{bmatrix}}_{C} \xi \tag{11.38}$$

where $e = [e_2, e_5]^T$ and

$$A_j = \begin{bmatrix} a_{2j} & 0 \\ 0 & a_{5j} \end{bmatrix}, \tag{11.39}$$

with $j = 0, 1, 2, 3$.

In the next few sections, we apply the iterative learning control techniques introduced in this book to design the applied FES levels (v_2, v_5) to track the desired trajectory for shoulder and elbow.

11.4 Norm Optimal Iterative Learning Control

In this section, simulation results for trajectory tracking using NOILC algorithm for 10 trials are presented. The trajectory used (as shown in Fig. 11.3) corresponds to lifting and extending the upper arm and forearm over a period of 6 s. The weight matrices used in the NOILC algorithm are $Q = I$, $R = 0.01I$, and $\epsilon^2 = 1$ for both muscles, and the weightings in the optimal control design (11.34) are $\boldsymbol{Q} = \boldsymbol{I}$ and $\boldsymbol{R} = 0.1\boldsymbol{I}$ (note the overloaded use of notation Q, R). The results are shown in Figs. 11.4,11.5,11.6, and 11.7.

As can be seen from Fig. 11.4, the NOILC algorithm achieves continuous improvement on the tracking performance. As expected, the tracking error norms for both shoulder and elbow converge monotonically. Note that a Gaussian measurement noise is included, due to which the final converged tracking error norm is non-zero. From Fig. 11.5 it can be seen that the shoulder and elbow positions on the last trial follow the trajectories very well (subject to the effect of measurement noise), and the maximum (filtered) tracking error is just 1.95×10^{-2} *rad* for shoulder and 4.06×10^{-2} *rad* for elbow, respectively, as shown in Fig. 11.6. The stimulation inputs to the muscles are also shown in Fig. 11.7.

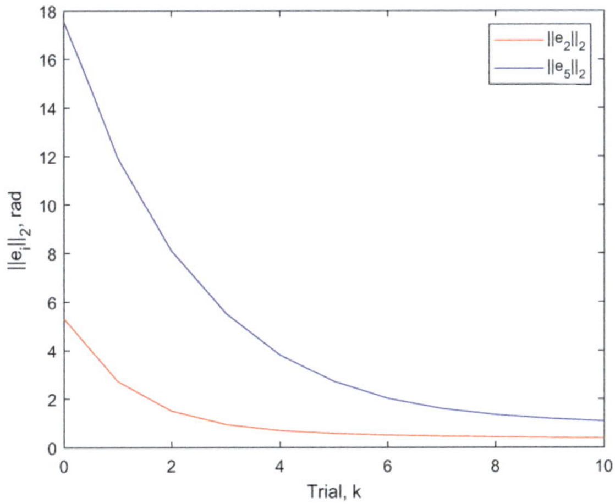

Fig. 11.4 Tracking error norms for shoulder and elbow using NOILC over 10 trials

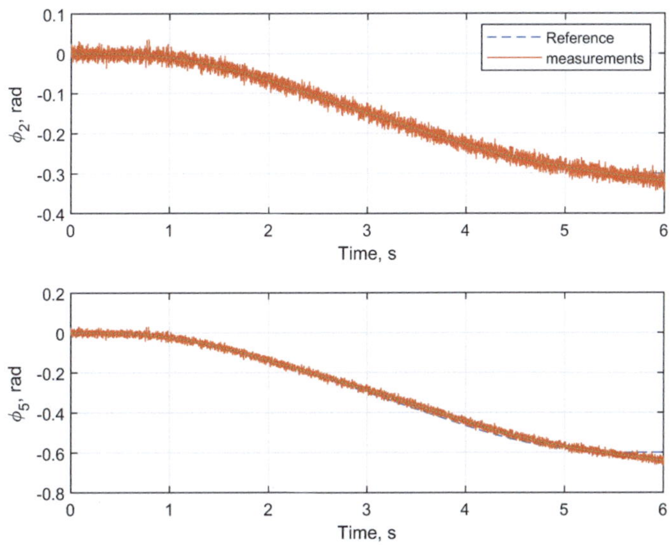

Fig. 11.5 Trajectory tracking for shoulder and elbow using NOILC (on 10th trial)

11.5 Accelerated NOILC with Error Corrected Reference

In this section, we demonstrate the performance of accelerated NOILC algorithm with error corrected reference (introduced in Sect. 7.4) with the expectation that the accelerated NOILC algorithm will lead to a faster convergence. We use one gradient step and a step size $\beta = 0.3097$. All other settings are the same as in the previous

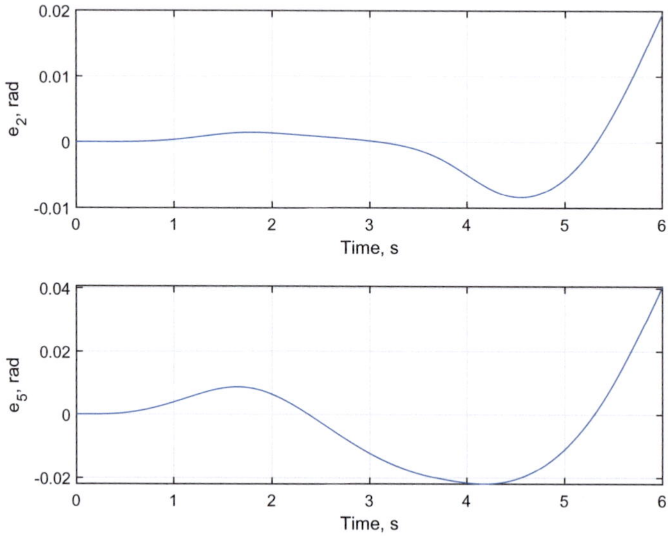

Fig. 11.6 Trajectory tracking error for shoulder and elbow using NOILC (on 10th trial)

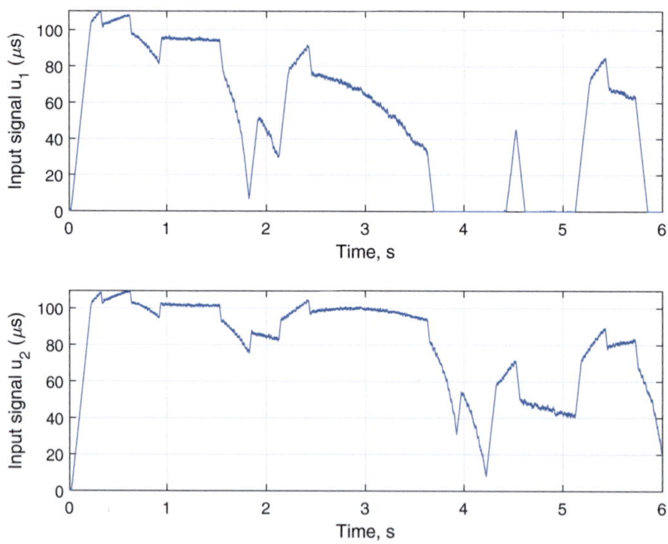

Fig. 11.7 Input signals for shoulder and elbow using NOILC (on 10th trial)

section. Simulation results for trajectory tracking using accelerated NOILC algorithm for 10 trials are presented in Figs. 11.8, 11.9, 11.10, and 11.11.

As can be seen from Fig. 11.8, the accelerated NOILC algorithm with error corrected reference improves the performance of the NOILC algorithm. To see this clearly, the tracking error norms are also shown in Table 11.1, from which it can be

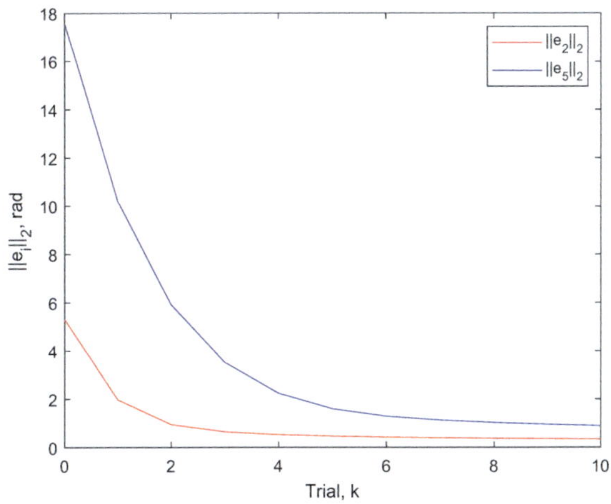

Fig. 11.8 Tracking error norms for shoulder and elbow using Accelerated NOILC over 10 trials

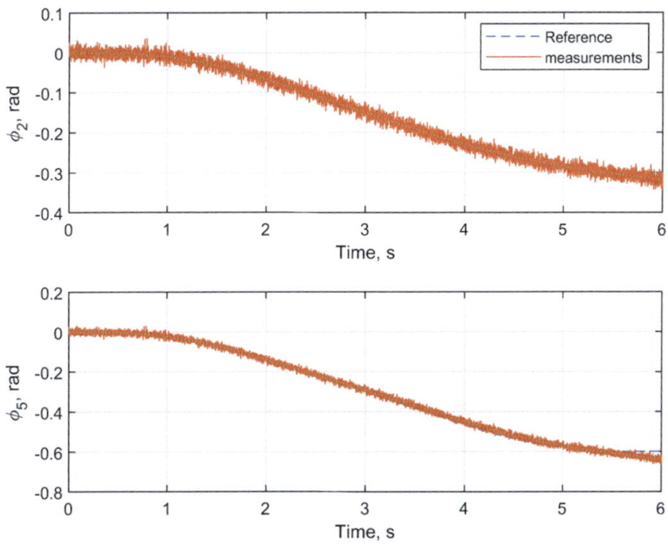

Fig. 11.9 Trajectory tracking for shoulder and elbow using Accelerated NOILC (on 10th trial)

seen that the improvement, especially during the first few iterations that is often of great practical importance, is substantial.

From Fig. 11.9 it can be seen that the shoulder and elbow positions on the last trial follow the trajectories very well (subject to the effect of measurement noise), and the maximum (filtered) tracking error is 1.88×10^{-2} *rad* for shoulder and 4.05×10^{-2} *rad* for elbow, respectively, as shown in Fig. 11.10. The stimulation inputs to the muscles are also shown in Fig. 11.11.

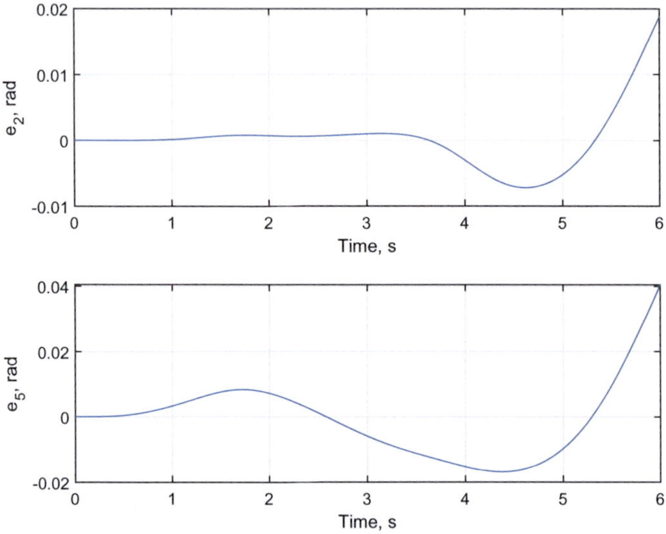

Fig. 11.10 Trajectory tracking error for shoulder and elbow using Accelerated NOILC (on 10th trial)

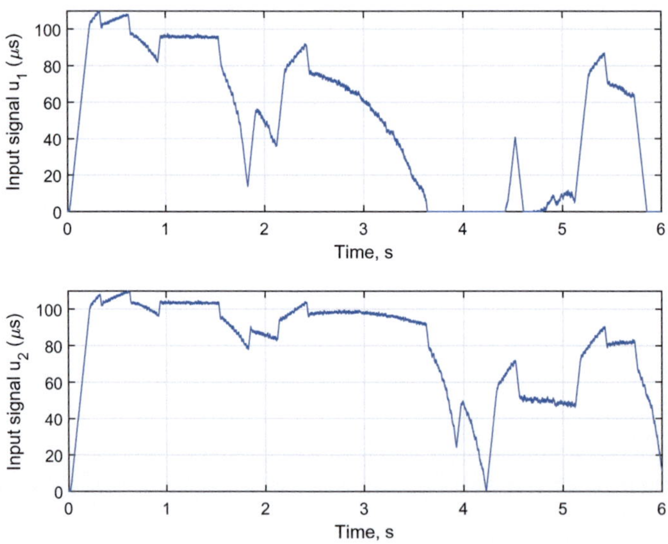

Fig. 11.11 Input signals for shoulder and elbow using Accelerated NOILC (on 10th trial)

Table 11.1 Tracking error norms for shoulder and elbow using NOILC and Accelerated NOILC

Iteration (k)	NOILC (e_2)	NOILC (e_5)	Accelerated NOILC with Error Corrected Reference (e_2)	Accelerated NOILC with Error Corrected Reference (e_5)
0	5.2986	17.5678	5.2986	17.5678
1	2.7222	11.9431	1.9688	10.1929
2	1.4962	8.0979	0.9437	5.9148
3	0.9429	5.5182	0.6435	3.5173
4	0.6939	3.8128	0.5271	2.2340
5	0.5722	2.7090	0.4634	1.5915
6	0.5028	2.0155	0.4218	1.2827
7	0.4575	1.5933	0.3917	1.1229
8	0.4250	1.3403	0.3688	1.0233
9	0.4001	1.1850	0.3517	0.9496
10	0.3803	1.0827	0.3396	0.8894

11.6 The Effect of Feedback Controller

In this section, we consider the effect of feedback controllers on the performance of the resulting iterative learning control design (as described in Sect. 7.1). The feedback controller used here is designed using optimal control techniques. To illustrate its effect, we use different weightings in the performance index to obtain the optimal feedback controller gain. In particular, in this section the weighting in the optimal control design on the input is $1/100$ of that in the previous two sections, resulting in a more aggressive feedback controller.

Using this controller, we compare the performance of the accelerated NOILC algorithm with error corrected reference (again, we use one gradient step and step size $\beta = 0.3097$—the same as in the previous section) with the previous section. All other settings remain the same. Simulation results for trajectory tracking for 10 trials are presented in Figs. 11.12, 11.13, 11.14, and 11.15.

As can be seen from Fig. 11.12, the NOILC algorithm with error corrected reference under the new feedback controller improves the performance as that described in the previous section. For ease of comparison, the tracking error norms are also shown in Table 11.2, from which it can be seen that the improvement, especially during the first few iterations that are often of great practical importance, again is substantial.

It is also worth noting that for the new (more aggressive) feedback controller used in this section, the final converged tracking error norms on the 10th trial are larger than that achieved in the previous section. This can also be seen from Fig. 11.13 that the maximum (filtered) tracking error is $2.55 \times 10^{-2}\,rad$ for shoulder and $8.78 \times$

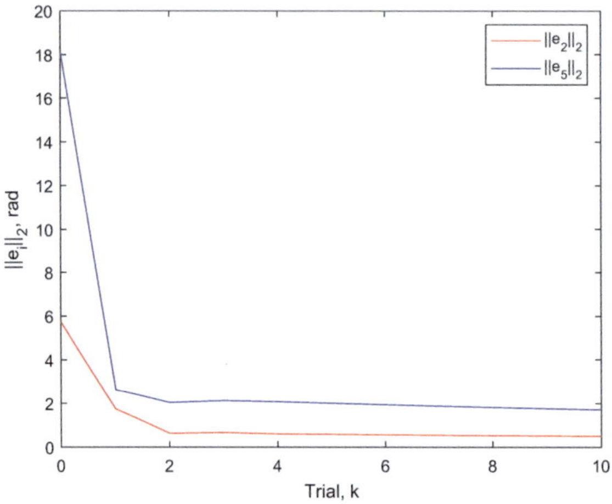

Fig. 11.12 Tracking error norms for shoulder and elbow using Accelerated NOILC with a new feedback controller over 10 trials

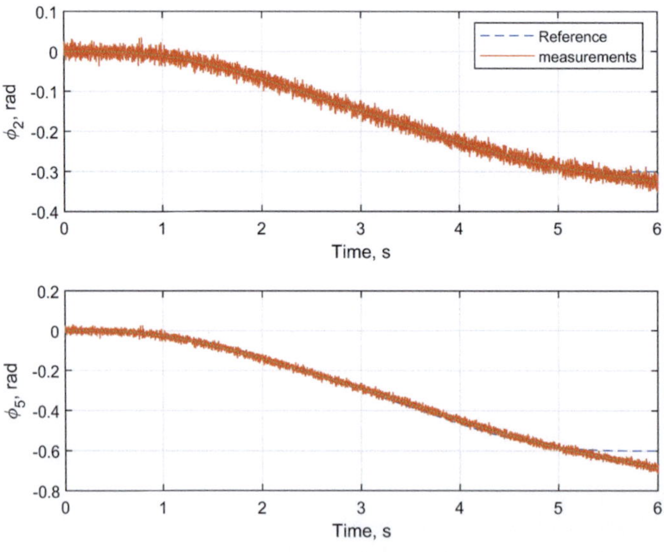

Fig. 11.13 Trajectory tracking for shoulder and elbow using Accelerated NOILC with a new feedback controller (on 10th trial)

10^{-2} *rad* for elbow, respectively, as shown in Fig. 11.14. The stimulation inputs to the muscles are also shown in Fig. 11.15.

A possible explanation for this is that a more aggressive feedback controller may be more sensitive to the non-repeated noise/disturbances (e.g., the random measurement noise in the elbow position)—a more detailed and rigorous investigation of this would be a topic for future research.

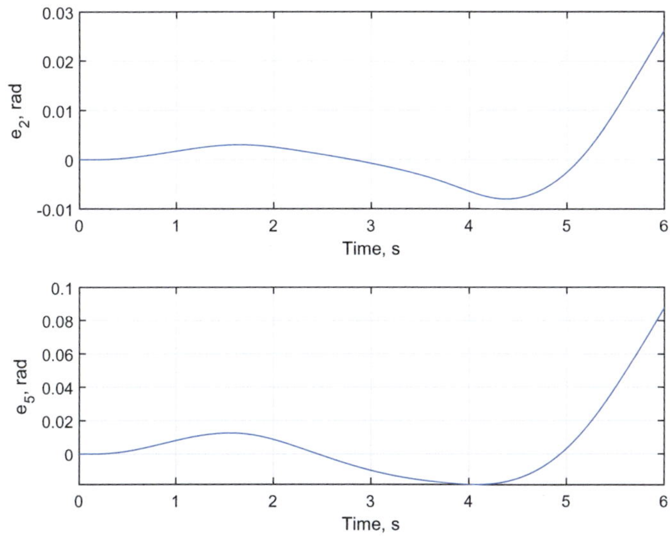

Fig. 11.14 Trajectory tracking error for shoulder and elbow using Accelerated NOILC with a new feedback controller (on 10th trial)

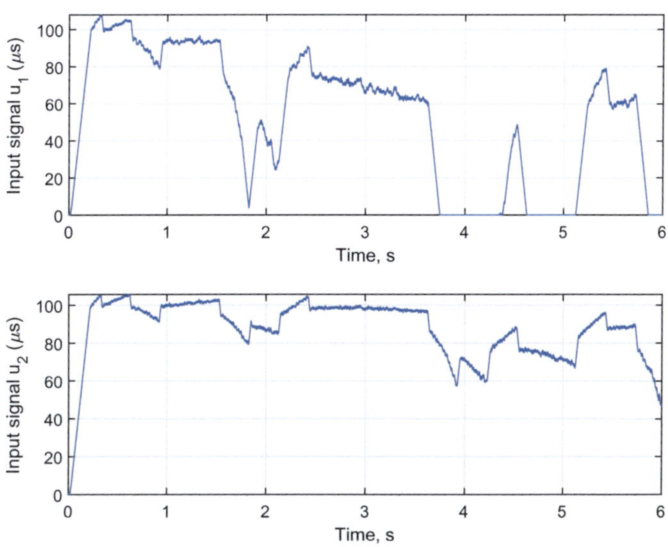

Fig. 11.15 Input signals for shoulder and elbow using Accelerated NOILC with a new feedback controller (on 10th trial)

Table 11.2 Tracking error norms for shoulder and elbow using Accelerated NOILC and with a new feedback controller

Iteration (k)	Accelerated NOILC with Error Corrected Reference (e_2)	Accelerated NOILC with Error Corrected Reference (e_5)	Accelerated NOILC with Error Corrected Reference and New Feedback Controller (e_2)	Accelerated NOILC with Error Corrected Reference and New Feedback Controller (e_5)
0	5.2986	17.5678	5.7262	18.0948
1	1.9688	10.1929	1.7424	2.6241
2	0.9437	5.9148	0.6270	2.0340
3	0.6435	3.5173	0.6664	2.1234
4	0.5271	2.2340	0.6030	2.0712
5	0.4634	1.5915	0.5854	2.0048
6	0.4218	1.2827	0.5628	1.9382
7	0.3917	1.1229	0.5445	1.8755
8	0.3688	1.0233	0.5281	1.8166
9	0.3517	0.9496	0.5137	1.7617
10	0.3396	0.8894	0.5011	1.7104

11.7 Summary and Discussion

In this chapter, a combined feedback (a feedback linearisation controller and an optimal state feedback controller) and iterative learning control scheme has been developed to perform upper arm trajectory tracking using FES upon a 3D exoskeleton robot for stroke rehabilitation. NOILC algorithm (and its acceleration) is then applied to further improve the trajectory tracking accuracy. Accelerated convergence performance is achieved by the use of an error corrected reference, as well as an improved feedback controller design.

Numerical simulations are presented to demonstrate the performance of the design. Future work will be needed to experimentally verify the proposed methods in upper arm trajectory tracking, and the successful outcomes will be used in further clinical trials to exploit the potential for stroke patient rehabilitation.

In this chapter it is assumed the dynamics of the human muscle are not changing over time. Human muscle dynamics actually can change over time due to fatigue—some investigation of this and incorporating it into design can be found in [23]. There are also other ILC design methods that can be applied, as well as other types of rehabilitation tasks—see, e.g., [7, 21] for more details.

References

1. Baratta R, Solomonow M (1990) The dynamic response model of nine different skeletal muscles. IEEE Trans Biomed Eng 37(3):243–251
2. Burridge JH, Ladouceur M (2001) Clinical and therapeutic applications of neuromuscular stimulation: a review of current use and speculation into future developments. Neuromodulation 4(4):147–154
3. Colombo R, Pisano F, Micera S, Mazzone A, Delconte C, Carrozza M, Dario P, Minuco G (2008) Assessing mechanisms of recovery during robot-aided neurorehabilitation of the upper extremity. Neurorehabil Neural Repair 22(1):50–63
4. Davoodi R, Andrews BJ (1998) Computer simulation of FES standing up in paraplegia: A self-adaptive fuzzy controller with reinforcement learning. IEEE Trans Rehabil Eng 6(2):151–161
5. De Kroon JR, Van der Lee JH, IJzerman MJ, Lankhorst GJ, (2002) Therapeutic electrical stimulation to improve motor control and functional abilities of the upper extremity after stroke: a systematic review. Clin Rehabil 16(4):350–360
6. Freeman CT (2015) Upper limb electrical stimulation using input-output linearization and iterative learning control. IEEE Trans Control Syst Technol 23(4):1546–1554
7. Freeman CT (2016) Control system design for electrical stimulation in upper limb rehabilitation. Springer International Publishing, Switzerland
8. Freeman CT, Hughes AM, Burridge JH, Chappell PH, Lewin PL, Rogers E (2009) Iterative learning control of FES applied to the upper extremity for rehabilitation. Control Eng Pract 17(3):368–381
9. Freeman CT, Tong D, Meadmore K, Cai Z, Rogers E, Hughes AM, Burridge JH (2011) Phase-lead iterative learning control algorithms for functional electrical stimulation based stroke rehabilitation. Proc Inst Mech Eng 225(6):850–859
10. Freeman CT, Rogers E, Hughes AM, Burridge JH, Meadmore KL (2012) Iterative learning control in healthcare: electrical stimulation and robotic-assisted upper-limb stroke rehabilitation. IEEE Control Syst Mag 32(1):18–43
11. Freeman CT, Tong D, Meadmore KL, Hughes AM, Rogers E, Burridge JH (2012) FES based rehabilitation of the upper limb using input/output linearization and ILC. In: American Control Conference, Montreal, pp 27–29
12. Hughes AM, Freeman CT, Burridge JH, Chappell PH, Lewin PL, Rogers E (2009) Feasibility of iterative learning control mediated by functional electrical stimulation for reaching after stroke. Neurorehabil Neural Rep 23(6):559–568
13. Hunt KJ, Munih M, Donaldson ND (1997) Feedback control of unsupported standing in paraplegia. I. Optimal control approach. IEEE Trans Rehabil Eng 5(4):331–340
14. Hunt KJ, Jaime RP, Gollee H (2001) Robust control of electrically stimulated muscle using polynomial H_∞ design. Control Eng Pract 9(3):313–328
15. Jezernik S, Wassink RGV, Keller T (2004) Sliding mode closed-loop control of FES: controlling the shank movement. IEEE Trans Rehabil Eng 51(2):263–272
16. Krebs HI, Palazzolo JJ, Dipietro L, Ferraro M, Krol J, Rannekleiv K, Volpe B, Hogan N (2003) Rehabilitation robotics: performance-based progressive robot-assisted therapy. Autonom Robot 15(1):7–20
17. Lum PS, Burgar CG, Shor PC (2004) Evidence for improved muscle activation patterns after retraining of reaching movements with the MIME robotic system in subjects with post-stroke hemiparesis. IEEE Trans Neural Syst Rehabil Eng 12(2):186–194
18. Meadmore KL, Hughes AM, Freeman CT, Cai Z, Tong D, Burridge JH, Rogers E (2012) Functional electrical stimulation mediated by iterative learning control and 3D robotics reduces motor impairment in chronic stroke. J Neuroeng Rehabil 32
19. Previdi F, Schauer T, Savaresi SM, Hunt KJ (2004) Data-driven control design for neuroprostheses: a virtual reference feedback tuning (VRFT) approach. IEEE Trans Control Syst Technol 12(1):176–182
20. Schmidt RA, Lee TD (1999) Motor learning. Motor control and learning: a behavioral emphasis, Part 3. Human Kinetics, Champaign, IL, pp 261–285

21. Seel T, Werner C, Raisch J, Schauer T (2016) Iterative learning control of a drop foot neuroprosthesis—generating physiological foot motion in paretic gait by automatic feedback control. Control Eng Pract 48:87–97
22. Watanabe T, Iibuchi K, Kurosawa K, Hoshimiya N (2003) A method of multichannel PID control of two-degree-of-freedom wrist joint movements by functional electrical stimulation. Syst Comput Jpn 34(5):25–36
23. Xu W, Chu B, Rogers E (2014) Iterative learning control for robotic-assisted upper limb stroke rehabilitation in the presence of muscle fatigue. Control Eng Pract 31:63–72
24. Zhang D, Guan TH, Widjaja F, Ang WT (2007) Functional electrical stimulation in rehabilitation engineering: a survey. In: 1st International Convention on Rehabilitation Engineering and Assistive Technology, Singapore, pp 221–226

Index

B. Chu and D. H. Owens, *Optimal Iterative Learning Control*, Advances in Industrial Control, https://doi.org/10.1007/978-3-031-80236-2

MIX
Papier aus verantwortungsvollen Quellen
Paper from responsible sources
FSC® C105338
FSC
www.fsc.org

If you have any concerns about our products,
you can contact us on
ProductSafety@springernature.com

In case Publisher is established outside the EU,
the EU authorized representative is:
Springer Nature Customer Service Center GmbH
Europaplatz 3, 69115 Heidelberg, Germany

Printed by Libri Plureos GmbH
in Hamburg, Germany